Mobile Information Systems: Concepts and Applications

Mobile Information Systems: Concepts and Applications

**Edited by
Dylan Glover**

WILLFORD PRESS

www.willfordpress.com

Published by Willford Press,
118-35 Queens Blvd., Suite 400,
Forest Hills, NY 11375, USA

ISBN: 978-1-68285-765-6

Cataloging-in-Publication Data

Mobile information systems : concepts and applications / edited by Dylan Glover.
    p. cm.
Includes bibliographical references and index.
ISBN 978-1-68285-765-6
1. Mobile communication systems. 2. Telecommunication systems. 3. Mobile computing.
4. Computer networks. I. Glover, Dylan.
TK5103.2 .M63 2020
621.384--dc23

For information on all Willford Press publications
visit our website at www.willfordpress.com

WILLFORD PRESS

Contents

Preface

Every book is a source of knowledge and this one is no exception. The idea that led to the conceptualization of this book was the fact that the world is advancing rapidly; which makes it crucial to document the progress in every field. I am aware that a lot of data is already available, yet, there is a lot more to learn. Hence, I accepted the responsibility of editing this book and contributing my knowledge to the community.

The immense popularity of mobile infrastructures and higher bandwidths of network has radically transformed the way people use information resources. In this context, mobile information systems (MIS) are a pertinent area of study. MIS are information systems that allow the access to information resources and services with the aid of end-user terminals. These terminals are movable in space and are provided with wireless communication. The advantage of MIS is that they provide new value-added services due to their mobility and flexibility. They also render other services such as user access across multiple devices and channels, and the ability to perform collaborative work either in synchrony or otherwise. The book presents researches and studies performed by experts across the globe. Also included in this book is a detailed explanation of the various concepts and applications of mobile information systems. The readers would gain knowledge that would broaden their perspective about this field.

While editing this book, I had multiple visions for it. Then I finally narrowed down to make every chapter a sole standing text explaining a particular topic, so that they can be used independently. However, the umbrella subject sinews them into a common theme. This makes the book a unique platform of knowledge.

I would like to give the major credit of this book to the experts from every corner of the world, who took the time to share their expertise with us. Also, I owe the completion of this book to the never-ending support of my family, who supported me throughout the project.

Editor

A Sun Path Observation System based on Augment Reality and Mobile Learning

Wernhuar Tarng ⓘ,[1] Kuo-Liang Ou ⓘ,[1] Yun-Chen Lu,[1] Yi-Syuan Shih,[2] and Hsin-Hun Liou[2]

[1]*Institute of Learning Sciences and Technologies, National Tsing Hua University, Hsinchu, Taiwan*
[2]*Department of Computer Science and Information Engineering, National Central University, Taoyuan, Taiwan*

Correspondence should be addressed to Kuo-Liang Ou; klou@mail.nd.nthu.edu.tw

Academic Editor: Jesus Fontecha

This study uses the augmented reality technology and sensor functions of GPS, electronic compass, and three-axis accelerometer on mobile devices to develop a Sun path observation system for applications in astronomy education. The orientation and elevation of the Sun can be calculated by the system according to the user's location and local time to simulate the Sun path. When holding the mobile device toward the sky, the screen will show the virtual Sun at the same position as that of the real Sun. The user can record the Sun path and the data of observation date, time, longitude, and latitude using the celestial hemisphere and the pole shadow on the system. By setting different observation times and locations, it can be seen that the Sun path changes with seasons and latitudes. The system provides contextual awareness of the Sun path concepts, and it can convert the observation data into organized and meaningful astronomical knowledge to enable combination of situated learning with spatial cognition. The system can solve the problem of being not able to record the Sun path due to a bad weather or topographical restrictions, and therefore it is helpful for elementary students when conducting observations. A teaching experiment has been conducted to analyze the learning achievement of students after using the system, and the results show that it is more effective than traditional teaching aids. The questionnaire results also reveal that the system is easy to operate and useful in recording the Sun path data. Therefore, it is an effective tool for astronomy education in elementary schools.

1. Introduction

Astronomy is the scientific study of celestial objects such as stars, planets, and galaxies as well as the phenomena that originate outside the Earth's atmosphere. It is an important field of study investigating the problems of immediate concern to human beings, for example, the Sun, Moon, stars, seasonal changes, day and night, lunar phase, and rising and setting of stars. These changes are closely related to our daily life. The astronomical observation is an important topic of natural science in elementary schools [1]. Students can think about the relative motion between celestial bodies and their interaction by observing the changes of astronomical phenomena and investigate the problems using scientific methods to obtain the answers. Consequently, it can enhance their critical-thinking and problem-solving abilities [2].

There are several learning units about astronomy education in K-12 Science Education Curriculum [3]. In the "Lunar phase" unit (the fourth grade of elementary school), students may find out that the change of lunar phase is regular by observing moonrise and moonset for a month. In the "Sun and four seasons" unit (the fifth grade of elementary school) and the "Sun, Earth, and Moon" unit (the third grade of junior high school), students may understand that the change of seasons, day and night, lunar phase, solar eclipse, and lunar eclipse are caused by the relative motion of celestial bodies, that is the Sun, Earth, and Moon, by observing sunrise, sunset, moonrise, and moonset and operating the celestial model. In the "Beautiful night sky" unit (the fifth grade of elementary school), students may conduct a long-term observation to discover that the change of star positions in the night sky is caused by the rotation of Earth and its revolution around the Sun. It may help them establish the concepts of time and space.

The Sun is closely related with human life because it provides the light and energy for all organisms on the Earth. The Earth's axis is tilted at 23.5°, making the Northern Hemisphere point at the Sun more directly half the year, and the Southern Hemisphere does the same the other half. In the Northern Hemisphere, days reach their maximum (minimum) lengths at the summer (winter) solstice when the upper half of the planet faces directly toward (away from) the Sun. In the spring and fall, days and nights are roughly equal at the two equinoxes. The change of day and night affects our lifestyle and daily routine, the variation of day (night) time impacts animal and plant adaptations and behaviors, and the change of seasons influences farming and harvest. Therefore, observing the Sun to know its impacts on human life is important in science education.

Due to the Earth's rotation, the position of the Sun moves in the sky from time to time. In traditional teaching activities of the Sun path observation, the students are required to record the pole shadow and the Sun's position using the teaching aids of celestial hemisphere, compass, and protractor for measuring the Sun orientation and elevation. The learning objective is to find out the change of the Sun's position in the sky. The teacher may also use pictures and animations to show the Earth's revolution around the Sun and the tilted angle of its rotational axis for students to understand the regular change of the Sun path in four seasons and the relationship between the latitude of the observer's location and elevation of the Sun.

The observation of Sun path is often restricted by time, weather, and topographical conditions. If the observational activity is scheduled at noon, it may be affected by the hot weather for which can result in heatstroke when a person has been exposed to a hot environment for an extended period of time. The activity may also be cancelled due to typhoons or rainfall for several days during summer and autumn. Besides, it is impossible to measure the pole shadow in cloudy days or if the Sun is obstructed by tall buildings. To compare the Sun paths in different seasons, the observation has to be continued for a whole year. Some students have no perseverance to complete the task, so they gradually lose learning interest and confidence.

In addition to the efforts required in observing and recording the Sun path, the astronomical concepts have to be developed by induction and imagination according to the recorded data, which is difficult especially when the data are not complete. For example, most students do not have the chance to conduct observations at different latitudes or the patience to continue recording for a whole year. In order to understand how the Sun path changes in different seasons or at different latitudes, the students have to imagine themselves standing at a certain latitude on the Earth and think about the Sun's position as the Earth is rotating by itself. Therefore, they may have misconceptions about the relative motion of celestial bodies when studying complex astronomical phenomena [4–6]. All these factors affect the development of astronomical concepts.

With the innovation of information and communication technology (ICT), mobile devices, such as personal digital assistant (PDA), smart phone, and tablet PC, are more powerful and have been applied widely in different fields of learning. As a result, learning activities are no longer limited to classroom teaching, and they can be conducted by using any device at anytime from anywhere to achieve the so-called ubiquitous learning [7, 8]. As the smart phone and tablet PC become more and more popular, many virtual reality (VR) and augmented reality (AR) apps have been developed on mobile devices for entertainment and educational applications. VR replaces the real world with a simulated one whereas AR enhances one's current perception of reality through superimposing VR objects onto real environments. AR emphasizes the combination of contexts in the real world with virtual or real situations to intensify its interaction with users. AR can integrate a real environment with virtual objects to enhance the comprehension of environmental context and the sense of reality in a more interactive way. Recent research has shown that applications of AR in learning can improve knowledge construction and engage learners in high-flow experience levels [9–11]. Therefore, AR can be used as an effective tool to promote science learning [12–15].

Many researchers believe that VR and AR technologies are becoming mature and will be broadly used in industries such as architecture and construction as well as science, technology, and engineering education within the next 10 years. The present time can be a turning point of the applications of VR and AR due to the increasing development of hardware and software. The applications of AR technology in industrial training and science education can enhance one's perception of reality through superimposing VR objects onto real environments to bridge the gap between the virtual and physical worlds. According to the review of augmented reality learning experiences (ARLEs) conducted by Drljevic, Wong, and Boticki [16], some ARLEs support one or more of active, constructive, authentic, intentional, and cooperative learning, but only two studies describe ARLEs that offer effective support for a teacher to guide the activities.

Recently, the AR technology has been widely used in creating learning experiences for educational settings in elementary and high schools. Santos et al. [17] conducted a review of AR applications intended to complement traditional curriculum materials, and their research results showed that ARLEs achieved a widely variable effect on student performance from a small negative effect to a large positive effect. A qualitative analysis was performed on the design aspects for ARLEs, including display hardware, software libraries, content authoring solutions, and evaluation techniques to support their finding that AR incurs three inherent advantages: real world annotation, contextual visualization, and vision-haptic visualization. The advantages were illustrated through the exemplifying prototypes supported by multimedia learning theory, experiential learning theory, and animate vision theory to inform the design of future ARLEs.

In our previous study [18], the AR technology has been applied to develop a lunar-phase observation system. The system enables the user to record the lunar phase, including its azimuth/elevation angles and the observation date/time.

In addition, the system can shorten the learning process by setting different dates and times for observation, so it can solve the problem of being unable to observe and record lunar phases due to bad weather or the Moon appearing late in the night. The experiment results showed that it is effective in learning the lunar concepts, and the questionnaire results revealed that students considered the system easy to operate and useful in locating the Moon and recording the lunar data.

In this study, the AR technology and sensor functions of GPS, electronic compass, and three-axis accelerometer on mobile devices are combined to develop a Sun path observation system for educational applications in elementary schools. A celestial model is developed to simulate the Earth's rotation and its revolution around the Sun. The orientation and elevation of the Sun can be calculated by this model according to the location (latitude and longitude) and local time of the user to simulate the Sun path in the sky. When holding the mobile device toward the Sun, the screen will show the virtual Sun in the same direction as that of the real Sun. The user can record the Sun path as well as the observation data such as the date, time, orientation, and elevation using the celestial hemisphere and pole shadow on the system and set different observation times and locations to understand that the Sun path changes with seasons and latitudes.

This study differs from the previous study because the celestial objects and the tools used for observation are different. The relative movement of the Sun to the Earth and the calculation of its path are different from those of the lunar-phase observation system since the Moon is revolving around the Earth which is also revolving around the Sun at the same time. When observing the Moon, the orientation and elevation are measured by the sensors on the mobile device and recorded as the data for calculating its position in the sky. To display the Sun path, the azimuth/elevation angles have to be mapped to the surface of the three-dimensional celestial hemisphere and match with the data calculated by the length and orientation of the pole shadow. Even with the same AR technology and the sensors on mobile devices, the coordinate systems of the targets for observation are different, and they have to be calculated according to the periods of the Earth's rotation and revolution.

In addition, the current study has to implement a celestial hemisphere and a pole stick for observing and recording the Sun path. The Sun path observation system enables the user to look from inside and outside of the celestial hemisphere to observe the Sun and its path by a simple motion-sensing operation. Also, the Sun paths of different seasons or different latitudes can be displayed together for comparison to help students understand that the Sun path changes with the seasons and latitudes. Finally, the function of autorecording is implemented to reduce the observation and recording time.

2. Research Method

Astronomy is an important topic in the curriculums of science education in elementary and high schools. However, students are often confused by the complicated spatial concepts involved in the relative motion of celestial bodies and therefore have misconceptions after learning. Due to the limited class time, some teachers can only demonstrate how to use the celestial hemisphere, compass, and protractor to record the Sun path and then require students to conduct observation by themselves after class. Without proper and timely supports of effective instructional scaffolding, students may have difficulties in obtaining the necessary experiences and knowledge for developing correct astronomical concepts. As a result, they may use incorrect methods to conduct observations and eventually lose their confidence and learning interest.

In practice, the teaching activities of astronomical observation in the curriculums of science education are designed based on the theory of "learning by doing" [19]. Therefore, it is better to encourage students to conduct outdoor observation and practical operation using the teaching aids, simulation models, and computer animation in real or virtual environments. To achieve the learning goals, the teacher can use some instructional strategies or assistant tools to enhance students' learning effectiveness. The objective is for the students to develop experimentation skills and conceptual knowledge by hands-on experiences.

Many students believe that summer is hotter because the distance between the Sun and the Earth is shorter. In fact, the Sun is farther from the Earth in the summer of the Northern Hemisphere, which is the winter of the Southern Hemisphere. Scientists have discovered that the distance between the Sun and the Earth is not the cause of seasonal change while it is caused by the revolution of the Earth around the Sun and the inclination of the Earth's axis. If the axis were perpendicular to the orbital plane, there would be no seasonal changes. However, the Earth's axis is tilted at 23.5°, causing the Earth's surface exposed to the direct sunlight to change from season to season. At the summer solstice (the summer of the Northern Hemisphere), Tropic of Cancer is exposed to the direct sunlight; at the winter solstice (the winter of the Northern Hemisphere), Tropic of Capricorn is exposed to the direct sunlight; and at the vernal equinox and the autumnal equinox (the spring and autumn of the Northern Hemisphere), the equator is exposed to the direct sunlight (Figure 1).

The Sun's position in the sky changes with time due to the Earth's rotation, and the Sun path varies with the Earth's position in its orbit as well as the observer's latitude on the Earth, which is caused by the inclination of the Earth' axis. At the summer solstice, the Sun is more northward in the sky in regions with higher latitudes in the Northern Hemisphere. In the Arctic Circle, the Sun path is always above the horizon in summer, and there are 24 hours of daylight everywhere north of 66.5° N latitude, dubbed "Arctic white nights." The Sun path also varies with seasons because the Earth's position keeps changing on its orbital plane. For example, Tropic of Cancer cuts across Taiwan at Chiayi, where the Sun is northward at the summer solstice. It is above the head at noon, and thus the shadow of a person can hardly be seen. At the winter solstice, the Sun's elevation is about 45° at noon, so the shadow is about the same length as one's height.

Because one cannot look at the Sun directly due to its dazzling light and the fact that the Sun's orientation and elevation are not marked by any signs in the sky, the four

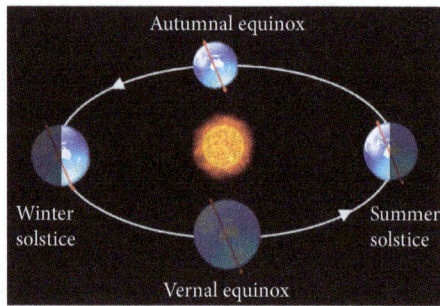

FIGURE 1: Seasonal changes by the inclination of the Earth's axis.

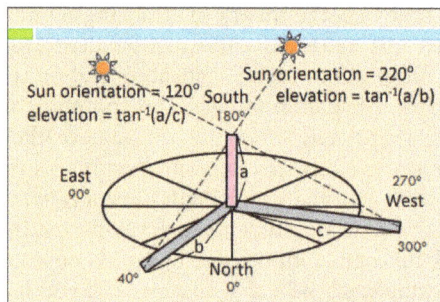

FIGURE 2: Measuring the Sun's orientation and elevation with the pole shadow.

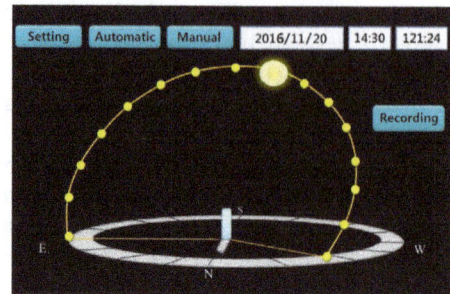

FIGURE 3: Proposed model of the Sun path observation system.

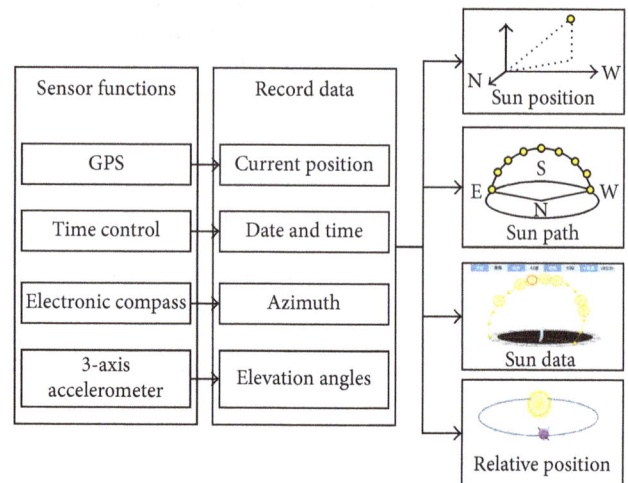

FIGURE 4: Main modules of the Sun path observation system.

seasons and 24 solar terms in ancient China were determined by setting up a sundial for measuring its shadow to tell the seasonal changes. In teaching the Sun path observation, the teachers often simulate the Sun with a flashlight for the students to know that the shadow of an object is in the direction opposite to the light source and the Sun's elevation can be calculated by the length of pole shadow using trigonometric functions (Figure 2). For example, the orientation of the Sun is 120° if the orientation of the pole shadow is 300°, and its elevation can be calculated by $\tan^{-1}(a/c)$. Finally, the measured orientation and elevation are marked on the celestial hemisphere to form the Sun path in a day. The observation results can also be used to induce the regular change of the Sun's position on a daily basis.

This study is motivated by the requirements of teaching the Sun path observation in the K-12 Science Education Curriculum in Taiwan. A celestial model containing the Sun and Earth is developed using the AR technology to simulate the Earth's rotation and its revolution around the Sun. A virtual celestial hemisphere with a pole at its center for measuring the shadow is placed on the Earth model for students to record the Sun's position in the sky. The students can observe and record the Sun's orientation and elevation as well as the related data such as the observation date, time, longitude, and latitude to understand that the Sun path changes with seasons and latitudes. They can also calculate the Sun's azimuth/elevation angles by measuring the orientation and length of the pole shadow for verifying correctness of the Sun's position.

To break through the limitations of observation time, location/ and weather conditions, this study aims at satisfying the teaching requirements by developing a Sun path

observation system for applications in science education in elementary schools. A preliminary study has been conducted on the research problems of (1) how to obtain the date, time, longitude, and latitude on the mobile device; (2) how to obtain the orientation and elevation of the mobile device with its sensors; (3) how to calculate the Sun path according to the Sun's position in its orbit as well as the observation date, time, longitude, and latitude; (4) how to develop a motion-sensing interface for the user to observe the Sun by holding the mobile device toward its direction; and (5) how to record data, generate the Sun path, and determine the relations between the Sun path and the Earth's position in its orbit as well as the user's latitude. The main modules of the Sun path observation system developed in this study include (1) a virtual Sun to simulate the real Sun's position, (2) a pole shadow for calculating the Sun orientation and elevation, and (3) a celestial hemisphere for recording the Sun path (Figure 3).

3. System Design and Operation

The operating procedure of the Sun path observation system is (1) obtaining the longitude and latitude coordinates from the GPS system and (2) calculating the orientation and elevation of the mobile device using the electronic compass and the three-axis accelerometer. The above data together

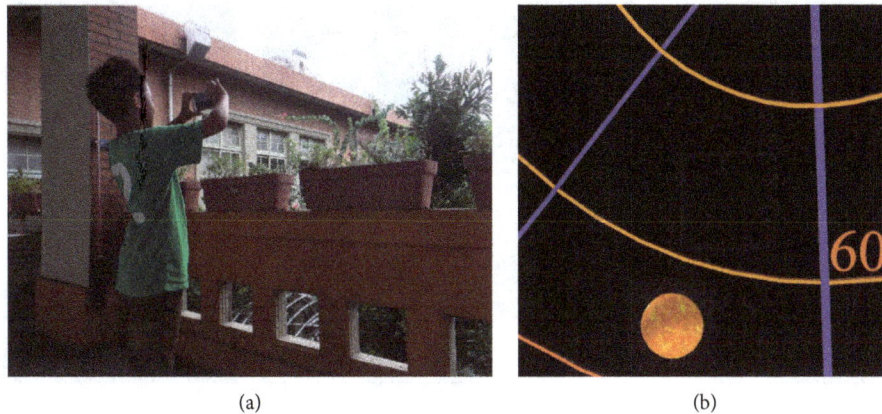

(a)   (b)

FIGURE 5: Holding the mobile device toward the Sun in the sky.

with the system date and time are input to the celestial model of the Earth's revolution around the Sun for (3) calculating the position of the virtual Sun in the sky. The system provides a motion-sensing interface for the user to conduct the Sun observation by holding the mobile device toward the sky to locate the Sun's position. The recorded data can be used for (4) displaying the Sun path on the hemisphere model and the Earth's position in its orbit.

The main modules of the Sun path observation system include the "GPS," "time control," "electronic compass," "three-axis accelerometer," "Sun position," "Sun path," "Sun data," "revolution model," "celestial hemisphere," and the modules for setting date, time, location, and operating speed (Figure 4).

The major functions of the Sun path observation system include (1) Sun observation, (2) recording the Sun path data, (3) setting observation date, time, and location, (4) displaying the Sun path, and (5) displaying the Earth's position in its orbit, and their user interface are described as follows.

3.1. Sun Observation.

After starting the Sun path observation system, the mobile device obtains the longitude and latitude coordinates from the GPS system and measures its orientation and elevation using the electronic compass and the three-axis accelerometer. To facilitate the Sun observation and displaying of the Sun data, the background color of the user interface is set to black. When the elevation of the mobile device is within 90° of the horizontal plane, the user can see a celestial hemisphere on the screen with a pole shadow located at its center. If the elevation of the mobile device is more than 90°, the system simulates the user looking at the sky from the center of the celestial hemisphere to observe the Sun (Figure 5). By holding the mobile device toward the Sun in the sky, the user can see the virtual Sun in the same direction as that of the real Sun. In addition, Sun's orientation and elevation, current date and time, longitude and latitude are displayed on the screen. The user can record the data by simply pressing the "Recording" button.

3.2. Recording the Sun Path Data.

When the user presses the "Recording" button, the observation data are stored as the

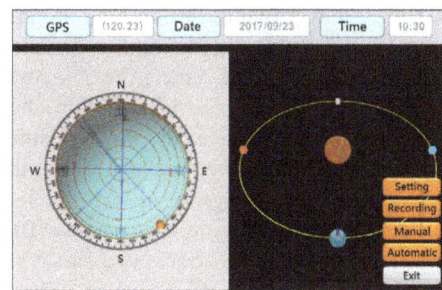

FIGURE 6: Recording Sun's position on the celestial hemisphere.

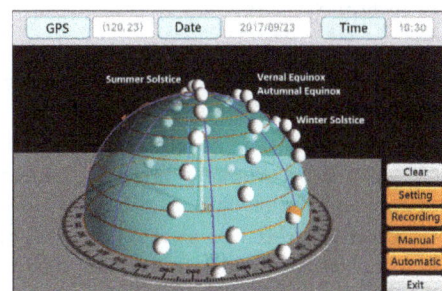

FIGURE 7: Displaying the Sun paths in different seasons at Tropic of Cancer.

environment variables which can be retrieved at any time for displaying the Sun path. Several sets of data, for example, in different seasons or at different latitudes, can be combined together and converted into the Sun paths on the celestial hemisphere for comparison. The user may set the recording mode as "Manual" to record the data one at a time or "Automatic" to record the data once per hour for the whole day. The operating speed may also be adjusted so that the user can obtain the Sun path in a short time (Figure 6).

3.3. Setting Observation Date, Time, and Location.

The Sun path observation system provides the function of setting observation date, time, and location. The user can press the "Setting" button to adjust the observation date and time. For

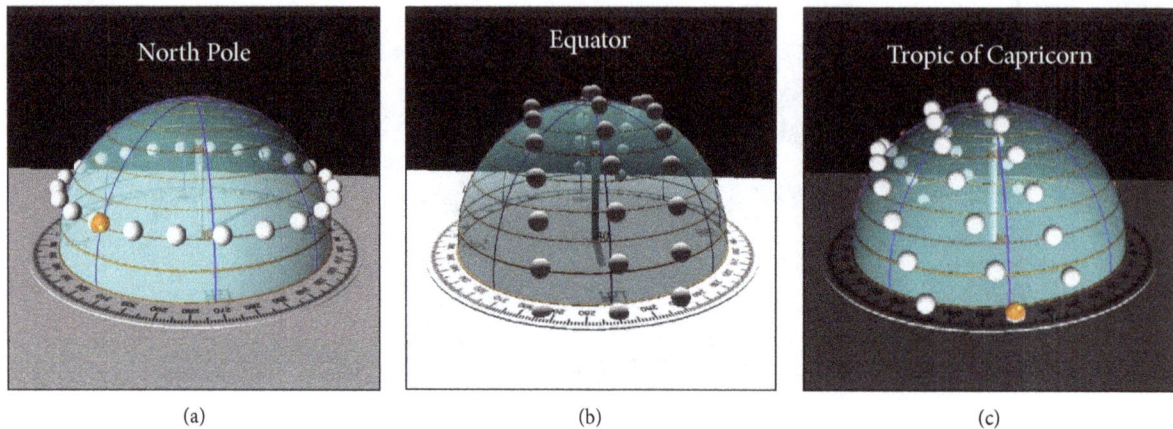

FIGURE 8: Comparing the difference of the Sun paths at different latitudes.

simplicity, the system provides the options of vernal equinox, summer solstice, autumnal equinox, and winter solstice for the teaching purpose. After setting the observation date, the user can see that the the Earth's position changes accordingly in its orbit and so does the Sun's position in the sky. The system date/time can be reset by pressing the "Now" button. The user may also change the observation location by setting its latitude and longitude coordinates. The system provides several representative coordinates, such as "Tropic of Cancer," "Tropic of Capricorn," "Equator," "North Pole," "South Pole," and the "Here" button for returning to the current location. The user may understand the relation between the Sun path and the latitude by setting different observation locations.

3.4. Displaying the Sun Path. The user can press the "Management" button to select the observation data for displaying the Sun path. The system will group the observation data recorded on the same day at the same location into a set of data and display Sun's positions on the celestial hemisphere according to the time sequence so that the user may see the Sun path during the observation day to achieve the teaching objectives of (1) observing sunrise and sunset by recording Sun's position in the sky, (2) understanding that Sun's position varies with time, and (3) knowing how to draw the Sun path on the celestial hemisphere. Also, the user may combine several sets of data (Figure 7) to compare the difference of the Sun paths at different latitudes (e.g., Tropic of Cancer, Tropic of Capricorn, Equator, North Pole, and South Pole) or in different seasons (e.g., the vernal equinox, the summer solstice, the autumnal equinox, and the winter solstice).

3.5. Displaying the Earth's Position in Its Orbit. In addition to the Sun path in the sky, the system can also display the Earth's revolution model to show the relation between the Earth's position in its orbit and the change of Sun path. The teacher may explain to students that the period of the Earth's revolution is one year, and the regions exposed to the direct sunlight change from season to season due to the inclination of the Earth's axis. After operating the Sun path observation

system, it is easier for students to understand that at the summer solstice, the Tropic of Cancer is exposed to the direct sunlight; at the vernal equinox and the autumnal equinox, the equator is exposed to the direct sunlight; and at the winter solstice, the Tropic of Capricorn is exposed to the direct sunlight (Figure 8). Using the system, the students can easily understand how the Earth rotates by itself at a tilted angle and revolves around the Sun along its orbit, and all these cause the Sun path to change with latitudes and seasons.

3.6. Longitudinal Effect on the Sun Path. Sun's position in the sky changes with time due to the Earth's rotation. To calculate the rotation angle of the Earth model per unit time, it is required to know the time for the Earth to rotate once (360°). Traditionally, a solar day (24 hours) is defined as the time between two successive transits of the Sun across the meridian. However, the angle that the Earth rotates for a solar day is slightly greater than 360° because the Earth is also revolving around the Sun during its rotation. The time for the Earth to rotate 360° (a sidereal day) can be calculated as follows. In each solar day, as the Earth makes a 360° revolution around the Sun in a year (about 365 days), it must turn about an additional degree to see the Sun on meridian. Therefore, it takes about $24 \times 60/361 = 4$ minutes for the Earth to rotate one degree, and therefore a sidereal day is only about 23 hours and 56 minutes. In the celestial model, the sidereal day is used to calculate the time required for the Earth to rotate 360° to obtain the accurate simulation results.

The Sun path varies with the Earth's position in its orbit as well as the observer's latitude on the Earth, and it is caused by the inclination of the Earth' axis. However, the Sun path at the same latitude remains the same even if the longitude changes. It can be seen in Figure 9 that the Sun paths are not changed if the longitude is moved from 120° to 210° while the latitude remains the same (23°). It is caused by the fact that the Earth surface at the same latitude exposed to the direct sunlight (at noon) always turns to the same position at about the same time. There are 360° of longitude and 24 hours in a day, so every 15° of longitude equals one time zone (an hour). A time zone is a region on the Earth with a uniform

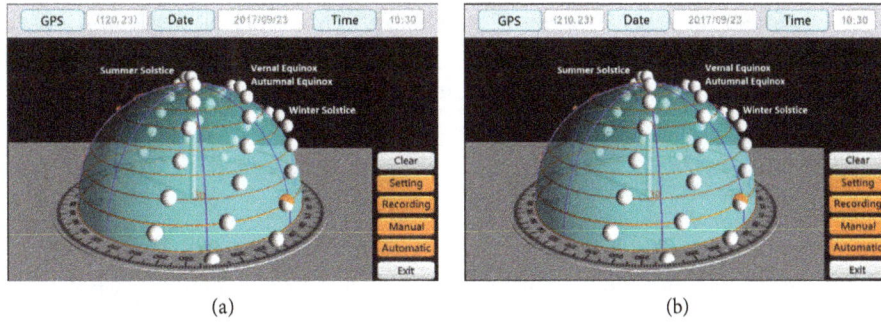

(a)   (b)

FIGURE 9: Comparing the difference of the Sun paths at different longitudes.

(a)   (b)

FIGURE 10: Conducting the teaching experiment. (a) control group; (b) experimental group.

TABLE 1: Variables of the teaching experiment.

	Control group	Traditional teaching aids
	Experimental group	The Sun path observation system
Covariance		Achievement test (pretest)
Dependent variable		Achievement test (posttest)
Control variable		Teacher, teaching time, learning contents

standard time for legal, commercial, and social purposes. Because the local time data obtained within a time zone are the same, the Sun position at the same latitude but different longitudes may be slightly different. However, it will not affect the Sun path since the Sun only appears earlier or later (within an hour) but following the same locus.

4. Teaching Experiment

A teaching experiment has been conducted to evaluate the learning effectiveness of students using traditional teaching aids (the celestial hemisphere, compass, and protractor) and the Sun path observation system. This study randomly selected two classes of 5th-grade students as the experimental samples, one class (28 students) as the control group and the other (28 students) as the experimental group, from an elementary school in northern Taiwan. The control group used traditional teaching aids and the experimental group used the Sun path observation system, respectively (Figure 10). A questionnaire survey was conducted by both groups to investigate the attitudes of students after using different tools for observation.

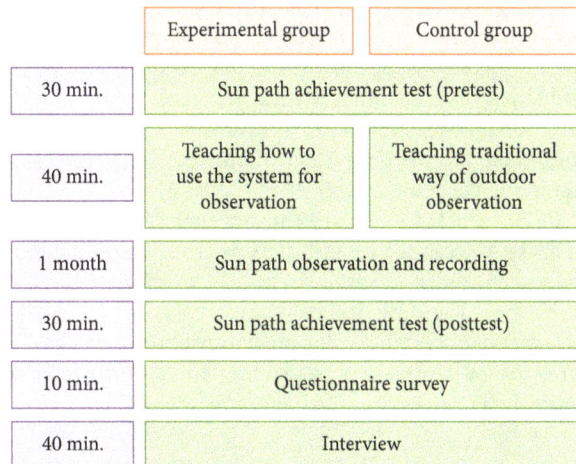

FIGURE 11: The flowchart of teaching experiment.

The achievement test used in this study contains 20 test items which were designed based on the teaching objectives, learning contents, and observation activities. The test items

TABLE 2: ANCOVA results of learning achievement for both groups.

| | Students | Average | | Progress | Standard deviation | f | Significance (p) |
		Pretest	Posttest				
Control group	28	40.18	78.75	38.57	7.77	10.83	0.002**
Experiment group	28	38.39	85.36	46.97	10.54		

** $p < 0.01$.

TABLE 3: Questionnaire results of using the teaching tool for the Sun path observation.

Question	Experimental group	Control group	Mean difference
(1) I am more interested in learning this subject than I did before by using the teaching tool	4.36	4.00	0.36*
(2) The teaching tool can help me find out new problems in learning this subject	3.82	3.61	0.21
(3) I would think about the observed phenomena in a new way by using the teaching tool	4.82	3.50	1.32***
(4) I like to use the teaching tool for learning this subject	4.36	3.71	0.65***
(5) I like to use the same way in learning other subjects	4.21	4.18	0.03
(6) I hope there is a chance to learn in this way in the future	4.89	3.68	1.21***
(7) I will recommend the teaching tool to other students	4.18	3.18	1.00***

* $p < 0.05$ and *** $p < 0.001$.

were revised according to the comments of two experts and two science teachers. The questionnaire used in this study is a system satisfaction survey according to the user's experiences after using the teaching tools for learning, and it was designed based on the discussion with experts and science teachers. This study adopted the 5-point Likert scale [20] (5 = strongly agree; 4 = agree; 3 = neutral; 2 = disagree; and 1 = strongly disagree).

This study adopted a nonequivalent pretest-posttest design to analyze if there is a significant difference in learning effectiveness between the two groups. Before the teaching activities, all students took a pretest on their background knowledge about the Sun path concepts. After conducting the teaching activities, they took the posttest and filled out the questionnaire. The variables of the teaching experiment are listed in Table 1, and the flowchart of the teaching experiment is shown in Figure 11.

In the teaching experiment, an analysis of covariance (ANCOVA) has been conducted to investigate if there is a significant difference in learning effectiveness between the two groups. The statistical software SPSS is used to analyze the test results (Table 2). For the control group, the average scores of pretest and posttest are 40.18 and 78.75, respectively. For the experimental group, the average scores of pretest and posttest are 38.39 and 85.36, respectively. The progress made by the experimental group (46.97) is higher than the progress made by the control group (38.57). The power size of the effect is the difference between the average outcomes in two different intervention groups. Cohen's d is the appropriate effect size measure if two groups have similar standard deviations and are of similar size, and it is calculated as

$$\text{Cohen's } d = \frac{(M_2 - M_1)}{\text{SD}} = \frac{(46.97 - 38.57)}{9.26} = 0.91, \quad (1)$$

where $\text{SD} = \sqrt{(\text{SD}_1^2 + \text{SD}_2^2)/2} = \sqrt{(7.77^2 + 10.54^2)/2} = 9.26$.

After excluding the effect of covariance (the background knowledge), the impact of independent variable (the teaching aids) is obtained as $f = 10.83$ and $p = 0.002 < 0.01$. A significant difference exists between the control group and the experimental group, and the learning achievement of the latter is higher than that of the former. In other words, the Sun path observation system is more effective than the traditional teaching aids for enhancing the learning achievement of students.

A statistical analysis is conducted on the questionnaire results by the two groups. The average scores in each selected item and their significance are computed for the seven questions to understand the attitudes of students after using the teaching aids to record the Sun path data (Table 3). The questionnaire results show that the experimental group was more satisfied with the tools they used than the control group.

The interview results from the teacher and the experimental group students are summarized in the following to support the validity of the questionnaire results.

(i) Most students considered the Sun path observation system very useful because it can be used to measure the orientation and elevation of the Sun easily without using the compass and protractor.

(ii) The function of automatic recording is helpful, and it can reduce the observation and recording time.

(iii) The functions of setting different observation dates and latitudes are very convenient because the students do not have to wait a long time or actually go there to record the Sun path data.

(iv) The function of combining several sets of data to compare the Sun paths in different seasons or at different latitudes can help them understand that the Sun path varies with latitudes and seasons.

(v) The system can be used in the classroom for observing and recording the Sun path to prevent the risk of heatstroke due to staying in the Sun for an extended period of time.

5. Conclusion

In this study, a Sun path observation system is developed using the AR technology and the sensor functions on mobile devices to help elementary school students observe and record the Sun path. The system has a built-in celestial model to simulate the Earth's rotation and its revolution around the Sun for calculating the Sun's position in the sky. When holding the mobile device toward the sky, the screen will show the virtual Sun in the same direction as that of the real Sun, and the students can record the Sun orientation and elevation as well as the relative data such as the observation date, time, longitude, and latitude. They can also set different observation times and locations to understand that the Sun path changes with seasons and latitudes.

A teaching experiment has been conducted to analyze the learning effectiveness of students using the Sun path observation system. A questionnaire survey has been conducted to understand the attitudes of students after using traditional teaching aids and the Sun path observation system. The experimental results show that the learning effectiveness of the experimental group is significantly higher than that of the control group. The questionnaire survey and interview results reveal that most students considered the system useful in recording the observation data, and they would recommend it to other students. They thought the system is easy to operate, and they would like to use it if there is a similar requirement in the future.

The system provides situated learning contents about the Sun path concepts, and it converts the observation data into organized and meaningful astronomical knowledge. It is useful for developing the spatial concepts required in learning the relative motion between the Sun and Earth. In addition, the system can solve the problem of being unable to observe the Sun path due to a bad weather or topographical restrictions. Therefore, it is a useful tool for astronomy education in elementary schools.

Acknowledgments

The authors wish to acknowledge the support by the Ministry of Science and Technology (MOST), Taiwan, under Contract nos. 103-2511-S-007-010-MY2 and 106B2096UB.

References

[1] J. Dunlop, "How children observe the Universe," *Publications of the Astronomical Society of Australia*, vol. 17, no. 2, pp. 194–206, 2000.

[2] R. W. Hollingworth and C. McLoughlin, "Developing science students' metacognitive problem solving skills online," *Australian Journal of Educational Technology*, vol. 17, no. 1, pp. 50–63, 2001.

[3] Kang Hsuan Educational Publishing Group, 2014, https://www.knsh.com.tw/Index.asp.

[4] K. J. Schoon, "Students' alternative conceptions of Earth and space," *Journal of Geological Education*, vol. 40, no. 3, pp. 209–214, 1992.

[5] J. Baxter, "Children's understanding of familiar astronomical events," *International Journal of Science Education*, vol. 11, no. 5, pp. 502–513, 1989.

[6] K. C. Trundle and R. L. Bell, "The use of a computer simulation to promote conceptual change: a quasi-experimental study," *Computers and Education*, vol. 54, no. 4, pp. 1078–1088, 2010.

[7] G. J. Hwang and C. C. Tsai, "Research trends in mobile and ubiquitous learning: a review of publications in selected journals from 2001 to 2010," *British Journal of Educational Technology*, vol. 42, no. 4, pp. E65–E70, 2011.

[8] C.-C. Chen and P.-H. Lin, "Development and evaluation of a context-aware ubiquitous learning environment for astronomy education," *Interactive Learning Environments*, vol. 24, no. 3, pp. 644–661, 2016.

[9] P. Sommerauer and O. Müller, "Augmented reality in informal learning environments: a field experiment in a mathematics exhibition," *Computers and Education*, vol. 79, pp. 59–68, 2014.

[10] M. B. Ibáñez, Á. Di Serio, D. Villarán, and C. D. Kloos, "Experimenting with electromagnetism using augmented reality: impact on flow student experience and educational effectiveness," *Computers and Education*, vol. 71, pp. 1–13, 2014.

[11] T. H. C. Chiang, S. J. H. Yang, and G. J. Hwang, "Students' online interactive patterns in augmented reality-based inquiry activities," *Computers and Education*, vol. 78, pp. 97–108, 2014.

[12] H. Sollervall, "Collaborative mathematical inquiry with augmented reality," *Research and Practice of Technology Enhanced Learning*, vol. 7, no. 3, pp. 153–173, 2012.

[13] D. M. Bressler and A. M. Bodzin, "A mixed methods assessment of students' flow experiences during a mobile augmented reality science game," *Journal of Computer Assisted Learning*, vol. 29, no. 6, pp. 505–517, 2013.

[14] H.-K. Wu, S. W.-Y. Lee, H.-Y. Chang, and J.-C. Liang, "Current status, opportunities and challenges of augmented reality in education," *Computers and Education*, vol. 62, pp. 41–49, 2013.

[15] H.-Y. Chang, H.-K. Wu, and Y.-S. Hsu, "Integrating a mobile augmented reality activity to contextualize student learning of a socioscientific issue," *British Journal of Educational Technology*, vol. 44, no. 3, pp. E95–E99, 2013.

[16] N. Drljevic, L. H. Wong, and I. Boticki, "Where does my augmented reality learning experience (ARLE) belong? A student and teacher perspective to positioning ARLEs," *IEEE Transactions on Learning Technologies*, vol. 10, no. 4, pp. 419–435, 2017.

[17] M. E. C. Santos, A. Chen, T. Taketomi, G. Yamamoto, J. Miyazaki, and H. Kato, "Augmented reality learning

experiences: survey of prototype design and evaluation," *IEEE Transactions on Learning Technologies*, vol. 7, no. 1, pp. 38–56, 2014.

[18] W. Tarng, Y.-S. Lin, C.-P. Lin, and K.-L. Ou, "Development of a lunar-phase observation system based on augmented reality and mobile learning technologies," *Mobile Information Systems*, vol. 2016, Article ID 8352791, 12 pages, 2016.

[19] L. Bot, P.-B. Gossiaux, C.-P. Rauch, and S. Tabiou, "Learning by doing: a teaching method for active learning in scientific graduate education," *European Journal of Engineering Education*, vol. 30, no. 1, pp. 105–119, 2005.

[20] R. Likert, "A technique for the measurement of attitudes," *Archives of Psychology*, vol. 22, no. 140, p. 55, 1932.

UWB/PDR Tightly Coupled Navigation with Robust Extended Kalman Filter for NLOS Environments

Xin Li [1,2] **Yan Wang** [1] **and Kourosh Khoshelham** [2]

[1]*School of Computer Science and Technology, China University of Mining and Technology, Xuzhou 221116, China*
[2]*Department of Infrastructure Engineering of the University of Melbourne, Melbourne 3010, Australia*

Correspondence should be addressed to Xin Li; linuxcumt@126.com

Academic Editor: Paolo Bellavista

The fusion of ultra-wideband (UWB) and inertial measurement unit (IMU) is an effective solution to overcome the challenges of UWB in nonline-of-sight (NLOS) conditions and error accumulation of inertial positioning in indoor environments. However, existing systems are based on foot-mounted or body-worn IMUs, which limit the application of the system to specific practical scenarios. In this paper, we propose the fusion of UWB and pedestrian dead reckoning (PDR) using smartphone IMU, which has the potential to provide a universal solution to indoor positioning. The PDR algorithm is based on low-pass filtering of acceleration data and time thresholding to estimate the step length. According to different movement patterns of pedestrians, such as walking and running, several step models are comparatively analyzed to determine the appropriate model and related parameters of the step length. For the PDR direction calculation, the Madgwick algorithm is adopted to improve the calculation accuracy of the heading algorithm. The proposed UWB/PDR fusion algorithm is based on the extended Kalman filter (EKF), in which the Mahalanobis distance from the observation to the prior distribution is used to suppress the influence of abnormal UWB data on the positioning results. Experimental results show that the algorithm is robust to the intermittent noise, continuous noise, signal interruption, and other abnormalities of the UWB data.

1. Introduction

Indoor positioning technology has a wide range of applications from supermarket shopping assistance to drone positioning and patient tracking in hospitals [1–3]. Among various approaches, the ultra-wideband (UWB) positioning technology is particularly attractive because of its decimeter-level positioning accuracy. However, in many practical scenarios, such as warehouse robot positioning and emergency response in crowded scenes, UWB signals may be blocked by people, cargos, or other obstacles, which will cause signal problems such as multipath effect, strength attenuation, and even signal loss, resulting in a sharp drop in the UWB positioning accuracy [4, 5]. The fusion of inertial measurement unit (IMU) in a pedestrian dead reckoning (PDR) method and UWB is an effective way to achieve high-precision positioning in nonline-of-sight (NLOS) environments. While UWB provides accurate absolute positioning

in line-of-sight conditions, the PDR ensures a continuous and smooth trajectory in periods of UWB signal loss.

There are a few works in the literature on combining UWB and inertial sensors. Fan et al. [6, 7] developed a loosely coupled fusion method based on the EKF to track the pedestrian movement. However, at least three effective UWB range measurements are required for this method, which may not be available under the NLOS conditions. Some scholars have presented a tightly coupled method which combines the UWB range measurement with the IMU measurements [8–10]. Although the positioning precision and stability are indeed improved by these methods, the signal noise and interruption of the beacons are not considered. Benini et al. [11] proposed an IMU/UWB/vision fusion method for the drone positioning, with two-dimensional positioning accuracy of 10 cm. Wang et al. [12] designed a tightly coupled GPS/UWB/IMU integrated system based on the adaptive robust Kalman filter.

González et al. [13] used particle filter algorithm to fuse the data of UWB, IMU, and odometer, achieving good positioning stability under the NLOS conditions. However, the particle filter algorithm is difficult to be realized in an embedded system due to its high computational complexity. The fusion of UWB and PDR has been proposed in a few other works [14–17]. However, existing UWB/PDR fusion methods are based on foot-mounted or body-worn IMUs. While the positioning accuracy of PDR with foot-mounted or body-worn IMU is higher than that of smartphone-based PDR, due to the use of zero-velocity update, these systems are generally not scalable and are limited to a few specific application scenarios.

There are many methods for outliers detection. The most direct method is to reject the observations with large residual directly. However, this method may lead to a lack of continuity in the estimates of the covariance [18]. Median-based filters can also be used to realize robust EKF, and this kind of filter may be highly robust, but not efficient. This filter is based on a block-based batch-reprocessing manner and is difficult for real-time applications [19]. By minimizing the estimation error of the worst case, H-infinity filter can control the influence of some uncertain observations on the system state, but it fails when large outliers exist [20]. M-estimation-based robust filters have been widely studied in recent years, such as the Huber-based unscented EKF is generalized to cope with the case that outliers in both the prediction and observation coexist simultaneously [21]. But M-estimation-based robust filters achieve robustness at the cost of reducing the accuracy of the nonlinear transformation itself. Essentially the M-estimation-based filters may be equivalent in some sense to the method of observation-noise inflating or observation-trimming, but with relatively computational complexities [22].

To address the above challenges, in this paper, we propose a tightly coupled UWB/PDR fusion method based on smartphone IMU for positioning in indoor environments. To the best of our knowledge, the proposed method is the first to address the UWB/PDR fusion using smartphone IMU. We also adopt the Madgwick algorithm for heading estimation in smartphone-based PDR and show that a significant improvement in heading estimation accuracy can be achieved. Finally, we propose the use of Mahalanobis distance from the observation to the prior distribution to suppress the influence of abnormal UWB data on the positioning results.

The remainder of the paper is organized as follows: In Section 2, the PDR algorithm based on smartphone inertial sensor is discussed and the theoretical analysis and experimental results of step detection, step length calculation, and heading calculation are given. In Section 3, the UWB/PDR fusion algorithm based on the EKF is studied, emphasizing the principle of the robust algorithm based on the Mahalanobis distance. In Section 4, several experiments are presented and the results are analyzed. In Section 5, the paper is summarized.

2. Overview of the Proposed Method

An overview of the proposed method for UWB/PDR fusion is shown in Figure 1. Using the IMU data, steps are detected

based on a combination of low-pass filtering, acceleration, and time threshold. Considering two different movement patterns, walking and running, the estimation of step length by using various models is compared and analyzed, and the optimal method for step length estimation and the corresponding parameters for different movement patterns are identified. Real-time heading estimation is performed by adopting the Madgwick algorithm [23], which improves the accuracy of heading estimation and enhances the positioning accuracy of the PDR algorithm.

The UWB/PDR fusion is done based on the EKF algorithm. To make the fusion robust against abnormal and outlying sensor readings, the use of the Mahalanobis distance from the observation to the prior distribution is proposed. As shown by experimental evaluation, the proposed method is robust to intermittent noise, continuous noise, signal interruption, and other abnormalities of the UWB data.

3. PDR Algorithm Based on Smartphone Inertial Sensors

The PDR algorithm consists of step detection, step length estimation, and heading estimation. Based on the movement characteristics of the pedestrian, the accelerometer data are used to measure the number of steps and estimate the step length. Then, in combination with the heading information obtained from the IMU, the current position of the pedestrian is calculated by using equation (1).

$$\begin{cases} N_{i+1} = N_i + s_t \cos \psi_t, \\ E_{t+1} = E_t + s_t \sin \psi_t, \end{cases} \quad (1)$$

where (N, E) are the position coordinates of the pedestrian, s is the step length, and ψ is the heading angle.

3.1. Step Detection. Walking is characterized by a periodic acceleration pattern. With the foot from lifting to landing, that is, with the gravity center from rising to falling, the acceleration data in the vertical direction follow a curve from peak to trough. Therefore, the detection of a step can be done based on acceleration measurements. Smooth region detection [24], zero-crossing detection [25], peak detection [26, 27], and autocorrelation method [28] are commonly used methods.

In this work, the method proposed in [29] is adopted, where the acquired vertical acceleration is first processed by low-pass filtering, so that spurious peaks are removed from the waveform of the acceleration data. Since the filtering order, the pass-band frequency, and the stop-band frequency are all set empirically, sporadic pseudopeaks and pseudotroughs might still exist in the waveform. Therefore, using the filtered data, the amplitude difference and the time difference between consecutive filtered values are used to identify the correct maxima and minima, thereby performing step detection, as described in [30]. The detection result is shown in Figure 2.

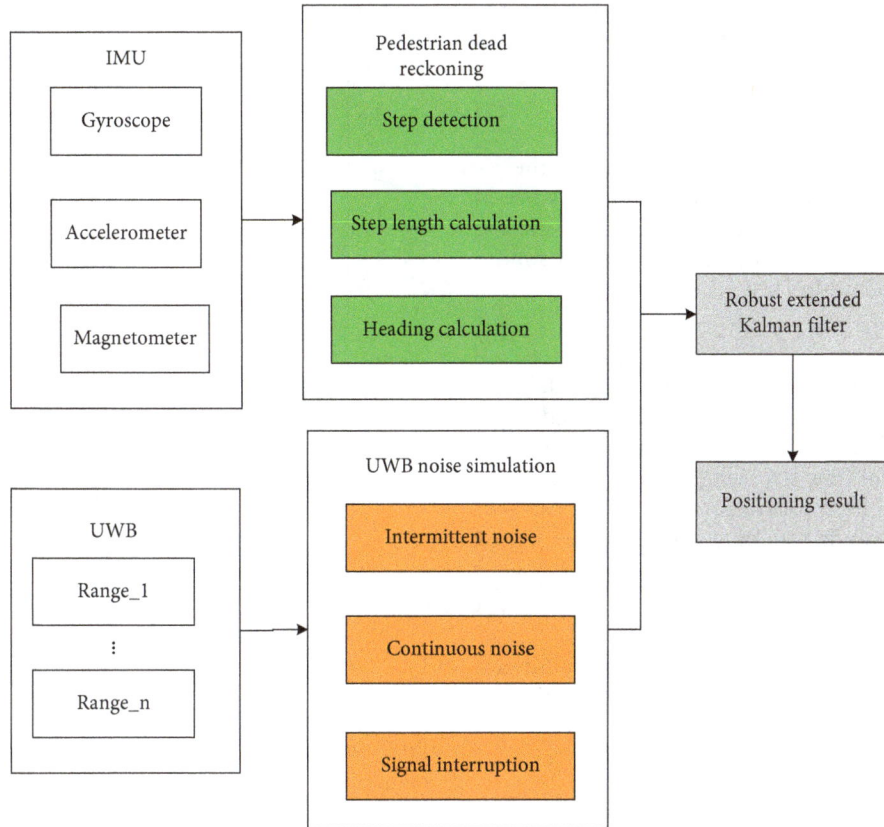

FIGURE 1: Framework of the research.

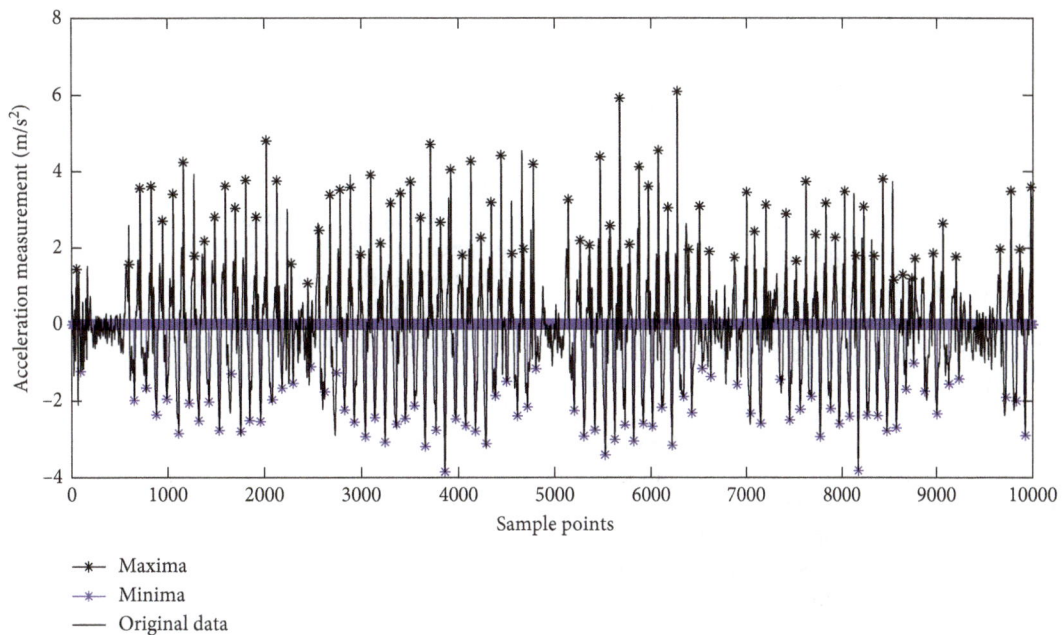

FIGURE 2: Identification of maxima and minima in the step detection.

3.2. Step Length Calculation. The commonly used step length calculation models can be roughly divided into three types: constant model, linear model, and nonlinear model. Among them, the constant model is the simplest one; however, it has the drawback that it cannot adapt to the real-time changes of the pedestrian velocity. A statistical table of step length at different levels is presented in [31], and the fixed step length can be selected from the table according to the movement pattern characteristics and the step frequency.

For the linear model, Levi and Judd [32] proposed the linear relationship between step length and step frequency as follows:

$$s = A + B \times f + C \times \text{var} + w, \tag{2}$$

where f is the step frequency of each step, that is, the reciprocal of the step cycle; var is the acceleration variance of each step; and w is the Gaussian noise. A and C are the regression coefficients obtained through training. A similar linear model is proposed in [33]:

$$s = A + B \times p + C \times \overline{a}_{\max}, \tag{3}$$

where p indicates the step cycle of each step, $\overline{a}_{\max}$ is the peak value of the acceleration after smoothing, and A and C are the regression coefficients obtained through training. The calculation model for the step length from a nonlinear perspective has also been analyzed, and a nonlinear step length calculation formula with only one parameter is used in [34]:

$$s = K \times \sqrt[4]{a_{\max} - a_{\min}}, \tag{4}$$

where $a_{\max}$ and $a_{\min}$ represent the maximum and minimum acceleration of each step, respectively, and K is the regression coefficient.

3.3. Heading Calculation by the Madgwick Algorithm. When the pedestrian walks with the smartphone held horizontally, the azimuth of the smartphone can be regarded as the heading angle of the pedestrian. The azimuth can be calculated by the built-in algorithm based on the gyroscope or the magnetometer in the smartphone. However, this heading calculation is largely influenced by the smartphone pose and the user's body movement. Figure 3 shows an example where a slight sway of the body causes a deviation of the calculated heading angle (the green line) from the actual heading angle (the blue line) reaching nearly 40 degrees.

In order to improve the accuracy of the heading calculation, the state-of-art Attitude and Heading Reference Systems (AHRS), i.e., the Madgwick algorithm, is introduced into the PDR algorithm in this paper. In this algorithm, in order to estimate the direction, the measurements of accelerometer, gyroscope, and magnetometer are combined with two absolute fields, the geomagnetic field and the gravity field, whose directions and intensities are known. The advantage of the AHRS algorithm is that the attitude error can be compensated continuously, improving the accuracy of the direction. As can be seen from Figure 3, the calculation result (the pink line) is basically consistent with the reference heading. The Madgwick algorithm consists of two steps as described below.

3.3.1. Initial Direction Calculation. Initially, it is assumed that the sensor is either in a stationary state or moves at a constant velocity, so that only the gravity vector is measured by the accelerometer. In addition, it is assumed that the magnetic field is nondisturbed, and therefore, only the geomagnetic field is measured by the magnetometer. The

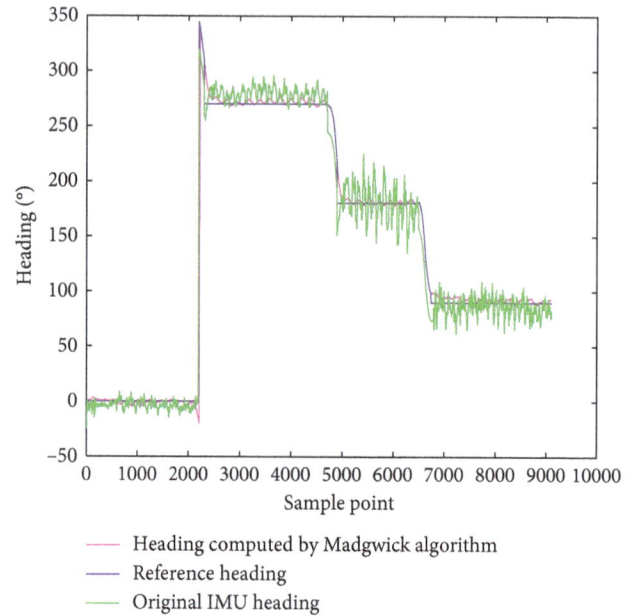

FIGURE 3: Comparison of the heading angles.

pitch (θ) and roll (ϕ) can be derived from $[f_x^b, f_y^b, f_z^b]$, the component of the gravity vector on the three axes.

$$\phi = \tan\left(\frac{f_y^b}{f_z^b}\right)^{-1},$$

$$\theta = -\sin\left(\frac{f_x^b}{g}\right)^{-1}. \tag{5}$$

The value of yaw (ψ) cannot be obtained by the gravity vector, but can be obtained by the magnetometer. The measurement of the magnetometer is recorded as $m^b = [m_x^b, m_y^b, m_z^b]$, representing the projection of the geomagnetic field on the three axes of the body frame (b-frame). Since pitch (θ) and roll (ϕ) are known, m^b can be converted onto the XOY plane of the navigation frame (n-frame), with its orthogonal component recorded as $m^n = [m_x^n, m_y^n, m_z^n]$. The conversion relationship between the two sets of magnetic data is

$$\begin{bmatrix} m_x^n \\ m_y^n \\ m_z^n \end{bmatrix} = \begin{bmatrix} \cos(\theta) & 0 & -\sin(\theta) \\ 0 & 1 & 0 \\ in(\theta) & 0 & \cos(\theta) \end{bmatrix}^{-1} \begin{bmatrix} 1 & 0 & 0 \\ 0 & \cos(\phi) & 0 \\ 0 & 0 & \cos(\phi) \end{bmatrix}^{-1} \begin{bmatrix} m_x^b \\ m_y^b \\ m_z^b \end{bmatrix}. \tag{6}$$

In the geographic frame, the b_x points to the north, and the b_y pointing to the east is equal to 0, so the geomagnetic field data are $b = [b_x, 0, b_z]$. Based on the angular transformation obtained by the accelerometer, the geomagnetic data in the n-frame and geographic frame are transformed onto the same XOY plane, and the angle difference between the two transformed data is the heading angle ψ, as shown in Figure 4. According to the trigonometric relationship, Formulas (7) and (8) are obtained.

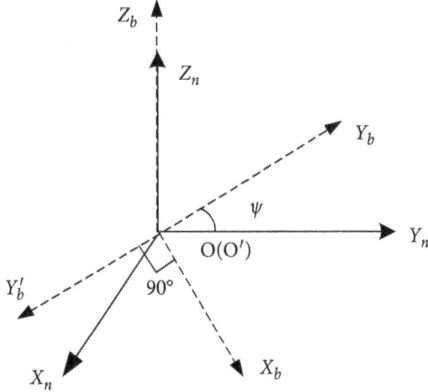

FIGURE 4: Navigation frame and geographic frame.

$$b_x = \text{sqrt}\left[\left(m_x^b\right)^2 + \left(m_y^b\right)^2\right], \qquad (7)$$

$$b_z = m_z^n. \qquad (8)$$

The relationship of the geomagnetic data between b-frame and geographic frame is

$$b = C_b^n m^b. \qquad (9)$$

The heading angle can be calculated by Formula (9):

$$\psi = \arctan\left(\frac{m_x^b \cos(\phi) + m_z^b \sin(\phi)}{[m_x^b \sin(\phi) - m_z^b \cos(\phi)]\sin(\theta) - m_y^b \cos(\theta)}\right). \qquad (10)$$

3.3.2. Attitude Calculation. For a specific vector, its magnitude and direction are the same when it is represented by different coordinate systems. This is the principle of the attitude calculation. However, error exists in the transformation matrix between the two coordinate systems. When a vector is transformed by a rotation matrix with error, a deviation between the transformed result and the original vector will appear. This deviation can be used to correct the rotation matrix, and the attitude is thus corrected. In the Madgwick algorithm, this rotation matrix is represented by a quaternion, and the attitude is calculated by modifying the quaternion. The accelerometer and the magnetometer are the main measured objects in the attitude calculation process. In the Madgwick algorithm, the current attitude is updated by the gyroscope:

$$q_{t+1} = q_t + 0.5 q_t \omega_t d_t, \qquad (11)$$

where q is the attitude of the sensor. The attitude error is corrected by calculating the difference between the accelerometer's measurement and the magnetometer's measurement under n-frame and b-frame. Assume that $e = [0, e_x, e_y, e_z]$ and $s = [0, s_x, s_y, s_z]$ are the vector coordinates under n-frame and b-frame, respectively. Then for the accelerometer, $e = [0, 0, 0, 1]$ and $s = [0, f_x^b, f_y^b, f_z^b]$; for the magnetometer

$e = [0, b_x, 0, b_z]$ and $s = [0, m_x^b, m_y^b, m_z^b]$. The transformation relationship is

$$s = q^{-1} e q. \qquad (12)$$

The error equation is defined as

$$E(q) = q^{-1} e q - s. \qquad (13)$$

The above formula is solved by the Gauss–Newton method:

$$q_{t+1} = q_t - \mu_t \frac{\nabla E}{||\nabla E||}, \qquad (14)$$

where μ_t is the step length and ∇ is the differential operator. Equations (11) and (14) are combined, and

$$q_{t+1} = q_t + 0.5 q_t \omega_t d_t - \beta \frac{\nabla E}{||\nabla E||} d_t, \qquad (15)$$

where β in equation (15) represents the weight, which is used to measure the error of the result calculated by the angular velocity.

4. Tightly Coupled UWB/PDR Fusion Algorithm

The UWB provides absolute positioning, but its performance is affected by the NLOS conditions. In contrast, the PDR method based on smartphone IMU data provides relative positioning characterized by error accumulation, but it is independent of the environmental conditions. Therefore, a UWB/PDR fusion algorithm is expected to outperform the individual techniques. In this section, a UWB/PDR fusion algorithm based on the robust EKF is presented.

4.1. UWB/PDR Fusion Algorithm. The acceleration data of the IMU are firstly used to determine whether the state is stationary or moving. At the beginning of the positioning, the pedestrian stands still for a few seconds. Gross error of the UWB positioning result in the stationary state is eliminated, and the average value is calculated as the starting position of the PDR algorithm. The initial orientation angle is obtained by using the Madgwick algorithm, and then, the PDR calculation is performed according to Formula (1). In the EKF fusion algorithm, the calculation position and the heading of the PDR algorithm are the state variables of the filter system, namely,

$$X_t = \left[N_t, E_t, \psi_t\right]^T. \qquad (16)$$

The state equation of the EKF algorithm is

$$\begin{cases} N_{t+1} = N_t + s_t \cos\psi_t, \\ E_{t+1} = E_t + s_t \sin\psi_t, \\ \psi_{t+1} = \psi_t + \omega_t d_t, \end{cases} \qquad (17)$$

where N_t and E_t are the north direction value and the east direction value of the terminal coordinates, respectively; s_t is the step length of each step obtained by step detection; ψ_t is the angle in the horizontal movement direction; and ω_t is the average angular velocity of the whole step. Derivative of Formula (17) is taken, and the state transition matrix is obtained:

$$F_t = \begin{bmatrix} 1 & 0 & -s_t \sin \psi_t \\ 0 & 1 & s_t \cos \psi_t \\ 0 & 0 & 1 \end{bmatrix}. \tag{18}$$

It is assumed that the position coordinate and the dynamic noise of the heading obey a Gaussian distribution with the mean of 0 and the covariance of $\sigma_N^2, \sigma_E^2, \sigma_\psi^2$. In actual applications, σ_ψ^2 is larger when turning and is smaller when going straight. Accordingly, by determining the pedestrian's movement pattern, such as going straight or turning, the dynamic noise of the variables related to the heading is adaptively determined. The dynamic noise matrix is

$$Q_t = \begin{bmatrix} \sigma_N^2 & 0 & 0 \\ 0 & \sigma_E^2 & 0 \\ 0 & 0 & \sigma_\psi^2 \end{bmatrix}, \tag{19}$$

where $\sigma_N^2 = \sigma_E^2 = 2$. When the pedestrian is going straight, $\sigma_\psi^2 = (2°)^2$; when the pedestrian is turning, $\sigma_\psi^2 = (15°)^2$.

The UWB ranging measurement $r_{t,i}$ and the heading angle $\tilde{\psi}_t$ calculated by the Madgwick algorithm are observations. Given n beacons with 2D coordinates denoted as $B_i = (B_{x,i}, B_{y,i})$, $i \in (1, n)$, the observation function is defined as

$$h(X_t) = \begin{bmatrix} \|B_1 - p_t\| \\ \vdots \\ \|B_n - p_t\| \\ \psi_t + \omega_t d_t \end{bmatrix}, \tag{20}$$

where $p_t = (x_t, y_t)$ is the position coordinates calculated by the PDR and $\|.\|$ represents the Euclidean distance. The Jacobian matrix of the observation equation is defined as

$$H_t = \begin{bmatrix} \dfrac{B_{x,1} - x_t}{\|B_1 - p_t\|} & \dfrac{B_{y,1} - y_t}{\|B_1 - p_t\|} & 0 \\ \vdots & \vdots & \vdots \\ \dfrac{B_{x,n} - x_t}{\|B_n - p_t\|} & \dfrac{B_{y,n} - y_t}{\|B_n - p_t\|} & 0 \\ 0 & 0 & 1 \end{bmatrix}. \tag{21}$$

The measurement noise matrix is

$$R = \begin{bmatrix} \sigma_{r_1}^2 & 0 & \cdots & 0 \\ 0 & \ddots & \cdots & 0 \\ \vdots & \cdots & \sigma_{r_n}^2 & 0 \\ 0 & \cdots & 0 & \sigma_\psi^2 \end{bmatrix}, \tag{22}$$

$$\sigma_{r_i}^2 = (0.4)^2.$$

4.2. Robustness to Abnormal UWB Range Measurements.

Suppose the UWB observation obeys Gaussian distribution, that is, $r_{t,i}$ obeys a Gaussian distribution with the mean of $h_i(X_t)$ and the variance of $H_{t,i} P_t^- H_{t,i}^T - R_{t,i}$, where P_t^- is the a priori covariance matrix of the state variables. Therefore, $\gamma_{t,i}$, the square of the Mahalanobis distance between $r_{t,i}$ and $h_i(X_t)$ obeys the χ^2 distribution [35], i.e.,

$$\gamma_{t,i} = (m_{t,i} - h_i(X_t))^T (H_{t,i} P_t^- H_{t,i}^T + R_{t,i})^{-1} (m_{t,i} - h_i(X_t))$$
$$\sim \chi_1^2, \tag{23}$$

where χ_1^2 represents a χ^2 distribution with the freedom degree of 1. That is to say, given the significance level α, then

$$\Pr(\gamma_{t,i} < \chi_{1,\alpha}^2) = 1 - \alpha, \tag{24}$$

where $\Pr()$ represents the probability of a random event and $\chi_{1,\alpha}^2$ is the α-quantile of the χ^2 distribution with the freedom degree of 1. In this paper, the significance level α is set to 0.001, and $\chi_{1,\alpha}^2$ is determined to be 6.2 according to the chi-square distribution table. Observations that do not satisfy the condition are considered as outliers. The effect of the abnormal observations on the posterior estimation can be attenuated by increasing the covariance of the abnormal values. Algorithm 1 is the robust EKF algorithm.

For observation with larger Mahalanobis distance, its covariance is required to be increased to reduce its influence on the posterior estimation. The covariance matrix for the new observation can be updated according to the following formula:

$$R'_{t,i} = \begin{cases} R_{t,i}, & \gamma_{t,i} < \chi_{1,\alpha}^2, \\ \left(\dfrac{\gamma_{t,i}}{\chi_{1,\alpha}^2}\right) \times R_{t,i}, & \gamma_{t,i} \geq \chi_{1,\alpha}^2, \end{cases} \tag{25}$$

where $\gamma_{t,i}/\chi_{1,\alpha}^2$ is the ratio of the Mahalanobis distance to the threshold of the observation at the current operating point and covariance. In each iteration process, the Mahalanobis distance between the observation and the current operating point is calculated according to the system state of the current operating point, and the observation covariance is updated accordingly. Before solving the posterior state, the observation covariance is firstly determined by the prior state, which is different from the standard Kalman filter.

For computational complexity, REKF is just added lines 2–7 for robust processing with respect to EKF. The problem here is to calculate the number of while loops. From the fourth line of the algorithm, we can see that each cycle $R_{t,i}$ is getting larger, and from Formula (23), the value of denominators is increasing, the value of numerator is unchanged, so each step $\gamma_{t,i}$ is getting smaller. That is, while loops can be finished within a limited number of loops (usually no more than 10 times in experiments), so they should be equivalent to EKF in terms of computational complexity.

Filtering: for $t = 1, 2, \ldots$
(1) State prediction
(2) **For** each measurement
(3)   **while** $\gamma_{t,i} \geq \chi^2_{1,\alpha}$
(4) $R_{t,i} = (\gamma_{t,i}/\chi^2_{1,\alpha}) \times R_{t,i}$
(5)   Recalculate $\gamma_{t,i}$ based on Formula (23)
(6)   **end**
(7) **end**
(8) State update

ALGORITHM 1: REKF algorithm Pseudocode.

5. Experiments

5.1. Experimental Setup. The test site is the underground garage of the University of Melbourne. As shown in Figure 5(a), 4 UWB beacons were placed at the four corners of the rectangular area. MATE9 of Huawei, China, is selected as the experimental mobile smartphone, as shown in Figure 5(b). The chip of BeSpoon, France, is used as the core chip of the UWB tag/beacon. The UWB tag and the smartphone are connected via a serial port. In order to keep the data of the IMU and the UWB synchronized, the ranging data from both the IMU and the UWB are simultaneously received by the smartphone. During the walking process, the smartphone is steady held in the hand of the pedestrian, basically keeping a uniform motion state. The azimuth of the smartphone is taken as the heading angle of the human body.

Two routes are set in the experiment: one is a rectangular route with fewer turnings and the other is an 8-shaped route with more turnings, as shown in Figure 6. In the experiment, two laps are taken along each route in a clockwise direction. The start point and the end point are marked in the figure. The rectangular points in the figure denote the location of the four beacons, and the four red circles represent the four columns of the underground garage.

5.2. Step Detection and Step Length Calculation

5.2.1. Step Detection. The PDR experiment was conducted by two boys and two girls. The height and weight of the two boys are 1.8 meters and 70 kilograms and 1.78 meters and 72 kilograms, respectively. The height and weight of the two girls are 1.65 meters and 55 kilograms and 1.64 meters and 50 kilograms, respectively. The step detection result is shown in Figure 7.

In the table, "Walk" represents the normal walking pattern and "Run" represents the running pattern, including split-step running, stride running, and normal running. It can be seen from the accuracy rate that when the user is walking forward at normal velocity, the accuracy rate of step detection is up to 98% or more, with an average of 99.2%, and the number of wrongly detected step is within 2; when the user is running forward at different velocities, the average accuracy rate is 97.2%, with the minimum accuracy rate of 94.2%, and the maximum wrongly detected step

number is 5. The results show that this method can be well applied to the normal walking pattern, while for the abnormal movement patterns, the parameters of step detection are required to be further studied and optimized.

5.2.2. Step Length Calculation. Formulas (2) and (3) of the linear model and Formula (4) of the nonlinear model based on acceleration statistics are verified in this paper. Four persons of different heights and different body sizes are selected to move in different patterns, including normal walking pattern and running pattern, in order to study the applicability and reliability of the three models in different movement patterns.

Seven sets of data for the normal walking pattern, including walking along the straight line, walking along the fold lines, etc., are firstly selected. All the step lengths are accumulated to calculate the pedestrians' movement distance, and the calculated value is compared to the actual distance. The least square method is used to solve the regression coefficients of the linear model and the coefficient of the nonlinear model. The regression coefficients of the linear model are $A_1 = 0.678$, $B_1 = -0.036$, $C_1 = 0.031$, $A_2 = 0.409$, $B_2 = 0.011$, $C_2 = 0.068$. The coefficient of the nonlinear model is $K = 0.425$. Then the step length is calculated, as shown in the following table.

Table 1 shows that the average absolute distance differences of the three models are 1.555 m, 0.931 m, and 2.032 m, respectively, with the maximum error no more than 5 meters. The variances of the absolute distance differences of the three models are 1.50, 0.61, and 2.04, respectively, indicating that the linear model is the most stable one with the smallest error. Similarly, four persons of different heights and different body sizes are selected to move in running patterns, including running along the straight line and running along the fold lines. Seven sets of running data are collected, and the corresponding movement distances are calculated, as shown in Table 2.

The average absolute distance differences obtained by the three models are 5.298 m, 4.973 m, and 4.713 m, respectively, and the one calculated by Formula (4) is 89% and 95% of that calculated by the other two models, respectively. The variances of the absolute distance differences of the three models are 19.74, 9.24, and 8.23, respectively, indicating that the nonlinear model is the most stable one with the highest accuracy.

In summary, when a pedestrian moves in a normal gait, the linear model corresponding to Formula (3) is preferred for calculating the step length; when the pedestrian moves in an abnormal gait, i.e., in the running pattern, the nonlinear model corresponding to Formula (4) is preferred. In practical applications, the movement pattern of the pedestrian should be firstly identified, and then the appropriate model is selected according to the identification result.

5.3. Fusion Positioning Using Original UWB Data

5.3.1. Analysis of Original UWB Data. As can be seen from Figure 8, some UWB ranging data of Route 1 and Route 2 are

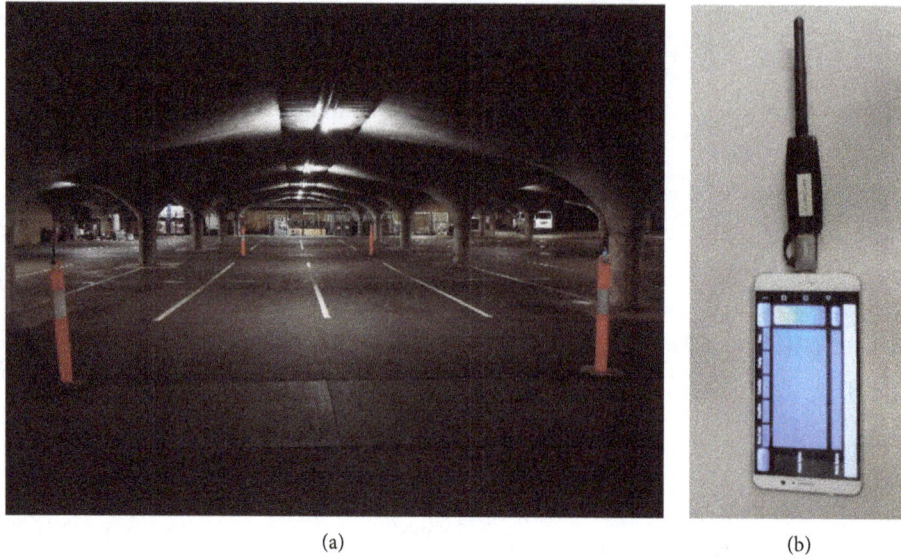

(a)   (b)

FIGURE 5: Experimental setup and equipment. (a) The underground garage and (b) UWB tag and smartphone.

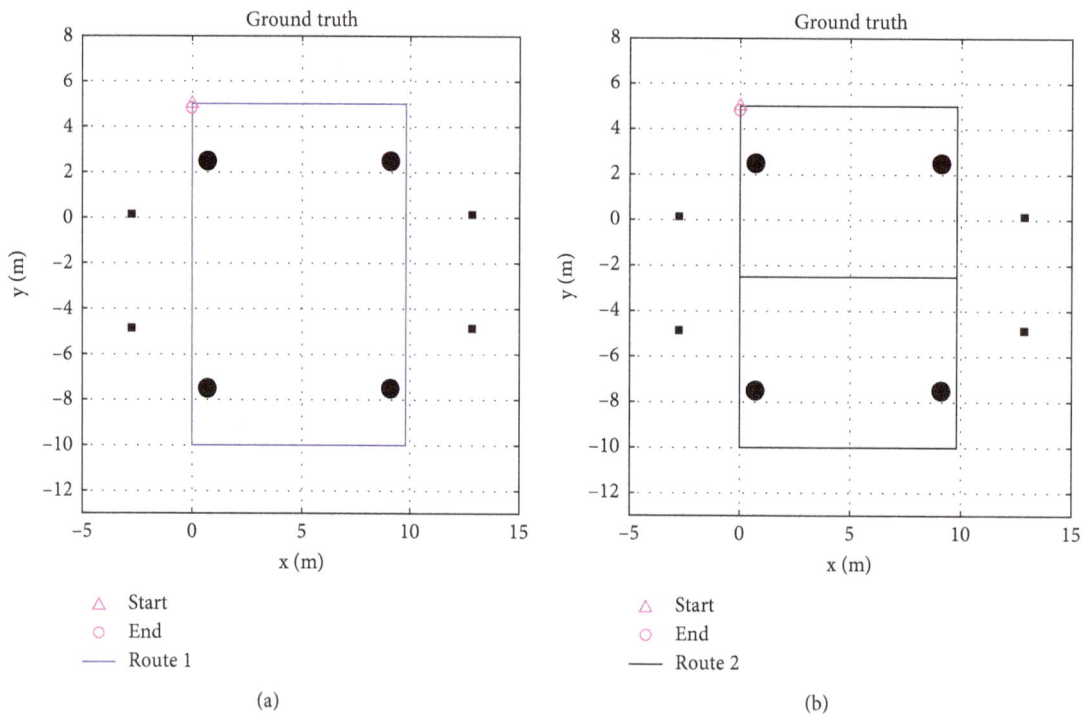

(a)   (b)

FIGURE 6: Experimental routes. (a) Route 1 and (b) Route 2.

blocked by the column, resulting in discontinuous and abrupt range measurement. More abnormal data appear on Route 2 due to the complexity of the route.

5.3.2. Analysis of the Estimated Trajectory. Figure 9 shows the positioning results of Route 1 and Route 2. In the figure, UWB represents the positioning result of the raw UWB data; PDR represents the positioning result of the smartphone IMU; EKF represents the UWB/PDR fusion positioning

result of the EKF algorithm; REKF represents the UWB/PDR fusion positioning result of the EKF algorithm based on the Mahalanobis distance.

In subgraph (a), for the UWB, most of the positioning points are consistent with the reference trajectory, but due to the blocking of the column and the pedestrian during walking, some abnormal points appear in the positioning result. For the PDR, the trajectory is relatively smooth, but due to the cumulative error of step length and direction, the overall positioning trajectory deviates from the reference

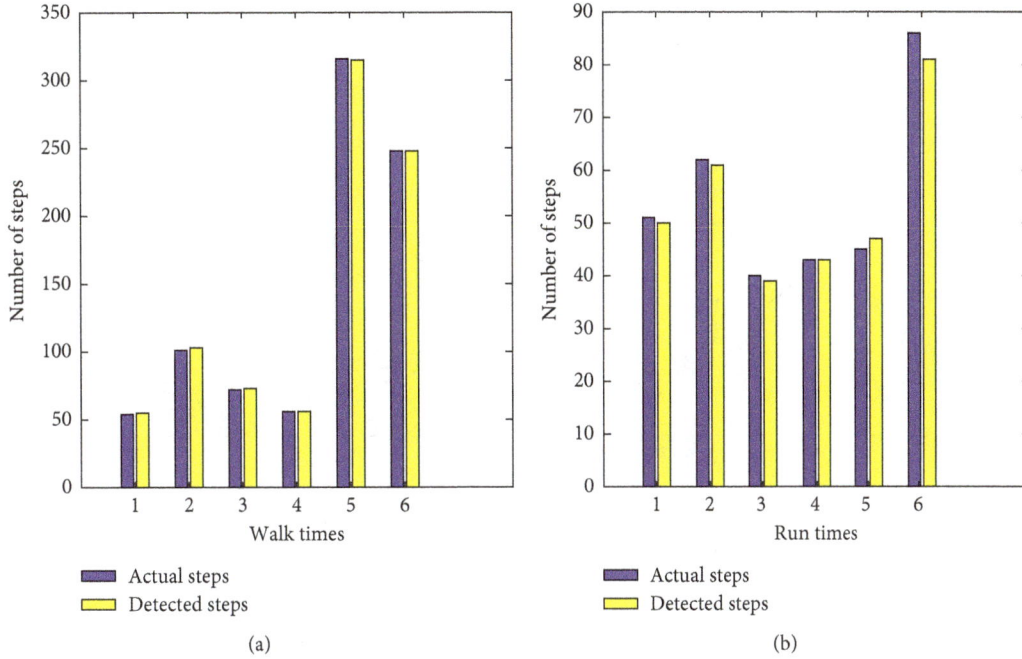

FIGURE 7: Step detection result.

TABLE 1: Normal gait step length calculation (unit: m).

Actual distance	40.5	71.4	43.2	39.5	81.6	211.68	179
Constant model	38.608	73.766	43.348	40.421	81.526	213.771	182.391
Linear model	41.411	71.372	44.799	38.894	82.381	209.381	178.707
Nonlinear model	38.427	73.065	46.227	39.895	81.520	215.612	175.950

TABLE 2: Abnormal gait step length calculation (unit: m).

Actual distance	55.2	52.5	40.5	40.8	31.8	39.5	81.6
Constant model	46.183	52.621	36.437	40.631	35.907	51.297	73.788
Linear model	46.865	54. 992	36.072	42.554	38.021	46.854	76.173
Nonlinear model	46.377	54.656	37.393	40.131	37.717	46.490	76.270

one. For the EKF, advantages of UWB and PDR algorithms are used and most of the positioning result is consistent with the reference trajectory, but several large jumps appear since the abnormal UWB cannot be processed by this algorithm. For the REKF, the positioning result almost coincides with the reference trajectory, presenting strong ability to process abnormal data.

Route 2 is more complex than Route 1. Therefore, in subgraph (b), for the UWB, more data are blocked by the pedestrian and the column, resulting in more abnormal points on the positioning result. For the PDR, the accumulation error of the direction grows with the increase of the number of turnings, resulting in a large shift in the overall trajectory. For the EKF, there are many jumps on the entire trajectory since the algorithm is unavoidably disturbed by the abnormal data. For the REKF, it still shows strong antinoise ability, with its positioning result basically consistent with the reference trajectory.

Table 3 shows the positioning results of different algorithms. From the perspective of the root mean square error (RMSE), the REKF algorithm presents the optimal performance. Subgraph (a) and subgraph (b) of Figure 10 show the values of γ corresponding to the four beacons of Route 1 and Route 2, respectively. Among them, γ is greater than the threshold of 6.2 for several times, indicating that abnormal UWB range measurement exist. The observation covariance is adjusted by γ, and the larger the abnormal ranging value, the larger the corresponding γ. For example, the maximum abnormal ranging value of Beacon 2 of Route 2 is 110m, and the corresponding γ is close to 5000. The effect of abnormal observations on the positioning result is inhibited by the dynamically changing γ.

5.4. Fusion Positioning Using UWB Data Contaminated with Noise. To further analyze the performance of the REKF algorithm, three kinds of noise are injected into the UWB

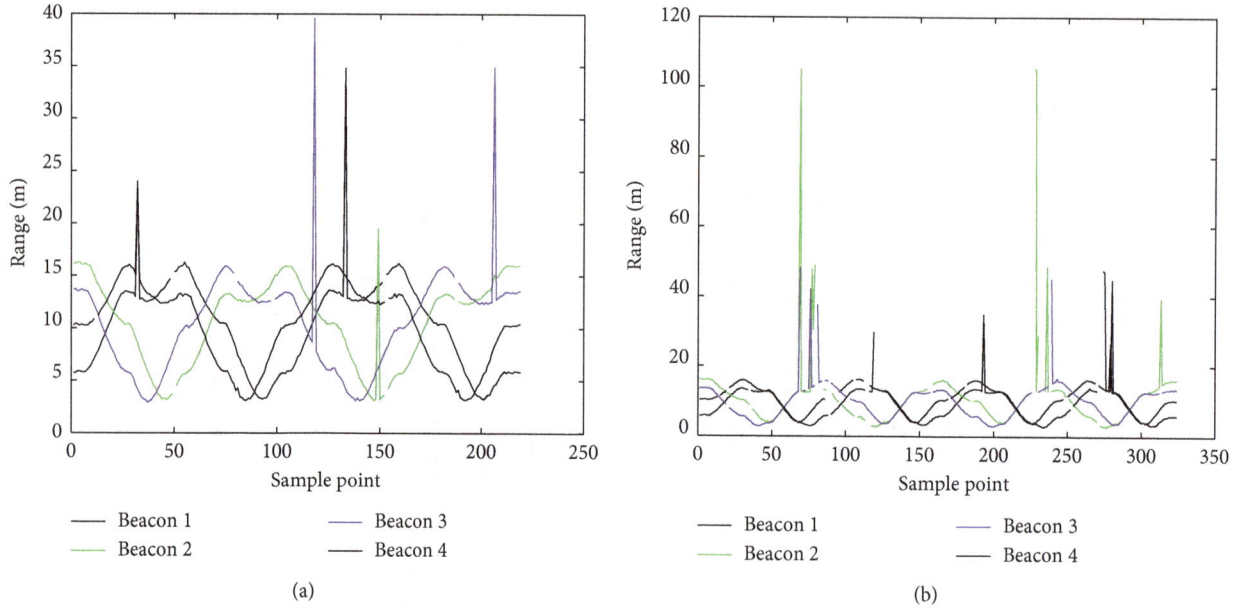

FIGURE 8: Range measurement of (a) Route 1 and (b) Route 2.

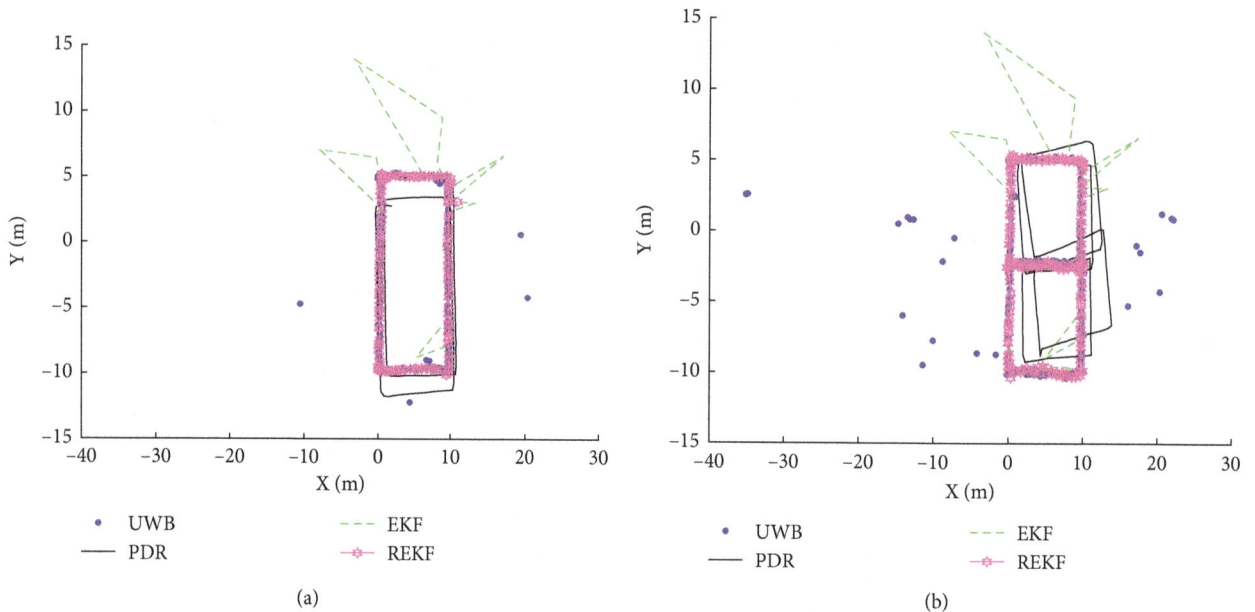

FIGURE 9: Positioning results of (a) Route 1 and (b) Route 2.

data of Route 2. The first one is uniformly increasing random noise of different ratios; the second one is continuous noise randomly injected; the third one is noise caused by the randomly blocked UWB beacons over a continuous time period.

5.4.1. Uniformly Increasing Random Noise of Different Ratios.
Gaussian white noise with an intensity of 30 dBW is used. The ratios of the injected noise account for 20%, 10%, and 7% of the total number of measurements, respectively.

Figure 11 is the result of the four beacons with 20% Gaussian white noise randomly and uniformly injected.

Figure 12 shows the positioning results of the REKF algorithm with 20%, 10%, and 7% noise injected, respectively. It can be seen that the REKF algorithm will be affected to a certain extent with the increase of the noise ratio, but compared to the continuous noise which will be explained in the next section, this uniformly injected noise results in a relatively small effect on the system state. The correct system state is used by the REKF algorithm to determine the quality of the current ranging value, thereby

TABLE 3: Positioning error analysis of Route 1 and Route 2.

	UWB	PDR	EKF	REKF
Route 1 (RMSE/m)	0.92	2.34	0.78	0.35
Route 2 (RMSE/m)	1.44	3.26	1.04	0.45

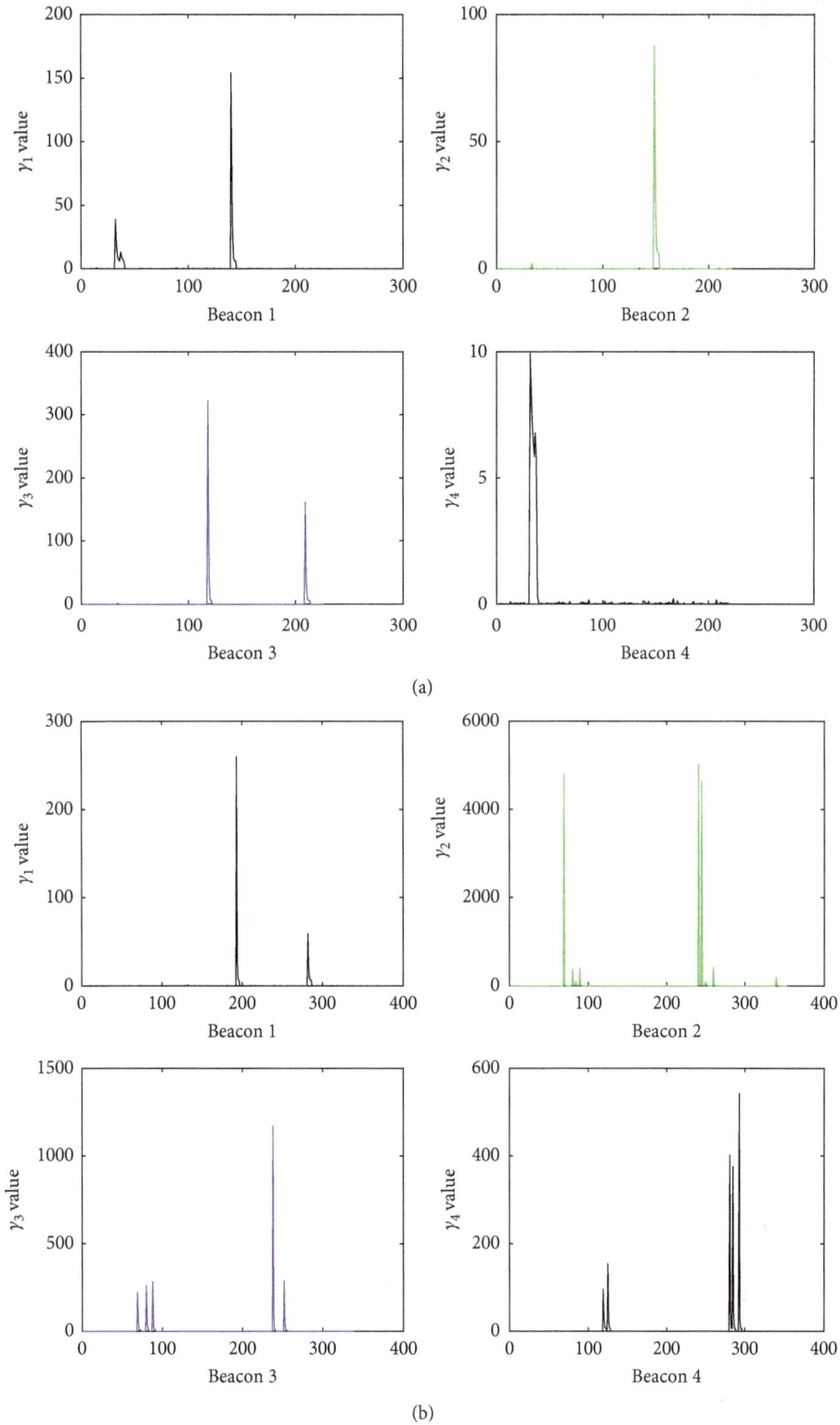

(a)

(b)

FIGURE 10: γ values of (a) Route 1 and (b) Route 2.

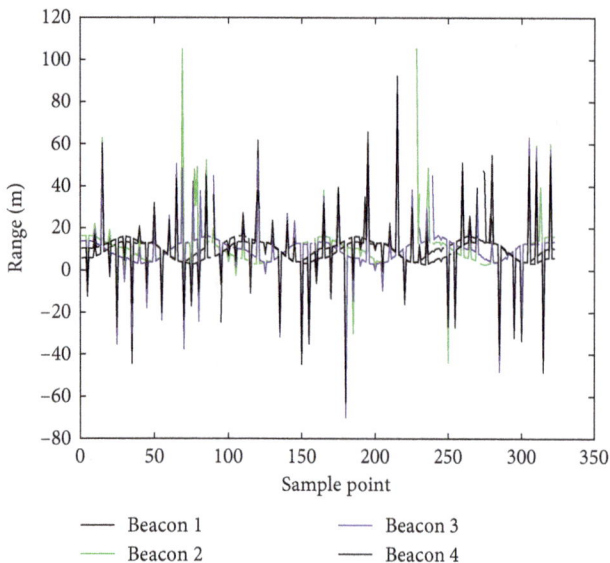

FIGURE 11: 4 beacons with 20% Gaussian white noise randomly and uniformly injected.

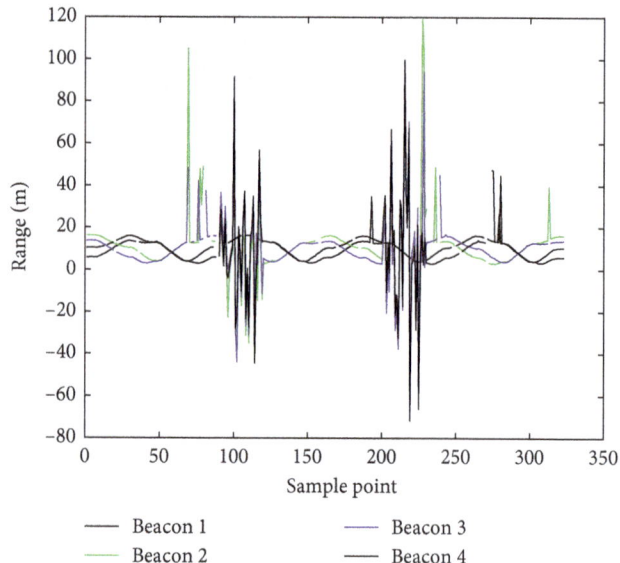

FIGURE 13: 3 beacons with continuous noise injected.

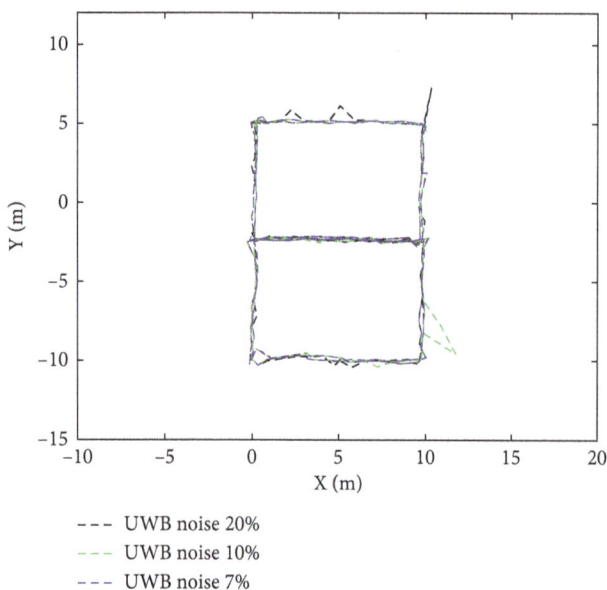

FIGURE 12: Positioning results of Route 2 with UWB noise of different ratios injected.

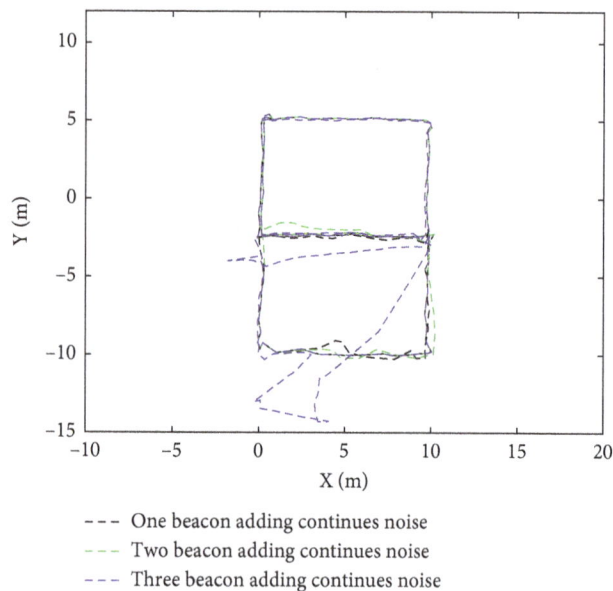

FIGURE 14: Positioning results of Route 2 with continuous noise injected into Beacons 1–3.

determining whether to rely more on the results of the PDR algorithm or not. In general, the trajectory obtained by using the REKF algorithm is basically consistent with the reference trajectory. The several jumps in the figure are caused by the adjacent relationship between the injected noise and the noise of the raw data, since such continuous error will cause instability of the system state.

5.4.2. Continuous Noise Randomly Injected. Figure 13 shows the results of Beacons 2–4 with 30 dBW Gaussian white noise continuously injected. The data segments with noise injected are 90~120 and 200~230.

Figure 14 shows the positioning results when different numbers of beacons are affected by such noise. It can be seen that when such noise is simultaneously injected into three beacons, the positioning result is seriously affected, as shown by the blue trajectory. This is because continuous abnormal data will cause large errors in the system state, and when the system state is inaccurate, the determination of the UWB observation's reliability will also be inaccurate, resulting in a wrong estimation of the observation covariance, and a system state which is difficult to be corrected. In other words, when the system state deviates from the real trajectory, the following correct observations will be defined as outliers. When such noise is injected into one or two beacons, the positioning result is less affected. In the robust EKF

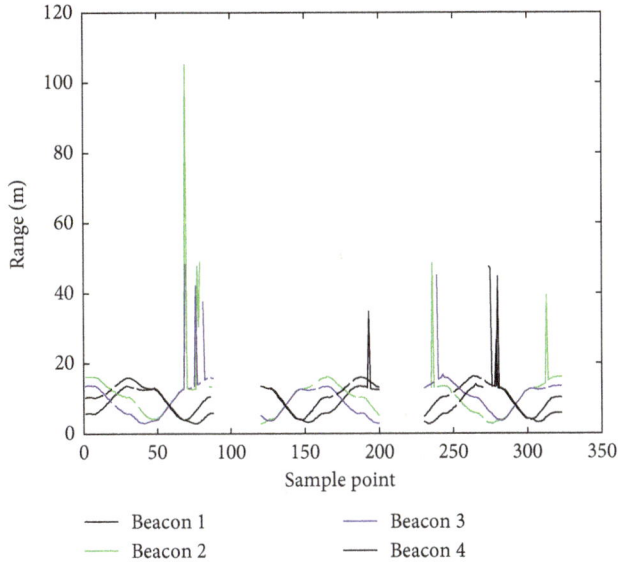

FIGURE 15: 4 beacons intermittently blocked.

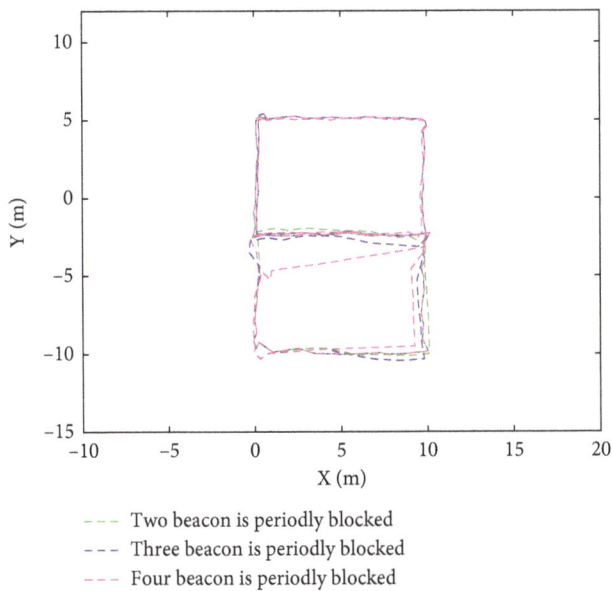

FIGURE 16: Positioning results of Route 2 with 2–4 beacons' data intermittently blocked.

algorithm, 2 to 3 reliable UWB observations are used to maintain the positioning accuracy of the system.

5.4.3. Noise Caused by the Blocked UWB Beacons over Continuous Time Period. Figure 15 shows the results of the four beacons being blocked on data segments of 90~120 and 200~230. Figure 16 shows the positioning results when different numbers of beacons are affected by such noise. It can be seen that when four beacons are simultaneously blocked, the positioning result is completely dependent on the PDR algorithm, and the positioning trajectory over the blocking time period is basically consistent with that of the PDR algorithm. When three or two beacons are blocked, the

positioning result is less affected. In other words, even if only one beacon's data are received by the smartphone, the positioning will also be assisted.

6. Conclusions

A UWB/PDR fusion positioning method based on the robust EKF is proposed in this paper. In this algorithm, the Mahalanobis distance between the observation and the system state is calculated to update the observation covariance, inhibiting the effect of abnormal observations on the positioning results. In addition, the Madgwick algorithm is introduced into the heading calculation of the PDR algorithm, effectively suppressing the cumulative error of the heading calculation. The experimental results show that in the case of intermittent or continuous UWB ranging noise and signal interruption, the proposed method exhibits strong robustness, with positioning accuracy higher than that of the EKF algorithm. However, how to improve the performance of the algorithm in the presence of stronger UWB ranging noise and much more signal interruption is required to be further tested.

Acknowledgments

This work was supported by the National Natural Science Foundation of China under grant number 41674030 and the China Postdoctoral Science Foundation under grant number 2016M601909 and a grant from the China Scholarship Council.

References

[1] X. Li, J. Wang, C. Liu, L. Zhang, and Z. Li, "Integrated WiFi/PDR/Smartphone using an adaptive system noise extended kalman filter algorithm for indoor localization," *ISPRS International Journal of Geo-Information*, vol. 5, no. 2, p. 8, 2016.

[2] X. Li, J. Wang, and C. Liu, "A bluetooth/PDR integration algorithm for an indoor positioning system," *Sensors*, vol. 15, no. 10, pp. 24862–24885, 2015.

[3] F. Santoso and S. J. Redmond, "Indoor location-aware medical systems for smart homecare and telehealth monitoring: state-of-the-art," *Physiological Measurement*, vol. 36, no. 10, pp. R53–R87, 2015.

[4] E. García, P. Poudereux, Á. Hernández, J. Ureña, and D. Gualda, "A robust UWB indoor positioning system for highly complex environments," in *Proceedings of 2015 IEEE International Conference on Industrial Technology*, pp. 3386–3391, IEEE, Seville, Spain, March 2015.

[5] J. Zhang and C. Shen, "Research on UWB indoor positioning in combination with TDOA improved algorithm and kalman filtering," *Modern Electronics Technique*, vol. 39, no. 13, pp. 1–5, 2016.

[6] Q. Fan, B. Sun, Y. Sun, and X. Zhuang, "Performance enhancement of MEMS-based INS/UWB integration for indoor navigation applications," *IEEE Sensors Journal*, vol. 17, no. 10, pp. 3116–3130, 2017.

[7] S. Sczyslo, J. Schroeder, S. Galler, and T. Kaiser, "Hybrid localization using UWB and inertial sensors," in *Proceedings of IEEE International Conference on Ultra-Wideband*, vol. 3, pp. 89–92, IEEE Xplore, Hannover, Germany, September 2008.

[8] Y. Xu and X. Chen, "Range-only UWB/INS tightly integrated navigation method for indoor pedestrian," *Chinese Journal of Scientific Instrument*, vol. 37, no. 8, pp. 142–148, 2016.

[9] Y. Xu, X. Chen, J. Cheng, Q. Zhao, and Y. Wang, "Improving tightly-coupled model for indoor pedestrian navigation using foot-mounted IMU and UWB measurements," in *Proceedings of IEEE Instrumentation and Measurement Technology Conference*, Taipei, Taiwan, May 2016.

[10] Q. Fan, B. Sun, Y. Sun, Y. Wu, and X. Zhuang, "Data fusion for indoor mobile robot positioning based on tightly coupled INS/UWB," *Journal of Navigation*, vol. 70, no. 5, pp. 1079–1097, 2017.

[11] A. Benini, A. Mancini, and S. Longhi, "An INS/UWB/vision-based extended kalman filter for MINI-UAV localization in indoor environment using 802.15.4a wireless sensor network," *Journal of Intelligent and Robotic Systems*, vol. 70, no. 1–4, pp. 461–476, 2013.

[12] J. Wang, Y. Gao, Z. Li, X. Meng, and C. M. Hancock, "A tightly-coupled GPS/INS/UWB cooperative positioning sensors system supported by v2i communication," *Sensors*, vol. 16, no. 7, 2016.

[13] J. González, J. L. Blanco, C. Galindo et al., "Mobile robot localization based on ultra-wide-band ranging: a particle filter approach," *Robotics and Autonomous Systems*, vol. 57, no. 5, pp. 496–507, 2009.

[14] L. Zwirello, C. Ascher, G. F. Trommer, and T. Zwick, "Study on UWB/IMU integration techniques," in *Proceedings of 2011 8th Workshop on Positioning, Navigation and Communication*, pp. 13–17, IEEE Xplore, Dresden, Germany, April 2011.

[15] L. Zwirello, X. Li, T. Zwick, C. Ascher, S. Werling, and G. F. Trommer, "Sensor data fusion in UWB-supported inertial navigation systems for indoor navigation," in *Proceedings of IEEE International Conference on Robotics and Automation*, pp. 3154–3159, Karlsruhe, Germany, May 2013.

[16] F. Zampella, A. De Angelis, I. Skog, D. Zachariah, and A. Jiménez, "A constraint approach for UWB and PDR fusion," in *Proceedings of International Conference on Indoor Positioning and Indoor Navigation (IPIN)*, pp. 1–9, Sydney, Australia, November 2012.

[17] P. Chen, Y. Kuang, and X. Chen, "A UWB/improved PDR integration algorithm applied to dynamic indoor positioning for pedestrians," *Sensors*, vol. 17, no. 9, p. 2065, 2017.

[18] H. E. Soken and C. Hajiyev, "Pico satellite attitude estimation via robust unscented kalman filter in the presence of measurement faults," *ISA Transactions*, vol. 49, no. 3, pp. 249–256, 2010.

[19] W. R. Wu and D. C. Chang, "Feedback median filter for robust preprocessing of glint noise," *IEEE Transactions on Aerospace and Electronic Systems*, vol. 36, no. 4, pp. 1026–1035, 2002.

[20] Z. Duan, J. Zhang, C. Zhang, and E. Mosca, "Robust H_2 and H_∞ filtering for uncertain linear systems," *Automatica*, vol. 42, no. 11, pp. 1919–1926, 2006.

[21] L. Chang, B. Hu, G. Chang, and A. Li, "Robust derivative-free Kalman filter based on Huber's M-estimation methodology," *Journal of Process Control*, vol. 23, no. 10, pp. 1555–1561, 2013.

[22] M. A. Gandhi and L. Mili, "Robust Kalman filter based on a generalized maximum-likelihood-type estimator," *IEEE Transactions on Signal Processing*, vol. 58, no. 5, pp. 2509–2520, 2010.

[23] S. O. H. Madgwick, A. J. L. Harrison, and R. Vaidyanathan, "Estimation of IMU and MARG orientation using a gradient descent algorithm," in *Proceedings of IEEE International Conference on Rehabilitation Robotics*, pp. 1–7, Zürich, Switzerland, June-July 2011.

[24] S. Y. Cho and C. G. Park, "MEMS based pedestrian navigation system," *Journal of Navigation*, vol. 59, no. 1, pp. 135–153, 2006.

[25] S. Beauregard and H. Haas, "Pedestrian dead reckoning: a basis for personal positioning," in *Proceedings of the Workshop on Positioning, Navigation and Communication (WPNC'06)*, Hannover, Germany, September 2006.

[26] S. H. Fang and T. N. Lin, "Indoor location system based on discriminant-adaptive neural network in IEEE 802.11 environments," *IEEE Transactions on Neural Networks*, vol. 19, no. 11, pp. 1973–1978, 2008.

[27] F. Gu, K. Khoshelham, J. Shang, F. Yu, and Z. Wei, "Robust and accurate smartphone-based step counting for indoor localization," *IEEE Sensors Journal*, vol. 17, no. 11, pp. 3453–3460, 2017.

[28] F. Weimann, G. Abwerzger, and B. Hofmannwellenhof, "A pedestrian navigation system for urban and indoor environments," in *Proceedings of International Technical Meeting of the Satellite Division of the Institute of Navigation*, pp. 1380–1389, Fort Worth, TX, USA, September 2007.

[29] F. Li, C. Zhao, G. Ding, J. Gong, C. Liu, and F. Zhao, "A reliable and accurate indoor localization method using phone inertial sensors," in *Proceedings of ACM Conference on Ubiquitous Computing*, pp. 421–430, Pittsburgh, PA, USA, September 2012.

[30] J. Teng, B. Zhang, J. Zhu, X. Li, D. Xuan, and Y. F. Zheng, "EV-LOC: integrating electronic and visual signals for accurate localization," *IEEE/ACM Transactions on Networking*, vol. 22, no. 4, pp. 1285–1296, 2014.

[31] O. Mezentsev and G. Lachapelle, "Pedestrian dead reckoning—a solution to navigation in GPS signal degraded areas," *Geomatica*, vol. 59, no. 2, pp. 5430–5434, 2005.

[32] R. W. Levi and T. Judd, "Dead reckoning navigational system using accelerometer to measure foot impacts," US Patent 5583776A, 1996.

[33] W. Kang, S. Nam, Y. Han, and S. Lee, "Improved heading estimation for smartphone-based indoor positioning systems," in *Proceedings of IEEE, International Symposium on Personal Indoor and Mobile Radio Communications*, pp. 2449–2453, IEEE, Sydney, Australia, September 2012.

[34] S. Moafipoor, D. A. Grejner-Brzezinska, and C. K. Toth, "A fuzzy dead reckoning algorithm for a personal navigator," *Navigation*, vol. 55, no. 4, pp. 241–254, 2008.

[35] G. Chang, "Robust Kalman filtering based on Mahalanobis distance as outlier judging criterion," *Journal of Geodesy*, vol. 88, no. 4, pp. 391–401, 2014.

An Analysis of Economic Impact on IoT Industry under GDPR

Junwoo Seo (iD),[1] **Kyoungmin Kim** (iD),[1] **Mookyu Park** (iD),[2] **Moosung Park**,[3] **and Kyungho Lee** (iD)[2]

[1]*Department of Cyber Defense (CYDF), Korea University, Seoul, Republic of Korea*
[2]*Center for Information Security Technologies (CIST), Korea University, Seoul, Republic of Korea*
[3]*Agency for Defense Development, Seoul, Republic of Korea*

Correspondence should be addressed to Kyungho Lee; kevinlee@korea.ac.kr

Academic Editor: Jeongyeup Paek

The EU GDPR comes into effect on May 25, 2018. Under this regulation, stronger legislation than the existing directive can be enforced. The IoT industry, especially among various industries, is expected to be heavily influenced by GDPR since it uses diverse and vast amounts of personal information. This paper first analyzes how the IoT industry handles personal information and summarizes why it is affected by GDPR. The paper then uses the cost definition of Gordon and Loeb model to estimate how GDPR affects the cost of IoT firms qualitatively and uses the statistical and legal bases to estimate quantitatively. From a qualitative point of view, GDPR impacted the preventative cost and legal cost of the Gordon and Loeb model. Quantitative view showed that the cost of IoT firms after GDPR could increase by three to four times on average and by 18 times if the most. The study finally can be applied to situational awareness of the economic impact on the certain industry.

1. Introduction

On April 14, 2016, the European parliament passed the General Data Protection Regulation (GDPR). This regulation strengthens the privacy rights of information entities and ensures that personal information is freely transferred among EU member states. European citizens are expected to control their personal information and create a high level of privacy protection in the European Union. With the introduction of GDPR, companies dealing with personal information are expected to be heavily influenced. Amidst a variety of industries, this paper focuses on the IoT industry, which collects and analyzes vast amounts of information from users.

The IoT industry is used in a variety of areas such as automotive, security and surveillance, medical, smart home, and T2T wireless networks. According to Gartner, there will be 11.2 billion Internet of "things" by 2018 [1]. This means that each of the "things" stores, reprocesses, and distributes more than 50 billion personal information. Statista, however, issued statistical data that 39% of European consumers completely denied that they were given sufficient information about personal information collected by IoT manufacturers [2]. According to this status quo, the IoT industry, which significantly transfers control of information use to individuals, is expected to be heavily influenced by GDPR and it is important to raise awareness of the situation of the economic impact on the certain industry.

This paper is an updated and revised version of previous study [3]. Section 2, which is a background of this paper, presents the characteristics of GDPR and security and privacy issues of IoT. Section 3 analyzes how IoT handles personal information, and Section 4 examines why GDPR affects the IoT industry. Section 5 analyzes the firms cost under GDPR to determine the economic impact of the IoT industry through the Gordon and Loeb model. Section 6 introduces related studies on the economic impact of GDPR and compare it with the paper. Finally, the conclusions are described in Section 7.

2. Background

This section first describes how the GDPR differs from the existing Directive. As a result, one can know which industries violate the regulation. Moreover, it describes the security issues of the IoT industry that is strongly relevant to GDPR.

2.1. General Data Protection Regulation.

2.1. General Data Protection Regulation. Basically, the EU legislation is divided into Directive and Regulations. While the Directive provides concrete results to be achieved, each member state has discretion over the transition to the national standard of the Directive. On the other hand, Regulations are legally binding on all member states and takes effect on the date set by all member states [4]. As a result of GDPR, Data Protection Directive 1995 (95/46/EC), which protects personal information in 1995, will be replaced. The following are key elements to discriminate between GDPR and Directive.

First, the definition of personal information has been further expanded [4]. According to Article 4 of GDPR, personal information is all information related to identified or identifiable information subjects. Location information, online identifiers, and genetic information that were not included in the definition of Data Protection Directive 1995 were included.

Second, there will be a two-tiered sanction regime on penalties [4]. Violations of the GDPR, which is considered a small-scale case, could result in a fine of either 10 million or 2% of a firm's global turnover (whichever is greater). The most serious violations may result in a fine of either 20 million or 4% of a firm's global turnover (whichever is greater). This is the maximum penalty that can be imposed for the most serious violation, such as if the customer's data processing consent is insufficient or if the design concept of the firm violates the core of personal information.

Third, the consent, requirement of the GDPR, is reinforced beyond the Directive [4]. If consent is required, clear and specific information should be provided, and a simple and easy language should be used. The subject also has the right to withdraw consent at any time.

2.2. IoT Security Issues. IoT has become an increasingly attractive target for global hackers. Recent studies have categorized and described various security issues that arise in IoT environments which have heterogeneity and large scale of objects. Zhi-Kai Zhang et al. categorized these issues into identification, authentication, lightweight cryptosystem, and software vulnerability analysis [5].

These technical security issues are a threat to IoT devices that contain private information. Even if these various vulnerabilities occur, the most important thing is the security update problem. According to HP News, one IoT device has an average of 25 vulnerabilities [6]. There have also been many studies that have confirmed that IoT devices have vulnerabilities exposed to attackers [7–9]. As mentioned earlier, the number of IoT devices will be over a billion, and of course, the number of vulnerabilities will be the much huge amount. Since there are security updates on various devices, users are inevitably overwhelmed with the update. This causes 1-day vulnerabilities to be steadily generated, and various hacking incidents are likely to occur. Even if the firm provides automated updates, they often stop when they focus on constructing the next device, leaving customers with slightly outdated hardware that can become a security risk.

Under these circumstances, various studies are underway to solve the security issue of IoT environment. Ping Zhang et al. investigated the potential privacy risk of mobile Internet users and proposed an extensible system based on a public cloud service to hide the mobile user's network location and the traffic of the other party [10]. Bugra Gedik and Ling Liu recognized that privacy-aware management of location information is a very important task in the widespread deployment of location-based services [11]. They used a flexible privacy personalization framework to support location anonymity for a wide range of mobile clients with context-sensitive privacy requirements. Tianyi Song et al. found that smart home inevitably causes security and privacy problems [12]. They propose an improved stability and privacy protection communication protocol for smart home systems. In the proposed scheme, data transmissions within smart home systems are protected by a symmetric encryption scheme with secret keys generated by a chaotic system.

2.3. IoT Privacy Issues. This section describes what privacy issues are in the IoT. The following sections describe how IoT handles personal information specifically, but this chapter explains the rough privacy issues occurred in IoT environment. Due to the close correlation between personal physical characteristics and status, the adoption of cyber-physical-social systems (CPSSs) has inevitably been challenged by users' privacy concerns [13]. The first step is about collecting data. In the case of the Internet, information about user behavior is collected while surfing the Internet. IoT will analyze not only search history but also the individual user's various life patterns. Thus, more diverse and sensitive information can be collected. There are also data collection policies for each IoT device, including the access control policies of each "things" and the types of data. This may conflict with the articles of the GDPR in establishing control policies.

In addition, the digital forgetting process that requires the deletion of personal information may be difficult [14]. One of the processes of processing personal information is that the current trend allows people the right to forget. If an individual requests to delete information, it is difficult to completely delete it technically because it has so many diverse and massive amounts of personal information in the IoT situation, and it is necessary to revise the information held by the company.

3. How IoT Handles Personal Information

This section analyzes how the IoT industry handles personal information and shows that the industry is under GDPR. The following subsections describe the personal information usage features of IoT devices that may conflict with the GDPR.

3.1. Usage and Exchange Information between IoT Devices. Each endpoint in the IoT environment, the "things", automatically transmits data, communicates with other endpoints, and works together. In IoT, "things" occasionally trade and operate on behalf of the user. For example, if a smart refrigerator perceives that food is scarce, it connects to the Internet and buys necessary items for the user. In this case, information utilization is automated and user information is exchanged to various subjects. GDPR, managing the use of personal information, can bring the result of diminishing advantages of IoT.

3.2. Analysis of Information Aggregated from IoT. Currently, IoT manufacturers are collecting huge amounts of information generated in IoT environments and research methods to analyze this huge amount of data to better understand the system and user behavior. To bring more value and revenue, IoT manufacturers analyze data that appear to be irrelevant and ascertain the relationship between a consumers behavioral and usage patterns. In other words, data delivered from one endpoint is less likely to cause a privacy issue, but data collected and aggregated at various endpoints can trigger privacy issues. Therefore, this aggregated information has a possibility to be included in the extended definition of the personal information to which the GDPR applies and corresponds to the personal information control domain.

For example, Vizio, electronic product development company, was recently fined $ 2.2 million for using content-aware software to track users pattern without consent. The company sold 11 million IoT TVs with a software program installed intentionally to track customers' detailed viewing habits. They linked the data to specific household statistics and sold that information to third-party marketers. Vizio insisted that they never paired TV data with personally identifiable information such as name or contact information. However, the data were collected as an analysis of personal TV habit information, so it is considered as sensitive information and a fine was imposed. If the GDPR is applied, the fine would have been $ 292 million, more than 100 times larger than the previous judgment [3].

4. Why the IoT Industry Is Affected by GDPR

This section analyzes the characteristics of IoT and the provisions of GDPR, which were introduced above, and explain why the IoT industry is strongly influenced by GDPR.

4.1. Consent. The personal information required in the IoT environment is not only sensitive information but also has enormous amounts. Although there is no privacy protection law specific to the IoT field, the Federal Trade Commission (FTC) is conducting policy discussions on security and privacy protection in the IoT environment [15]. The FTC provided comments on the concerns of privacy breaches of IoT devices and the direction of information protection activities related to IoT through the "Benefits, Challenges,

and Potential Roles for the Government in Fostering the Advantage of the Internet of Things [16]."

In this statement, the FTC said that IoT devices that can collect, transmit, and share sensitive consumer information about their physical and lifestyle habits are dangerous when combined with information collected from other devices. Due to the characteristics of collection and sharing of personal information of IoT, there is a possibility of more difficulties in the consent process. In addition, when considering the sharing of personal information among T2T and the personal information utilization of companies, we can see that the consent is much difficult. According to the GDPR, there are conditions in which it is necessary to inform the purpose of collecting personal information in easy-to-understand terms and to simplify the consent process. By interfering with the consent process, consumers will be more hesitant to collect sensitive and diverse personal information, and these regulations will have a major economic impact.

4.2. Right to Compensation. Right to Compensation is an article that is directly related to all companies as well as IoT firms, but the reason why IoT industry has a big influence is that one attack vector can lead to various types of personal information leakage. From a business perspective, if the firm complies with the provisions of the GDPR, the firm can acquire rights to Right not to Compensate. However, IoT companies find it difficult to obtain these rights. There is a high probability that it will not be able to keep up with the security updates of a vast amount of devices, and the conditions for protecting the personal information that each device collects are much harder than for other industries' firms.

5. Situational Awareness for Measuring Economic Impact in IoT Industry under GDPR

This section analyzes the economic impact of the IoT industry. The paper uses the cost definition of the Gordon and Loeb model to estimate how GDPR affects the cost of IoT firms qualitatively and uses statistical and legal basis to estimate quantitatively.

According to the Gordon and Loeb model, the amount of damage can be calculated as follows: direct costs, indirect costs, explicit costs, and implicit costs [17]. First, the direct cost refers to damage cost that is directly caused by a specific infringement event. In other words, it is the amount of hardware or software lost due to an accident. On the other hand, indirect cost refers to prevention cost that is incurred to inhibit information security breaches. Furthermore, the explicit cost is the cost that is explicitly visible due to a specific infringement violation. The explicit cost includes a preinvestment cost to avoid damages and damage cost and a cost to recover damages caused by infringement. On the contrary, implicit cost does not include damage cost caused by an infringement, but the damage cost for situations that may arise thereafter. This involves, for example, the cost of legal liability for an infringement event, including falling stock prices or reduced sales due to a declining reputation of the affected company. Using the Gordon and Loeb model, this paper analyzes and indicates the defined costs that are expected to alter due to GDPR.

According to Article 82 (Right to Compensation and Liability), any person who has suffered material or non-material damage due to a violation of GDPR rules has the right to demand compensation for damages [4]. Especially, the Directive mentions only damages; however, GDPR can be compensated for pecuniary and nonpecuniary losses.

Article 83 explains the general conditions for imposing administrative fines, which are not automatically applied and are imposed on each individual case. Therefore, it is beyond the bounds of possibility to measure the fines accurately, due to characteristics of ones imposition and absence of judgment. However, by utilizing the definition of the Gordon and Loeb model, the paper presents the increment of certain cost elements. In this regard, the Article 82 and 83 result in an increase in the *legal cost* from the Gordon and Loeb model.

In addition, Articles 37, 38, and 39 describe the designation, status, and obligations of the Data Protection Officer (DPO), respectively. Controllers and processors must designate a DPO in following three cases: (1) public authorities, (2) large-scale regular and systematic monitoring of intelligence entities, and (3) large-scale processing of sensitive information or criminal records. Moreover, DPO should also have an in-depth understanding of GDPR and personal information processing tasks and expertise in national privacy laws. Since the designation of the DPOs is mandatory and their qualities must be proven, Articles 37, 38, and 39 will affect the *preventative cost*.

To sum up, GDPR affects two costs (legal cost and preventative cost) of the Gordon and Loeb model as written above. In order to analyze the economic impact of GDPR, the estimated cost of damages before and after GDPR should be compared. According to the Ponemon Institute, in 2016, the average number of breached records reached 24,089 [18]. Based on the research, the paper selects four average personal data breach cases to analyze the economic impact of the GDPR on the IoT industry.

Table 1 provides information on how much data breach has occurred and how much annual turnover the company has for each case. The paper firstly estimates how much four cases, regardless of GDPR, resulted in the cost loss of the firm. According to the Ponemon Institute, the average cost per data leakage (capita) of a data breach for the past four years is $150 [18]. The average cost per capita of a data breach includes both the direct and indirect costs incurred by the breach. Based on this research, estimates of the four cases' costs can be derived by multiplying the number of breached personal data by the average data breach cost per capita. In order to analyze how GDPR affects, the paper then examines the cost assuming GDPR application to the cases. As shown in Figure 1, two component costs of the Gordon and Loeb model that GDPR affects were derived, the legal cost and the preventative cost. The paper assumes that each violation of GDPR was considered to be a case of lesser

TABLE 1: Four personal data breach cases in the IoT industry.

	Number of breached personal information	Annual turnover of a firm EUR (millions)
A	43,000	10.0
B	28,000	3,133.8
C	23,200	386.6
D	20,000	17.5

infringement, due to an averageness of the cases, resulting in a fine 10 million or 2% of a firm's global turnover (whichever is greater). Taking into account each of these costs, Figure 2 shows how disastrous the prospective cost of a firm is.

Unlike the rest of the cases where the cost of a firm increases three to four times, the cost of firm B is expected to increase by about 18.6 times because of the provisions of GDPR. GDPR determines fine based on the annual turnover of a firm for the previous year, which is based on total sales of the firm, not the sales of the single branch or corporation that violates the regulations. This is a measure to secure the punitive penalties by preventing bogus companies or affiliated companies from using the methods to circumvent regulations. Therefore, firms with large annual turnover can be fined well exceeding 10 million.

The economic impact of data breach incidents can be an important indicator of decision making. "Situational Awareness (SA)" is an essential concept in this decision-making process. SA means a process that recognizes the time and environmental factors that an event occurs and responds to future threats. This process is used as a framework to respond to threats such as disaster, financial, and cybersecurity. SA's representative framework is based on Endsley's model. Endsley's model consists of recognition of the elements of the environment within the volume of time and space, an understanding of the meaning, and a process of projecting the state in the near future conditions [19]. The economic impact of GDPR, which can arise from data breach incidents, can be projected from SA to future conditions and used to predict damage.

6. Related Works

There were several studies on the economic impact of GDPR. Hofheinz Paul and Michael Mande argued that the GDPR is a "regulatory wall" to expand privacy [20]. They said the GDPR will have a major impact on the opportunities for US companies which provide digital services in the EU. Several studies have concluded that GDPR would be detrimental not only for foreign companies but also for economies in Europe. Christensen Laurits et al. argue that the GDPR will raise the production costs of companies in the EU by 20%. This could result in a 0.3% employment decline and a 3% decline in the company [21]. Avi Goldfarb and Catherine E. Tucker said the GDPR could hurt the efficiency of digital advertising and hinder its ability to generate revenue through digital content [22]. This paper first explains why a specific industry, IoT, is affected by GDPR, and it is meaningful to measure the loss magnitude to measure the risk of IoT industry under GDPR.

	Explicit cost		Implicit cost
Direct cost	Cost of lost profit	Cost of lost productivity	Legal cost
	Cost of recovery	Cost of lost data	
Indirect cost	Preventative cost		Cost of reputation effect

FIGURE 1: The Cost of Gordon and Loeb model that GDPR affects.

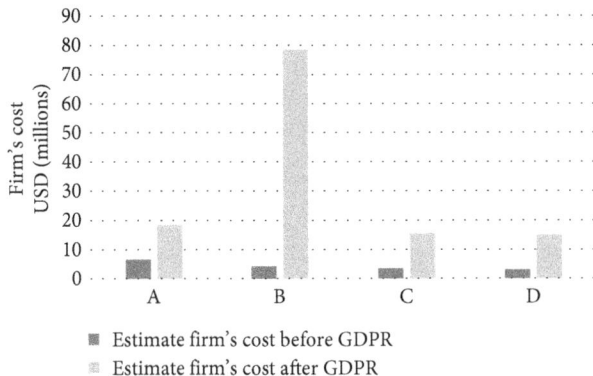

FIGURE 2: Comparison of estimate firm's cost before and after GDPR.

7. Conclusion

The impact of GDPR and its regulatory scope have heightened the tension of the firms. In this tension and concern, this paper first shows that by application of GDPR, the IoT industry is affected by GDPR in Sections 3 and 4. Section 5 uses the cost definition of the Gordon and Loeb model to estimate how GDPR affects the cost of IoT firms qualitatively and uses the statistical and legal bases to estimate quantitatively. Although there is a constraint to progress with limited data, the paper analyzes the economic impact of the certain industry from legal changes in two aspects (qualitative and quantitative), thus identifying industries that are vulnerable to legal changes.

The 2015 Icontrol State of the Smart Home study found that 44% of all Americans were "very concerned" about the possibility of their information getting stolen from their smart home and 27% were "somewhat concerned [23]." These public perceptions show the psychological concerns of people using IoT. In the presence of these concerns, the GDPR will add to the security of the user's personal information. However, companies will have to prepare and respond in order to find a way to fully utilize the IoT's functions under the GDPR.

Finally, situational awareness is an important step in the decision-making process. This study can be applied to situational awareness of the economic impact on the certain industry. Our next study is to develop the study to build an IoT risk management framework to anticipate and reduce risks based on the FAIR methodology, rather than simply measuring postaccident losses.

Acknowledgments

This work was supported by Defense Acquisition Program Administration and Agency for Defense Development under the contract UD060048AD.

References

[1] R. van der Meulen, "Gartner says 8.4 billion connected things will be in use in 2017, up 31 percent from 2016," 2017, http://www.gartner.com/newsroom/id/3598917.

[2] Statista, "Level of agreement regarding internet of things (IoT) manufacturers sufficiently informing consumers about information the devices can collect in europe in 2016," 2016, https://www.statista.com/statistics/609021/trust-in-iot-device-manufacturers-eu/.

[3] J. Seo, K. Kim, M. Park, M. Park, and K. Lee, "An analysis of economic impact on IoT under GDPR," in Proceedings of 2017 International Conference on Information and Communication Technology Convergence (ICTC), pp. 879–881, IEEE, Jeju Island, Korea, October 2017.

[4] C. of the European Union, "Regulation (EU) 2016/679 of the european parliament and of the council," 2016, http://data.consilium.europa.eu/doc/document/ST-5419-2016-INIT/en/pdf.

[5] Z.-K. Zhang, M. C. Y. Cho, C.-W. Wang, C.-W. Hsu, C.-K. Chen, and S. Shieh, "IoT security: ongoing challenges and research opportunities," in Proceedings of 2014 IEEE 7th International Conference on Service-Oriented Computing and Applications (SOCA), pp. 230–234, IEEE, Matsue, Japan, November 2014.

[6] K. Rawlinson, "Hp study reveals 70 percent of internet of things devices vulnerable to attack," 2014, http://www8.hp.com/us/en/hp-news/press-release.html?id=1744676#.WjJ_ld9l_BV.

[7] A. Costin, J. Zaddach, A. Francillon, D. Balzarotti, and S. Antipolis, "A large-scale analysis of the security of embedded firmwares," in Proceedings of USENIX Security Symposium, pp. 95–110, San Diego, CA, USA, August 2014.

[8] P.-Y. Chen, S.-M. Cheng, and K.-C. Chen, "Information fusion to defend intentional attack in internet of things," IEEE Internet of Things Journal, vol. 1, no. 4, pp. 337–348, 2014.

[9] V. Sharma, K. Lee, S. Kwon et al., "A consensus framework for reliability and mitigation of zero-day attacks in IoT," Security and Communication Networks, vol. 2017, Article ID 4749085, 24 pages, 2017.

[10] P. Zhang, M. Durresi, and A. Durresi, "Enhanced internet mobility and privacy using public cloud," Mobile Information Systems, vol. 2017, Article ID 4725858, 11 pages, 2017.

[11] B. Gedik and L. Liu, "Protecting location privacy with personalized k-anonymity: architecture and algorithms," IEEE Transactions on Mobile Computing, vol. 7, no. 1, pp. 1–18, 2008.

[12] T. Song, R. Li, B. Mei, J. Yu, X. Xing, and X. Cheng, "A privacy preserving communication protocol for IoT applications in smart homes," IEEE Internet of Things Journal, vol. 4, no. 6, pp. 1844–1852, 2017.

[13] X. Zheng, Z. Cai, J. Yu, C. Wang, and Y. Li, "Follow but no track: privacy preserved profile publishing in cyber-physical social systems," IEEE Internet of Things Journal, vol. 4, no. 6, pp. 1868–1878, 2017.

[14] T. Xu, J. B. Wendt, and M. Potkonjak, "Security of IoT systems: design challenges and opportunities," in Proceedings of 2014 IEEE/ACM International Conference on Computer-Aided

Design, pp. 417–423, IEEE Press, San Jose, CA, USA, November 2014.

[15] FTC, *The Internet of Things: Privacy and Security in a Connected World*, FTC, Washington, DC, USA, 2015.

[16] R. Hagemann, "The benefits, challenges, and potential roles for the government in fostering the advancement of the internet of things," 2016.

[17] L. A. Gordon and M. P. Loeb, "The economics of information security investment," *ACM Transactions on Information and System Security (TISSEC)*, vol. 5, no. 4, pp. 438–457, 2002.

[18] Ponemon Institute, *2017 Cost of Data Breach Study*, Ponemon Institute, Traverse City, MI, USA, 2017.

[19] M. R. Endsley, "Toward a theory of situation awareness in dynamic systems," *Human Factors*, vol. 37, no. 1, pp. 32–64, 1995.

[20] P. Hofheinz and M. Mandel, "Bridging the data gap: how digital innovation can drive growth and create jobs," *Lisbon Council-Progressive Policy Institute Policy Brief*, vol. 15, 2014.

[21] L. Christensen, A. Colciago, F. Etro, and G. Rafert, *The impact of the data protection regulation in the EU*, Intertic Policy Paper, Intertic, USA, 2013.

[22] A. Goldfarb and C. E. Tucker, "Privacy regulation and online advertising," *Management Science*, vol. 57, no. 1, pp. 57–71, 2011.

[23] Icontrol Networks, *Icontrol State of the Smart Home Report*, Icontrol Networks, Austin, TX, USA, 2015.

4

Nonparametric Blind Signal Detection based on Logarithmic Moments in very Impulsive Noise

Jinjun Luo [ID],[1] **Shilian Wang** [ID],[1] **Eryang Zhang,**[1] **and Xin Man** [ID][2]

[1]*School of Electronic Science and Engineering, National University of Defense Technology, Changsha 410073, China*
[2]*College of Electronic Engineering, Naval University of Engineering, Wuhan 430033, China*

Correspondence should be addressed to Jinjun Luo; jinjunluo@outlook.com

Academic Editor: Jinglan Zhang

The detection problem in impulsive noise modeled by the symmetric alpha stable (SαS) distribution is studied. The traditional detectors based on the second or higher order moments fail in SαS noise, and the method based on the fractional lower order moments (FLOMs) performs poorly when the noise distribution has small values of characteristic exponent. In this paper, a detector based on the logarithmic moments is investigated. The analytical expressions of the false alarm and detection probabilities are derived in nonfading channels as well as Rayleigh fading channels. The effect of noise uncertainty on the performance is discussed. Simulation results show that the logarithmic detector performs better than the FLOM and Cauchy detectors in very impulsive noise. In addition, the logarithmic detector is a nonparametric method and avoids estimating the parameter of the noise distribution, which makes the logarithmic detector easier to implement than the FLOM detector.

1. Introduction

Signal detection in noise is a classical problem in signal processing, and it is a key technology for mobile information systems. Usually, mobile information systems are designed under the assumption of Gaussian noise for mathematical convenience. However, in many applications, the background noise is non-Gaussian and shows impulsive behavior, such as man-made noise, interference from adjacent channel, and radio frequency interference (RFI) [1–3]. In impulsive noise, the conventional detection methods optimized under the Gaussian noise assumption suffer from a drastic performance degradation, which degrades the overall performance of the mobile information systems. Thus, more accurate models for impulsive noise are developed. The generalized Gaussian (GG) distribution, Gaussian mixture (GM) model, Middleton Class-A (MCA) model, and symmetric alpha stable (SαS) distribution are alternative models for impulsive noise.

The SαS distribution has a solid theoretical foundation that it satisfies the generalized central limit theorem and has an advantage of stability. In addition, empirical data prove that the SαS distribution is a successful model for impulsive noise in mobile information systems. For example, radio frequency interference (RFI) for the embedded wireless data transceivers can be modeled using symmetric alpha stable (SαS) distributions with a characteristic exponent of 1.43 [2]. The co-channel interference in a communication link in a Poison field of interferers shows very impulsive behavior and can be modeled using SαS distributions with small characteristic exponent values [4]. In medical diagnosis or mobile health data processing, evoked potential (EP) detection is an important problem. Experimental analysis showed that the noised EP signal could be successfully modeled using SαS distributions with characteristic exponent values between 1.06 and 1.94 [5]. In [6], the noise fitting procedure indicated that the SαS distribution could fit the noise from GSM-R antennas much better than the Middleton Class-A model and Gaussian distributions. Due to these advantages, the SαS distribution has received an increasing attention for modeling the impulsive noise in recent years. In this paper, we consider the signal detection problem in very impulsive SαS distributed noise.

In the early literature, the proposed detectors for SαS noise usually require prior knowledge of the transmitted signal. This kind of detectors mainly includes the optimal detector and the locally optimal detector under a weak signal assumption. As the probability density function (PDF) of

$S\alpha S$ has no closed form expression, the optimal and locally optimal detectors should be implemented through numerical methods, which results in a heavy computational burden. Thus, some locally suboptimal detectors are proposed, such as the soft limiter, hole puncher, and local Cauchy detector [3]. Some detectors based on the approximate expression for the PDF of the $S\alpha S$ distribution are also investigated [7, 8].

In some practical applications, such as the communication countermeasure, and cognitive radio, the transmitted signal is not available. The aforementioned detectors cannot be used directly, and thus blind or semiblind detection techniques are required. The generalized likelihood ratio test (GLRT) is an optimal solution, but it is extremely time-consuming [9]. The blind Cauchy detector is a special case of the GLRT and only performs well for some characteristic exponent in $S\alpha S$ noise. Recently, a scheme based on the fractional lower order moment (FLOM) is investigated in [10]. It performs better than the blind Cauchy detector. However, when the noise distribution has small values of the characteristic exponent, the FLOM detector performs poorly. In addition, the moment order of the FLOM should be determined, whose choice depends on the estimated characteristic exponent.

The logarithmic process was first introduced in [11]. Then the logarithmic moment (LM) was utilized for parameter estimation of the $S\alpha S$ distribution [12–14]. However, its application to signal detection has not yet received much attention. In this paper, we apply the logarithmic moment to signal detection and propose a detector based on it. The performance of the logarithmic detector is investigated in detail. The main contributions include the following: (i) We derive the analytical expressions of the false alarm and detection probabilities in nonfading channels as well as Rayleigh fading channels. (ii) The effect of the noise uncertainty on the detection probability is discussed, and the detection performance with different characteristic exponents and sample sizes is also investigated. (iii) The simulation results illustrate that the logarithmic detector performs better than the Cauchy and FLOM detectors when the characteristic exponent is smaller than 1.5. (iv) The logarithmic detector is a nonparametric method and avoids estimating the parameter of the noise distribution, which make the logarithmic detector easier to implement than the FLOM detector.

The remainder of this paper is organized as follows. In Section 2, the system and noise models are presented and the conventional detectors are introduced. In Section 3, the logarithmic detector is proposed and its performance is analyzed. In Section 4, extensive simulations are carried out. Finally, conclusions are drawn in Section 5.

2. System Model and Conventional Detectors

2.1. System Model. In this paper, the word "signal detection" refers to deciding whether a signal is present or not in the received data. It can be formulated as a binary hypothesis testing problem, described by H_0: signal absent and H_1: signal present. It can be written as

$$
\begin{aligned}
H_0 &: z(n) = w(n), \\
H_1 &: z(n) = hs(n) + w(n),
\end{aligned}
\tag{1}
$$

where $z(n)$ is the received data sample at time $n \in \{1, 2, \ldots, N\}$, $w(n)$ represents the impulsive noise, $s(n)$ is the transmitted signal, and h is the channel gain between the transmitter and receiver.

The impulsive noise $w(n)$ is modeled by an independent and identically distributed (i.i.d.) $S\alpha S$ distribution. Since the received signal is usually a sum of multiple nonline of sight (NLOS) signals, $s(n)$ can be modeled as a Gaussian random variable with zero mean according to the central limit theorem and is assumed to be independent of the noise $w(n)$.

The $S\alpha S$ distribution is described by its characteristic function:

$$
\varphi_\alpha(t) = \exp\left(-\gamma |t|^\alpha\right),
\tag{2}
$$

where $0 < \alpha \le 2$ and $\gamma > 0$. α is the characteristic exponent and determines the thickness of the distribution tail. When α is smaller, the tail is more thick and the pulse characteristic is more noticeable. γ is the scale parameter, also known as the dispersion parameter. It indicates the degree of the dispersion spread which is similar to the variance in a Gaussian distribution.

The PDF of the $S\alpha S$ distribution is the inverse Fourier transform of the characteristic function, namely,

$$
f_\alpha(x) = \frac{1}{2\pi} \int_{-\infty}^{\infty} \varphi_\alpha(t) e^{-jtx} \, dx.
\tag{3}
$$

The PDF does not have a closed form expression except for $\alpha = 2$ and $\alpha = 1$. In the two special cases, the $S\alpha S$ distribution reduces to Gaussian distribution and Cauchy distribution, respectively.

2.2. Conventional Detectors. In blind signal detection, the transmitted signal and channel gain are usually not available, so the likelihood ratio test (LRT) cannot be used directly. We can first estimate the unknown parameters and then apply the likelihood ratio test, which is known as the generalized likelihood ratio test, defined as [9]

$$
\Lambda_{\text{GLRT}} = \sum_{n=1}^{N} \log \frac{f_\alpha(z(n) - \hat{h}\hat{s}(n))}{f_\alpha(z(n))},
\tag{4}
$$

where $\hat{h}$ and $\hat{s}(n)$ are the maximum likelihood estimates of h and $s(n)$. Due to the numerical evaluation of $f_\alpha(x)$ and the parameter estimation, the GLRT detector is too complex to be applied for real-time applications. In the special case of $\alpha = 1$, the GLRT detector becomes a Cauchy detector [9], namely,

$$
\Lambda_{\text{Cauchy}} = \sum_{n=1}^{N} \log \left\{ 1 + \frac{|z(n)|^2}{\gamma^2} \right\}.
\tag{5}
$$

In the case of $0 < \alpha < 2$, the second and higher order moments do not exist, so the traditional detector based on these moments cannot be used. However, the fractional lower moments (FLOMs) of any order less than α do exist. Then, the FLOM detector is proposed [10], which is defined as

$$\Lambda_{\text{FLOM}} = \frac{1}{N} \sum_{n=1}^{N} |z(n)|^p, \tag{6}$$

where p is the order of the fractional moment. In order to ensure the existence of the FLOM statistic variance, the allowable value of p should be limited between 0 and $\alpha/2$. The choice of p depends on α which could be estimated by some methods [12, 13].

3. Nonparametric Logarithmic Detector and Performance Analysis

3.1. Logarithmic Moment-Based Detector. The logarithmic moment has been successfully applied to parameter estimation of the SαS distribution, but few studies investigate its application to signal detection. In this paper, we proposed a detector based on the logarithmic moment (LM). The LM detector is defined as

$$\Lambda_{\text{LM}} = \frac{1}{N} \sum_{i=1}^{N} \log|z(n)| \underset{H_0}{\overset{H_1}{\gtrless}} \eta, \tag{7}$$

where η is the detection threshold which could be determined by a given P_f. Compared with the FLOM detector, this detector does not need to decide the moment order p, which is convenient in practical applications.

3.2. Performance Analysis of the Logarithmic Detector in Nonfading Channels. The statistic Λ_{LM} is the sum of N independent random variables. When N is large enough, the statistic follows an asymptotic Gaussian distribution under either hypothesis due to the central limit theorem. To calculate the false alarm and detection probabilities, the mean and variance of the statistic should be calculated first.

Under hypothesis H_0, the mean and variance of Λ_{LM} can be calculated as (see Appendix for detail)

$$u_0 = E\{\Lambda_{\text{LM}}|H_0\} = C_e\left(\frac{1}{\alpha} - 1\right) + \frac{1}{\alpha}\log\gamma, \tag{8}$$

$$\sigma_0^2 = E\{\Lambda_{\text{LM}}^2|H_0\} - E^2\{\Lambda_{\text{LM}}|H_0\} = \frac{\pi^2}{6N}\left(\frac{1}{\alpha^2} + \frac{1}{2}\right). \tag{9}$$

Under hypothesis H_1, the mean of Λ_{LM} can be expressed as

$$u_1 = E\{\Lambda_{\text{LM}}|H_1\} = -\int_0^\infty E(u)\ln(u)F(u)\,du - C_e * \int_0^\infty E(u)F(u)\,du, \tag{10}$$

where

$$E(u) = \exp\left(-\gamma u^\alpha - \frac{h^2\sigma_s^2}{2}u^2\right), \tag{11}$$

$$F(u) = \gamma\alpha u^{\alpha-1} + h^2\sigma_s^2 u, \tag{12}$$

$$C_e = 0.577215\ldots. \tag{13}$$

The variance of Λ_{LM} can be computed as

$$\sigma_1^2 = \frac{1}{N}\left\{E\left[(\log|z|)^2\right] - (E(\log|z|))^2\right\}$$

$$= \frac{1}{N}\left\{E\left[(\log|z|)^2\right] - u_1^2\right\}, \tag{14}$$

where

$$E\left[(\log|z|)^2\right] = \int_0^\infty E(u)(\ln(u))^2 F(u)\,du$$
$$+ 2C_e \int_0^\infty E(u)\ln(u)F(u)\,du$$
$$+ \left(C_e^2 + \frac{\pi^2}{12}\right)\int_0^\infty E(u)F(u)\,du. \tag{15}$$

Then the false alarm and detection probabilities can be calculated as

$$P_f = \{\Lambda_{\text{LM}} > \eta|H_0\} = Q\left(\frac{\eta - \mu_0}{\sigma_0}\right), \tag{16}$$

$$P_d = \{\Lambda_{\text{LM}} > \eta|H_1\} = Q\left(\frac{\eta - \mu_1}{\sigma_1}\right). \tag{17}$$

The expression of P_f has a closed form, so the threshold η can be easily calculated from this expression, where the parameters α and γ can be estimated by many methods [12, 13]. Although the expression of P_d has no closed form, it can be evaluated by numerical methods. Through evaluating the expression of P_d, we can conveniently obtain the relationship between the detection probability and the sample size under different GSNRs, which gives us useful guideline for the choice of the sample size in practical applications.

In the following, we consider the effect of noise uncertainty on the logarithmic detector. Noise uncertainty is mainly caused by temperature variation, nonlinearity of components, initial calibration error, and interference [15, 16]. In general, a detector with a fixed threshold is designed based on the nominal noise power which is estimated in a finite time, so the performance will degrade when the actual noise power does not equal the nominal noise power.

From (8), (9), and (16), the threshold for logarithmic detector depends on the knowledge of the parameters γ and α which should be estimated by taking a large number of samples but some residual uncertainty still exists. Next, we only analyze the effect of uncertainty of γ on the detection performance; a similar procedure can be carried out for the parameter α.

We denote the nominal noise dispersion or the estimated noise dispersion by γ_n and the actual noise dispersion by γ_a. In the presence of noise uncertainty, $\gamma_a = \rho\gamma_n$, where ρ is the noise uncertainty coefficient, and its value in dB is denoted by $\beta = 10\log_{10}\rho$. Usually, β is uniformly distributed in the interval $[-B, B]$ [10, 17]. Then, the PDF of β can be expressed as

$$f(\beta) = \begin{cases} \dfrac{1}{2B}, & -B < \beta < B \\ 0, & \text{otherwise.} \end{cases} \tag{18}$$

According to the relationship between γ_a and β, we can easily obtain the PDF of γ_a as follows:

$$f(\gamma_a) = \begin{cases} \dfrac{10}{2B\gamma_a \ln 10}, & \gamma_n 10^{-B/10} < \gamma_a < \gamma_n 10^{B/10} \\[2mm] 0, & \text{otherwise.} \end{cases} \tag{19}$$

In nonpresence of noise uncertainty, the false alarm and detection probabilities can be calculated using (16) and (17). When the noise uncertainty does exist, the noise dispersion γ is no longer a constant value. In this case, we view (16) and (17) as conditional probabilities, where the condition is that γ is fixed. Then P_f and P_d are functions of the actual noise dispersion γ_a, and they can be rewritten as $P_f(\gamma_a)$ and $P_d(\gamma_a)$, respectively. By averaging (16) and (17) over (19), the probabilities of false alarm and detection under noise uncertainty can be obtained as

$$\begin{aligned} \overline{P}_f &= \int_b^c P_f(\gamma_a) f(\gamma_a)\, d\gamma_a, \\ \overline{P}_d &= \int_b^c P_d(\gamma_a) f(\gamma_a)\, d\gamma_a, \end{aligned} \tag{20}$$

where $b = \gamma_n 10^{-B/10}$ and $c = \gamma_n 10^{B/10}$. $\overline{P}_f$ and $\overline{P}_d$ do not have a closed form expression, but they can be evaluated by numerical methods.

3.3. Performance Analysis of the Logarithmic Detector in Rayleigh Fading Channels.

In Rayleigh fading channels, P_f remains the same as that of (16) in nonfading channels. The average detection probability can be evaluated by averaging the conditional P_d in the nonfading case as given in (17) over the channel gain. In this case, P_d in (17) can be regarded as a function of h, rewritten as $P_d(D|h)$. The average detection probability $\overline{P}_d$ in Rayleigh fading channels can be obtained as follows:

$$\overline{P}_d = E_h[P_d(D|h)]. \tag{21}$$

Since the channel gain h is Rayleigh distributed, its PDF is shown as

$$f(h) = \frac{h}{\sigma^2} \exp\left(-\frac{h^2}{2\sigma^2}\right), \quad h \geq 0, \tag{22}$$

where $\sigma^2 = \sqrt{2/\pi} E[h]$. So, $\overline{P}_d$ can be calculated as

$$\begin{aligned} \overline{P}_d &= \int_0^\infty P_d(D|h) f(h)\, dh \\ &= \int_0^\infty Q\left(\frac{\eta - \mu_1}{\sigma_1}\right) f(h)\, dh. \end{aligned} \tag{23}$$

It is difficult to find a closed form expression for $\overline{P}_d$, but we can evaluate it by numerical methods. Specially, when the GSNR is low enough where $\sigma_s^2 \approx 0$ and $\sigma_1^2 \approx \sigma_0^2$, $\overline{P}_d$ can be simplified as

$$\overline{P}_d = \int_0^\infty Q\left(\frac{\eta - \mu_1}{\sigma_0}\right) f(h)\, dh. \tag{24}$$

Because σ_0^2 is not a function of h, the integral will be greatly simplified.

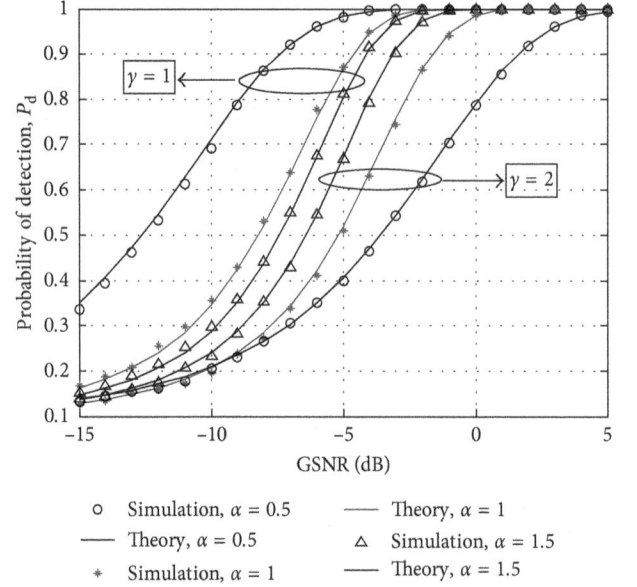

FIGURE 1: Detection probability versus GSNR for the logarithmic detector, $P_f = 0.1$, and $N = 1000$.

4. Simulation Results and Analysis

In this section, simulation results are presented to show the performance of the logarithmic detector. The channel is assumed fading free, unless otherwise specified. The transmitted signal is a Gaussian process with zero-mean and variance $\sigma_s^2 = E[|s(n)|^2]$. The results are based on 10000 Monte–Carlo simulations with a generalized signal-to-noise ratio (GSNR) defined as $\text{GSNR} = 10\log_{10}(\sigma_s^2 \sigma_h^2/\gamma)$ [10].

Figure 1 shows the detection probability versus the GSNR for the logarithmic detector in nonfading channels, where $P_f = 0.1$ and $N = 1000$. The curves are obtained from the simulation using (7) and by evaluating the analytical expressions in (16) and (17). The theoretical results are consistent with the Monte–Carlo simulation results very well for different α and γ, which proves the accuracy of the analytical expressions of the false alarm and detection probabilities. In addition, for a fixed false alarm probability, the detection probability increases with the increment of α when $\gamma = 2$, whereas it decreases with the increment of α when $\gamma = 1$. Therefore, the relationship between the detection probability and α depends on the values of γ.

Figure 2 shows the detection probability versus characteristic exponent α for different values of the GSNR and dispersion parameter γ, where $P_f = 0.1$ and $N = 1000$. The relationship between the detection probability and α is not definite, and it depends on the values of the GSNR and γ. For a fixed α and GSNR, the detection probability increases with the decrement of γ. When $\text{GSNR} = -8$ dB and $\gamma = 1$, the detection probability decreases with the increment of α. This result indicates that the detection performance is better in the presence of increased impulsive noise, which opposites researchers' intuition that more impulsive noise results in worse detection performance. This intuition is one-sided and based on an assumption that the detection performance is determined by the background noise. In fact, the detection

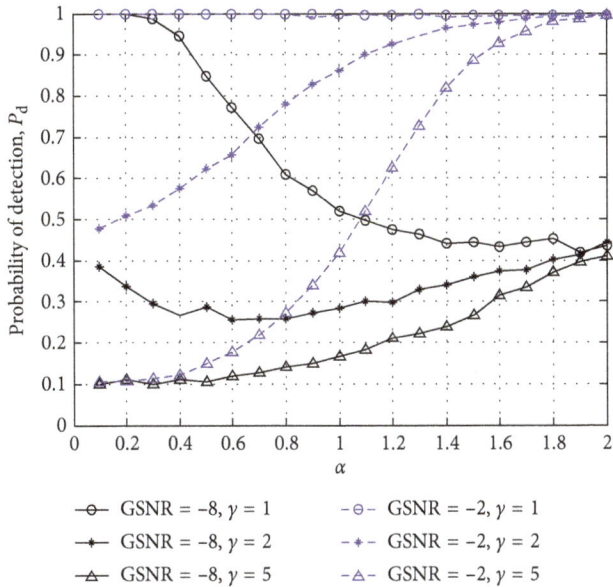

FIGURE 2: Detection probability versus α of the logarithmic detector, $P_f = 0.1$, and $N = 1000$.

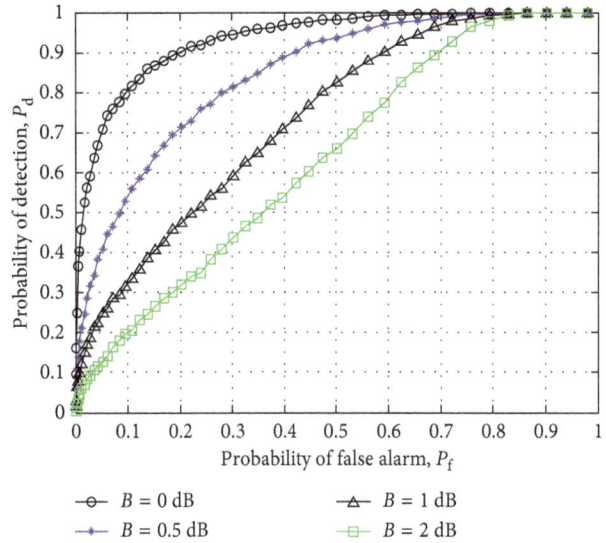

FIGURE 4: ROC of the logarithmic detector with noise uncertainty, $\alpha = 1.5$, $\gamma = 1$, GSNR $= -5$ dB, and $N = 1000$.

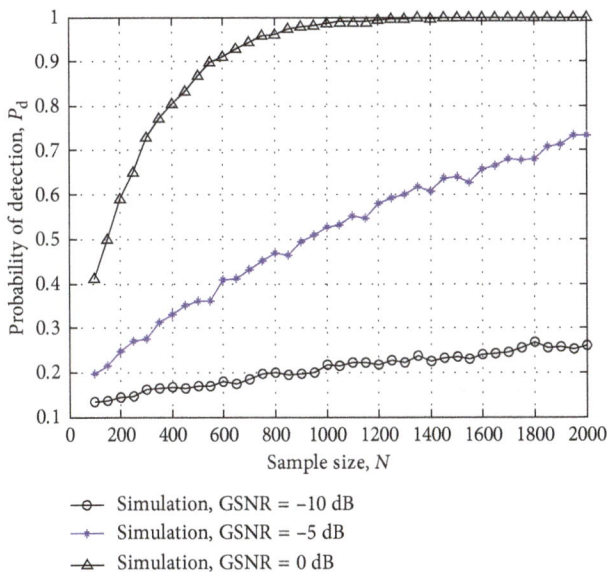

FIGURE 3: Detection probability versus number of samples for the logarithmic detector, $\alpha = 1$, $\gamma = 2$, and $P_f = 0.1$.

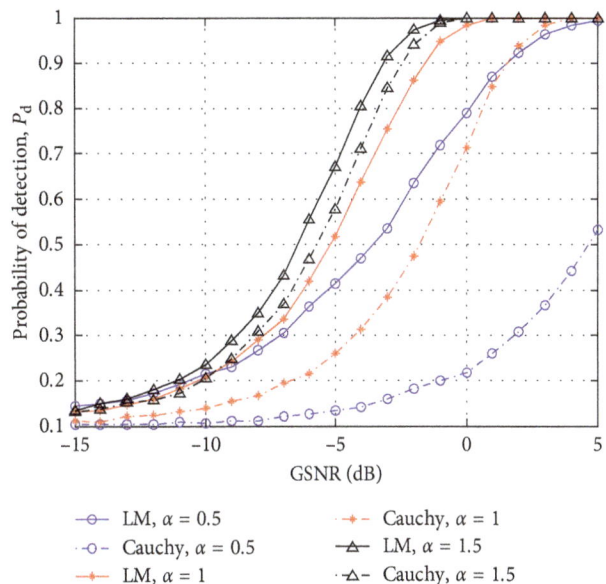

FIGURE 5: Detection probability versus GSNR for the logarithmic and Cauchy detectors in nonfading channels, $\gamma = 2$, $P_f = 0.1$, and $N = 1000$.

performance depends not only on the background noise but also on the detector. This special result indicates that the logarithmic detector has a good ability to deal with the impulsive noise.

Figure 3 shows the relationship between the detection probability and the sample size N, where $\alpha = 1$, $\gamma = 2$, and $P_f = 0.1$. The detection probability increases rapidly with the increment of N under a high GSNR, whereas it increases slowly with the increment of N under a low GSNR. Thus, when the GSNR is low, we should collect much more sample data to achieve a desired detection probability.

Figure 4 shows the receiver operation characteristic (ROC) curves of the logarithmic detector with different values of

noise uncertainty, where $\alpha = 1.5$, $\gamma = 1$, GSNR $= -5$ dB, and $N = 1000$. From this figure, we can see that the detection probability decreases with the increment of the noise uncertainty. When the value of B increases from 0 dB to 2 dB, the detection probability at $P_f = 0.1$ decreases from 0.8 to 0.2. To mitigate the effect of noise uncertainty, co-operative detection techniques are presented in the literature [18]. In addition, more accurate methods should be adopted to estimate the parameters α and γ based on more sample data.

Figure 5 compares the detection probability between the logarithmic and Cauchy detectors in nonfading channels, where $\gamma = 2$, $P_f = 0.1$, and $N = 1000$. We can see that the

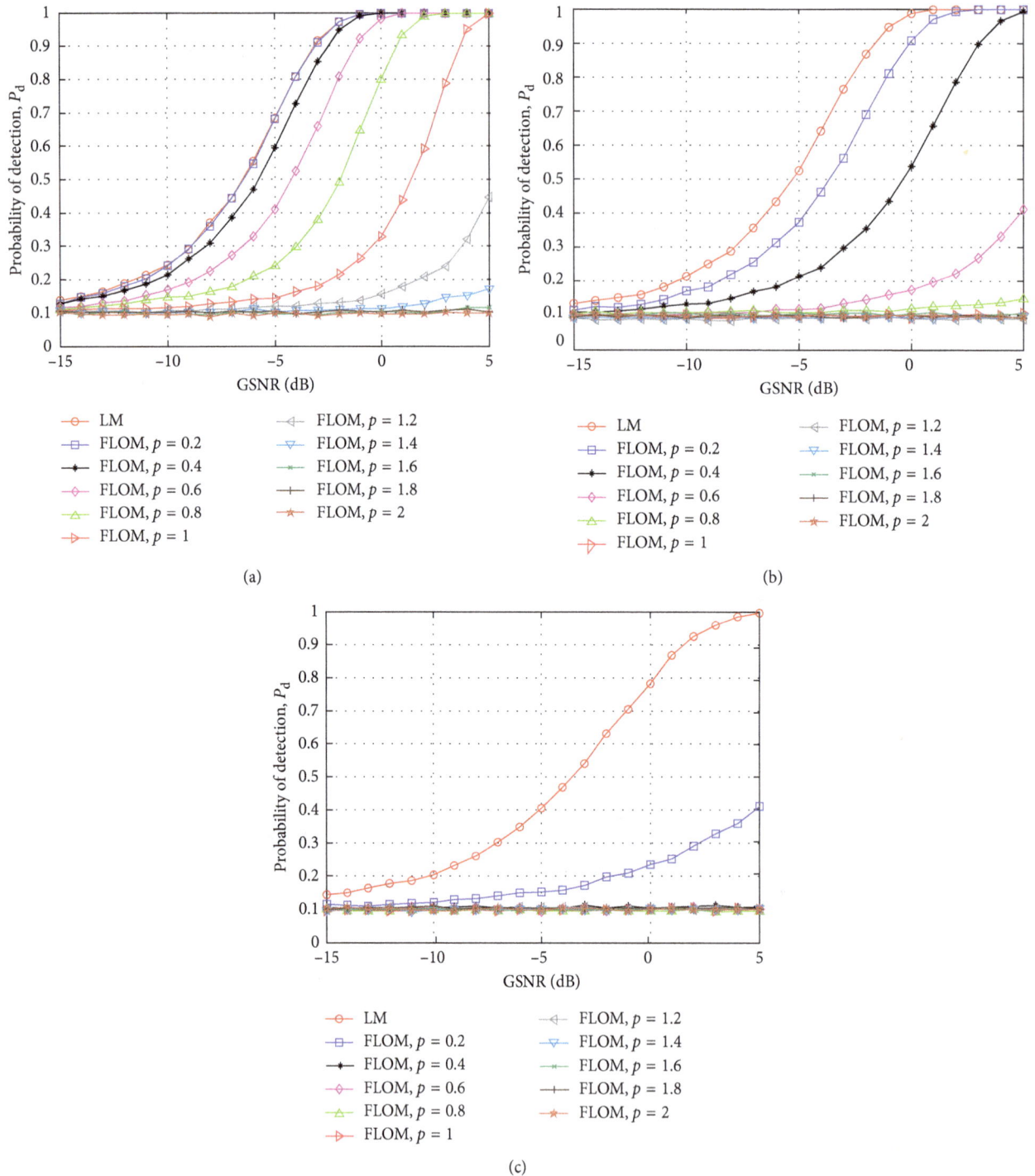

Figure 6: Detection probability versus GSNR for the logarithmic and FLOM detectors in nonfading channels, $\gamma = 2$, $P_f = 0.1$, and $N = 1000$. (a) $\alpha = 1.5$; (b) $\alpha = 1$; (c) $\alpha = 0.5$.

logarithmic detector outperforms the Cauchy detector for different α. In addition, the advantage of the logarithmic detector over the Cauchy detector becomes more obvious when α turns smaller. It indicates that the logarithmic detector is more suitable for detection in very impulsive noise than the Cauchy detector.

In Figure 6, the performance of the logarithmic detector is compared with the FLOM detector in nonfading channels,

where $\gamma = 2$, $P_f = 0.1$, $N = 1000$, and $p \in \{0.2 : 0.2 : 2\}$. In Figure 6(a), when the order p decreases from 2 to 0.2, the detection probability of the FLOM detector becomes higher and approaches the performance of the logarithmic detector; the same phenomenon can be seen in Figure 6(b) and Figure 6(c). When α decreases, the FLOM detector shows a more serious performance degradation than the logarithmic detector, which proves that the logarithmic

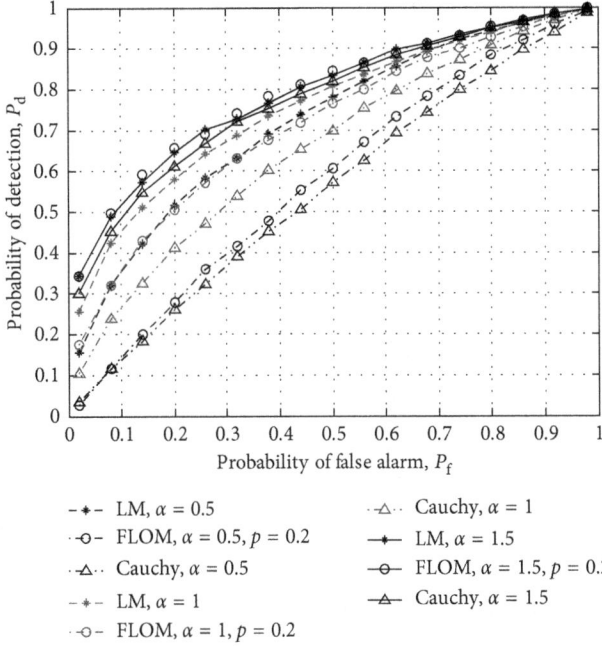

FIGURE 7: ROC of the logarithmic, FLOM, and Cauchy detectors in Rayleigh channels, $\gamma = 2$, $N = 1000$, and GSNR $= -5$ dB.

detector is more robust than the FLOM detector. Thus, the logarithmic detector performs better than the FLOM detector, and it is more suitable for signal detection in very impulsive noise.

Figure 7 shows the ROC curves of the logarithmic, FLOM, and Cauchy detectors for different values of α in Rayleigh channels. We can see that the logarithmic detector performs better than the FLOM and Cauchy detectors when $\alpha < 1.5$. In addition, the advantage of the logarithmic detector over the FLOM and Cauchy detectors is much greater when α is smaller. Thus, whether the channel is nonfading or Rayleigh fading, the logarithmic detector can achieve better performance than the FLOM and Cauchy detectors, especially for small values of α.

5. Conclusion

This paper studies the signal detection in impulsive noise modeled by SαS distribution. The logarithmic detector was proposed and the performance was analyzed. Simulation results show that the logarithmic detector performs better than the FLOM and Cauchy detectors in SαS noise for small values of α. In addition, the proposed detector is a nonparametric method, which makes it convenient in practical applications. Hence the logarithmic detector is a good alternative to the FLOM detector for signal detection in SαS impulsive noise.

Appendix

A.1 The Mean and Variance of Λ_{LM} under H_0

If Y is the logarithmic moment of a SαS random variable Z, namely, $Y = \log|Z|$, the mean and variance of Y can be expressed as [12]

$$E(Y) = C_e\left(\frac{1}{\alpha} - 1\right) + \frac{1}{\alpha}\log\gamma,$$

$$\mathrm{Var}(Y) = E\left\{(Y - E\{Y\})^2\right\} = \frac{\pi^2}{6}\left(\frac{1}{\alpha^2} + \frac{1}{2}\right), \tag{A.1}$$

where $C_e = 0.57721566\ldots$ is the Euler constant.

Under H_0, the observation signal is

$$z(n) = w(n). \tag{A.2}$$

So, $z(n)$ is a SαS random variable. Let $Y_n = \log[|z(n)|]$, then the mean of Λ_{LM} can be calculated as

$$u_0 = E\left[\frac{1}{N}\sum_{n=1}^{N} Y_n\right] = E[Y_n] = C_e\left(\frac{1}{\alpha} - 1\right) + \frac{1}{\alpha}\log\gamma. \tag{A.3}$$

And the variance can be obtained as

$$\sigma_0^2 = E\left[\left(\frac{1}{N}\sum_{n=1}^{N} Y_n\right)^2\right] - \left(E\left[\frac{1}{N}\sum_{n=1}^{N} Y_n\right]\right)^2$$

$$= \frac{1}{N^2}\sum_{n=1}^{N}\left\{E[Y_n^2] - (E[Y_n])^2\right\} = \frac{1}{N} * \frac{\pi^2}{6}\left(\frac{1}{\alpha^2} + \frac{1}{2}\right). \tag{A.4}$$

A.2 The Mean and Variance of Λ_{LM} under H_1

Under H_1, the observation signal is

$$z(n) = hs(n) + w(n). \tag{A.5}$$

In order to calculate the mean and variance of Λ_{LM}, we should firstly compute the first and second order moments of $Y = \log|Z|$. The logarithmic moments can be calculated from the FLOM as follows [13]:

$$E\left[(\log|z|)^n\right] = \lim_{p \to 0}\frac{d}{dz^n}E[|z|^p]. \tag{A.6}$$

To calculate $E[(\log|z|)^n]$, we should firstly compute $E[|z|^p]$ under H_1 as follows:

$$E[|z|^p] = E[|z(n)|^p] = E[|hs(n) + w(n)|^p]. \tag{A.7}$$

Note that h remains constant during the detection period, so $hs(n)$ still obeys Gaussian distribution, namely, $hs(n) \sim N(0, h^2\sigma_s^2)$. Its characteristic function is $\Phi_{hs}(u) = \exp(-(h^2\sigma_s^2/2)|u|^2)$. Since the characteristic function of the sum of n statistically independent random variables is equal to the product of the characteristic functions of the individual random variables [19], the characteristic function of $z(n)$ is obtained as

$$\Phi_z(u) = \Phi_{hs}(u)\Phi_w(u) = \exp\left(-\gamma|u|^\alpha - \frac{h^2\sigma_s^2}{2}|u|^2\right). \tag{A.8}$$

Because $s(n)$ and $w(n)$ have symmetric probability density functions (PDFs), it is easy to prove that $z(n)$ also has symmetric PDF.

Then with a similar procedure to calculate $E|X|^p$ of a SαS random variable X in [20], we can obtain

$$E\left[|z|^p\right] = \frac{1}{\Gamma(1-p)\cos((\pi/2)p)} \int_0^\infty \exp\left(-\gamma u^\alpha - \frac{h^2\sigma_s^2}{2}u^2\right)$$
$$\times (\gamma\alpha u^{\alpha-p-1} + h^2\sigma_s^2 u^{1-p})du.$$

(A.9)

Substituting (A.10) into (A.7), we can obtain the first and second order moments of $Y = \log|Z|$. Let $E[|z|^p] = A(p)/D(p)$, where $A(p) = \int_0^\infty \exp(-\gamma u^\alpha - (h^2\sigma_s^2/2)u^2)(\gamma\alpha u^{\alpha-p-1} + h^2\sigma_s^2 u^{1-p})du$ and $D(p) = \Gamma(1-p)\cos((\pi/2)p)$. The first order moment of Y is obtained as

$$E[\log|z|] = \lim_{p\to 0}\frac{dE[|z|^p]}{dp} = \lim_{p\to 0}\frac{A'(p)D(p) - A(p)D'(p)}{D^2(p)},$$

(A.10)

where

$$A'(p) = -\int_0^\infty E(u)\ln(u)\left(\gamma\alpha u^{\alpha-p-1} + h^2\sigma_s^2 u^{1-p}\right)du,$$

$$D'(p) = \Gamma'(1-p)\cos\left(\frac{\pi}{2}p\right) - \frac{\pi}{2}\Gamma(1-p)\sin\left(\frac{\pi}{2}p\right).$$

(A.11)

Using the integral in [21], we can get

$$\Gamma'(1-p) = -\int_0^\infty x^{-p}e^{-x}\ln(x)dx = -\Gamma(1-p)\psi(1-p).$$

(A.12)

From the equation in [21], we have

$$\psi(1) = -C_e.$$

(A.13)

So,

$$\lim_{p\to 0}\Gamma'(1-p) = C_e.$$

(A.14)

Then, the first order logarithmic moment is

$$E[\log|z|] = \lim_{p\to 0}\frac{dE[|z|^p]}{dp}$$
$$= -\int_0^\infty E(u)\ln(u)F(u)du - C_e\int_0^\infty E(u)F(u)du,$$

(A.15)

where $E(u)$ and $F(u)$ are defined in (11) and (12).

The second order moment of Y is obtained as

$$E\left[(\log|z|)^2\right] = \lim_{p\to 0}\frac{d^2E[|z|^p]}{dp^2}$$

(A.16)

$$= \lim_{p\to 0}\frac{d}{dp}\left[\frac{A'(p)D(p) - A(p)D'(p)}{D^2(p)}\right].$$

To solve the equation above, the key issue is to calculate $\lim_{p\to 0}D''(p)$ which should compute $\lim_{p\to 0}\Gamma''(1-p)$. From the equation in [21], we can obtain

$$\Gamma''(1-p) = \int_0^\infty t^{-p}e^{-t}(\ln t)^2 dt$$

(A.17)

$$= \Gamma(1-p)\{\psi^2(1-p) + \zeta(2,1-p)\},$$

where $\zeta(z, q)$ is defined as [21]

$$\zeta(z, q) = \sum_{n=0}^\infty \frac{1}{(q+n)^z}, \quad [\text{Re } z > 1, \ q \neq 0, -1, -2, \ldots].$$

(A.18)

From the equation in [22], we know

$$\zeta(2) = 1 + \frac{1}{2^2} + \frac{1}{3^2} + \cdots = \frac{\pi^2}{6}.$$

(A.19)

So,

$$\lim_{p\to 0}\Gamma''(1-p) = \lim_{p\to 0}\Gamma(1-p)\{\psi^2(1-p) + \zeta(2,1-p)\}$$

$$= C^2 + \frac{\pi^2}{6},$$

$$\lim_{p\to 0}D''(p) = C^2 - \frac{\pi^2}{12}.$$

(A.20)

Using the quotient rule, we can obtain

$$E\left[(\log|z|)^2\right] = \lim_{p\to 0}\frac{d^2E[|z|^p]}{dp^2}$$

$$= \int_0^\infty E(u)(\ln u)^2 F(u)du$$

$$+ 2C\int_0^\infty E(u)\ln uF(u)du$$

$$+ \left(C^2 + \frac{\pi^2}{12}\right)\int_0^\infty E(u)F(u)du.$$

(A.21)

After obtaining the first and second order moments of $Y = \log|Z|$, the mean and variance of Λ_{LM} can be easily obtained as follows:

$$u_1 = E\left\{\frac{1}{N}\sum_{n=1}^N Y_n\right\} = E[\log|z|],$$

$$\sigma_1^2 = E[\Lambda_{LM}^2] - (E[\Lambda_{LM}])^2$$

(A.22)

$$= \frac{1}{N^2}\left\{\sum_{n=1}^N E[Y_n^2] - \sum_{n=1}^N (E[Y_n])^2\right\}$$

$$= \frac{1}{N}\left\{E\left[(\log|z|)^2\right] - (E[\log|z|])^2\right\},$$

where $E[\log|z|]$ and $E[(\log|z|)^2]$ are shown as (A.15) and (A.21), respectively.

Acknowledgments

This work is supported by the National Natural Science foundation of China (Project no. 61501484).

References

[1] G. A. Tsihrintzis and C. L. Nikias, "Performance of optimum and suboptimum receivers in the presence of impulsive noise modeled as an alpha-stable process," *IEEE Transactions on Communications*, vol. 43, no. 2, pp. 904–914, 1995.

[2] M. Nassar, K. Gulati, A. K. Sujeeth, N. Aghasadeghi, B. L. Evans, and K. R. Tinsley, "Mitigating near-field interference in laptop embedded wireless transceivers," *Journal of Signal Processing Systems*, vol. 63, no. 1, pp. 1–12, 2011.

[3] C. L. Nikias and M. Shao, *Signal Processing with Alpha-Stable Distributions and Applications*, Wiley, Hoboken, NJ, USA, 1995.

[4] X. Yang and A. P. Petropulu, "Co-channel interference modeling and analysis in a Poisson field of interferers in wireless communications," *IEEE Transactions on Signal Processing*, vol. 51, no. 1, pp. 64–76, 2003.

[5] X. Kong and T. Qiu, "Adaptive estimation of latency change in evoked potentials by direct least mean p-norm time-delay estimation," *IEEE Transactions on Biomedical Engineering*, vol. 46, no. 8, pp. 994–1003, 1999.

[6] K. Hassan, R. Gautier, I. Dayoub, M. Berbineau, and E. Radoi, "Multiple-antenna-based blind spectrum sensing in the presence of impulsive noise," *IEEE Transactions on Vehicular Technology*, vol. 63, no. 5, pp. 2248–2257, 2014.

[7] E. E. Kuruoglu, W. Fitzgerald, and P. Rayner, "Near optimal detection of signals in impulsive noise modeled with a symmetric alpha-stable distribution," *IEEE Communications Letters*, vol. 2, no. 10, pp. 282–284, 1998.

[8] X. Li, J. Sun, S. Wang, L. Fan, and L. Chen, "Near-optimal detection with constant false alarm ratio in varying impulsive interference," *IET Signal Processing*, vol. 7, no. 9, pp. 824–832, 2013.

[9] H. G. Kang, I. Song, S. Yoon, and Y. H. Kim, "A class of spectrum-sensing schemes for cognitive radio under impulsive noise circumstances: structure and performance in nonfading and fading environments," *IEEE Transactions on Vehicular Technology*, vol. 59, no. 9, pp. 4322–4339, 2010.

[10] X. Zhu, W.-P. Zhu, and B. Champagne, "Spectrum sensing based on fractional lower order moments for cognitive radios in α-stable distributed noise," *Signal Processing*, vol. 111, pp. 94–105, 2015.

[11] V. M. Zolotarev, "Integral transfomations of distributions and estimates of parameters of multidimensional spherically symmetric stable laws," in *Contribution to Probability: A Collection of Papers*, E. Lukacs, J. Gani, and V. K. Rohatgi, Eds., pp. 283–305, New York Academic, New York, NY, USA, 1981.

[12] X. Ma and C. L. Nikias, "Parameter estimation and blind channel identification in impulsive signal environments," *IEEE Transactions on Signal Processing*, vol. 43, no. 12, pp. 2884–2897, 1995.

[13] E. E. Kuruoglu, "Density parameter estimation of skewed-stable distributions," *IEEE Transactions on Signal Processing*, vol. 49, no. 10, pp. 2192–2201, 2001.

[14] J. Gonzalez, J. Paredes, and G. Arce, "Zero-order statistics: a mathematical framework for the processing and characterization of very impulsive signals," *IEEE Transactions on Signal Processing*, vol. 54, no. 10, pp. 3839–3851, 2006.

[15] R. Tandra and A. Sahai, "SNR walls for signal detection," *IEEE Journal of Selected Topics in Signal Processing*, vol. 2, no. 1, pp. 4–17, 2008.

[16] A. Mariani, A. Giorgetti, and M. Chiani, "Effects of noise power estimation on energy detection for cognitive radio applications," *IEEE Transactions on Communications*, vol. 59, no. 12, pp. 3410–3420, 2011.

[17] W. Yin, P. Ren, J. Cai, and Z. Su, "Performance of energy detector in the presence of noise uncertainty in cognitive radio networks," *Wireless Networks*, vol. 19, no. 5, pp. 629–638, 2013.

[18] A. Mohammadi, M. R. Taban, J. Abouei, and H. Torabi, "Cooperative spectrum sensing against noise uncertainty using Neyman-Pearson lemma on fuzzy hypothesis test," *Applied Soft Computing*, vol. 13, no. 7, pp. 3307–3313, 2013.

[19] J. G. Proakis, *Digital Communications*, The McGraw-Hill Companies Inc., New York, NY, USA, 4th edition, 2001.

[20] M. Shao and C. L. Nikias, "Signal processing with fractional lower order moments: stable processes and their applications," *Proceedings of the IEEE*, vol. 81, no. 7, pp. 986–1010, 1993.

[21] S. Gradshteyn and I. M. Ryzhik, *Table of Integrals, Series, and Products*, USA Academic, New York, NY, USA, 7th edition, 2007.

[22] M. Abramowitz and I. A. Stegun, *Handbook of Mathematical Functions: with Formulas, Graphs, and Mathematical Tables*, vol. 55, National Bureau of Standards, Washington, DC, USA, 1965.

5

What Makes People Actually Embrace or Shun Mobile Payment

Yali Zhang [iD],[1] **Jun Sun** [iD],[2] **Zhaojun Yang** [iD],[3] **and Ying Wang** [iD][2]

[1]*School of Management, Northwestern Polytechnical University, Xi'an 710072, China*
[2]*College of Business and Entrepreneurship, University of Texas Rio Grande Valley, Edinburg, TX 78539-2999, USA*
[3]*School of Economics and Management, Xidian University, Xi'an 710126, China*

Correspondence should be addressed to Yali Zhang; zhangyl@nwpu.edu.cn

Academic Editor: Floriano Scioscia

Mobile payment is becoming increasingly popular, but it encounters the resistance from certain user groups. This study examines the factors that influence both the technology acceptance and actual usage aspects of mobile payment adoption from the perspective of the general systems theory. Based on a literature review, it conceptualizes the embedding relationships among relevant behavioral processes, personal characteristics, and extrinsic factors and develops a research model. Together, the extrinsic factors in terms of culture, subjective norm, and socioeconomic status and main personal characteristics including demographics, personality traits, and past behavior are hypothesized to have direct and moderating effects on mobile payment acceptance and usage. The observations collected from China and the USA support most of the hypothesized relationships and reveal interesting cross-culture differences. Whereas users in the USA appear to be more rational and risk-averse, people in China seem more subject to social influence. The findings contribute to the mobile payment literature by deepening the understanding of adoption stages and expanding the scope of explanatory variables beyond technology acceptance.

1. Introduction

Mobile payment, by its name, refers to the use of mobile devices and wireless technologies to make payments for goods, services, and bills [1]. Along with the fast pace of smartphone diffusion in the recent years, mobile payment has become increasingly popular [2]. Yet mobile payment requires mobile data services, and there has been resistance from a large proportion of people due to concerns such as security and privacy [3, 4]. Due to the inertia, many mobile payment services fail to reach intended customers.

Quite a few studies investigate why people use mobile payment, yet there still exist several research gaps, the most prominent of which concern the overemphasis on technology adoption and the lack of multination analyses [5]. Most studies on mobile payment adoption just predict behavioral intention with technology-related perceptions based on theoretical frameworks like the technology acceptance model (TAM) and the unified theory of acceptance and use of technology (UTAUT), but actual usage is what really matters [6]. Single-country samples further weaken the generalizability of findings as each market is unique, especially in the cultural influence on user behavior [7].

The main goal of this study, therefore, is to understand why people actually use or avoid mobile payment in distinct cultural contexts. It is essential to include other relevant variables than just technology-related perceptions and collect empirical observations from different countries. As an effort, this study identifies personal characteristics and extrinsic factors pertinent to mobile payment and examines their relationships with both technology acceptance and actual usage. For cross-culture comparisons, it draws samples from China and the USA where people are familiar with mobile payment development but cultures are different. In this way, this study responds to the call for more meaningful research on mobile payment user behavior.

The remaining of this article is organized as follows: First, it conducts a literature review that leads to a systems conceptualization of mobile payment adoption. Based on it, a research model is developed to hypothesize the relationships involved in the phenomenon. Then, it describes the methodology to collect observations for hypothesis testing and cross-culture comparison. Based on the results, theoretical and practical implications are discussed.

2. Theoretical Background

Mobile payment is a complex sociotechnical phenomenon that requires a holistic understanding. The general systems theory views such a complex phenomenon as a cohesive conglomeration of interdependent parts for adaptation to the environment [8]. From such a perspective, mobile payment adoption can be viewed as a system that comprises behavioral processes, personal characteristics, and extrinsic factors, the interactions among which shape the behavioral tendency of an individual on whether to use mobile payment or not. Figure 1 categorizes the elements involved in such a system as identified from the literature. At the center are the behavioral processes of mobile payment adoption, which includes two stages: technology acceptance and actual usage. Behavioral processes are affected by personal characteristics and extrinsic factors. Specific to each individual, personal characteristics include demographics, personality traits, and experience/habit associated with the mobile payment. On the other hand, extrinsic factors exert influences through social mechanisms including culture, social influence, and socioeconomic status. Depending on their "closeness" to behavioral processes, personal characteristics and extrinsic factors have different layers: internal-layer past behavior and socioeconomic status at the bottom, midlayer personality traits and social influence in the middle, and external-layer demographics and culture at the top.

Technology acceptance is the most studied aspect of mobile payment research. The main predicting variables including perceived ease-of-use and perceived usefulness from TAM and similar constructs like effort expectancy and performance expectancy from UTAUT are used by the majority of mobile payment adoption studies [6]. Yet technology acceptance is just a necessary condition of actual usage, which has rarely been included in empirical analyses. The original TAM included the path from behavioral intention to usage behavior [9], yet most studies on mobile payment adoption based on TAM and related frameworks stop at behavioral intention. Actual usage of mobile payment is more than just a yes-or-no decision as typically operationalized for traditional system use. Rather, it concerns how much money, to which extent, and for what purposes a person makes payments with the use of mobile technologies.

Regulating the behavioral processes of mobile payment adoption are personal characteristics of users. Among all, gender, age, personal innovativeness, risk aversion, experience, and habit are identified as the most relevant to mobile payment [5]. Gender and age describe user demographics that concern not only mobile payment but also general information technology adoption [10, 11]. On the other hand, personal innovativeness and risk aversion pertain to

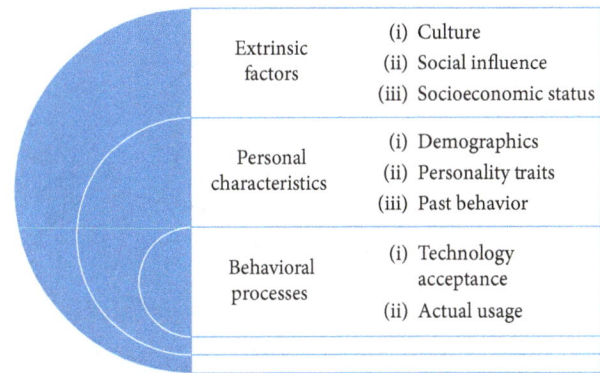

FIGURE 1: A systems view of mobile payment adoption.

the personality traits that are closely related to mobile technology adoption and financial transactions [12, 13]. Mobile payment is a technological innovation that dramatically changes the way in how people make financial transactions, especially in the countries where cash is still the main method, and brings some uncertainties. Thus personal innovativeness and risk aversion represent two sides of a coin related to mobile payment adoption [14]. Finally, both experience and habit are related past behavior that influences future usage of general information technologies as well as mobile payment in specific [1, 15].

Compared with personal characteristics, extrinsic factors are less studied in the extant mobile payment literature. In particular, culture has rarely been included in adoption studies, but it is supposed to have an impact on user behavior. On the other hand, social influence has drawn more attention, as it and the similar construct of the subjective norm are frequently included in the general information technology adoption studies based on TAM and UTAUT [11]. Meanwhile, the variables associated with socioeconomic status including income and education are occasionally included but still underrepresented in mobile payment adoption studies [6].

3. Research Model

The systems view of mobile payment adoption leads to the development of a research model to examine the relationships among relevant variables. As shown in Figure 2, the model includes two stages of behavioral processes: technology acceptance on the left side and actual usage on the right side. The first stage's core comprises TAM constructs, which are affected by personality traits and social influence. Compared with other reflective psychological constructs in the model, the final dependent variable in the second stage is a formative construct to capture actual usage with mobile payment frequency, scope, and amount. User demographics, socioeconomic status, past behavior, and culture play different moderating roles.

Compared with traditional payment methods, mobile payment enables users to make financial transactions anywhere and anytime with great convenience [16]. From the perspective of TAM [9], individual perception of the technology in terms of perceived usefulness and perceived

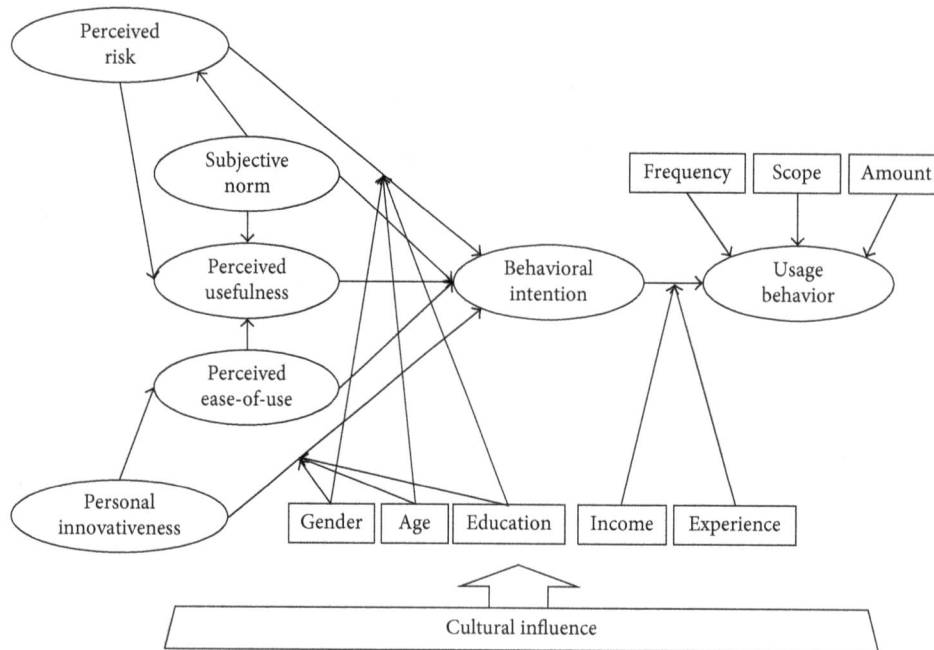

FIGURE 2: Research model.

ease-of-use directly affect the intention to use it. As for mobile payment, it is found that such user cognitions have direct impacts on behavioral intention [17]. Moreover, existing studies also found that users' evaluation of ease-of-use positively affects their belief in usefulness for general information technologies as well as mobile payment [18].

H1: Perceived Usefulness has a positive linear relationship with Behavioral Intention.

H2a: Perceived Ease-of-Use has a positive linear relationship with Behavioral Intention.

H2b: Perceived Ease-of-Use has a positive linear relationship with Perceived Usefulness.

As a personality trait, personal innovativeness describes the tendency of an individual to try out new technologies and innovations [19]. Empirical evidence suggests that personal innovativeness affects user acceptance of IT-based innovation [20]. Mobile payment is an IT-based innovation in mobile commerce, and personal innovativeness has an impact on user evaluation of technology usability [16, 21]. Highly innovative users are likely to have a more positive attitude towards new technologies in terms of the desire to acquire new skills than less innovative users [17, 22]. For mobile payment users, therefore, personal innovativeness may have a partial mediating relationship with behavioral intention through perceived ease-of-use.

H3a: Personal Innovativeness has a positive linear relationship with Behavioral Intention.

H3b: Personal Innovativeness has a positive linear relationship with Perceived Ease-of-Use.

On the other side of the coin to personal innovativeness, risk aversion is a personality trait that has a negative implication for information technology adoption [23, 24]. Perceived risk is the manifestation of risk aversion pertaining to the use of specific technologies/innovations that may expose individuals to certain loss or harm [25]. For mobile payment, in particular, perceived risk is the biggest concern that prevents users from accepting the new technology [26]. Mobile payment users mainly worry about unauthorized use, concerns on device and network reliability, privacy leaks, and transactions errors [27]. When people are aware of the potential loss or harm from the use of a system, they tend to downgrade its value and usefulness and hesitate to use it [4, 16].

H4a: Perceived Risk has a negative linear relationship with Behavioral Intention.

H4b: Perceived Risk has a negative linear relationship with Perceived Usefulness.

Subjective norm captures the social influence on the use of a new system from the relevant views and actions of the peers who have direct or indirect experiences with it [28]. Compared with traditional methods, mobile payment brings obvious advantages as well as potential risks. Facing the dilemma, an individual usually observes the behavior of surrounding people and seeks advice from peers to get more convinced [21, 29]. Therefore, the subjective norm is found to affect people's willingness to use mobile payment [16]. The technology not only supports consumer-business transactions but also integrates seamlessly with social media for personal transfer (e.g., digital "hongbao" or red envelope). The more the people around use mobile payment, the more likely a person is to perceive its value due to network externality [12, 27]. Meanwhile, others' positive view and active use of mobile payment may mitigate the individual's fear of uncertainty.

H5a: Subjective Norm has a positive linear relationship with Behavioral Intention.

H5b: Subjective Norm has a positive linear relationship with Perceived Usefulness.

H5c: Subjective Norm has a negative linear relationship with Perceived Risk.

Behavioral intention is widely regarded as the antecedent to actual technology usage at the individual level [30]. Yet actual usage is rarely included in the empirical analyses of mobile payment adoption. One study on mobile wallet adoption found that behavioral intention explained a large portion of the variance in usage behavior, the operationalization of which was oversimplified though [10]. More clearly defined and measured, actual usage will be included in this study to test its relationship with behavioral intention.

H6: Behavioral Intention has a positive linear relationship with Usage Behavior.

Many existing studies based on TAM and UTAUT examine how user demographics moderate the relationships between behavioral intention and its predictors. In this study, personal innovativeness and perceived risk reflect the personality traits that help explain perceived ease-of-use and perceived usefulness. If user demographics serve as the moderators for all of them, their effects are likely to be confounded. Rather, it makes more theoretical and practical sense to investigate the interactions among personal characteristics in terms of user demographics and personality traits.

Men and women vary in their overall attitude toward computers and associated usage behavior [31]. Two genders exhibit different perceptions and behaviors due to their different socially constructed cognitive structures to encode and process information [32]. They have distinct perceptions of innovative technologies: males care more about usefulness and relative advantage of systems [33, 34], and females are more concerned about ease-of-use and subjective norm [33–35]. Gender differences are also noticed in the studies of web-based shopping and mobile banking adoption as women are generally more risk-averse than men [36, 37]. For user adoption of mobile payment, therefore, gender is likely to interact with personal innovativeness and perceived risk on their effects on behavioral intention.

H7a: Gender moderates the relationship between Personal Innovativeness and Behavioral Intention.

H7b: Gender moderates the relationship between Perceived Risk and Behavioral Intention.

User perceptions and attitudes toward computer technologies also vary across age groups [11]. Mobile payment involves smartphone usage, and the learning curve becomes steeper when age increases. Thus, age is found to moderate the effects of effort expectancy and social influence on user intention in mobile learning [38]. As an innovation involving financial transactions, mobile payment is likely to follow a similar pattern of adoption to that of online shopping, which is also subject to age disparity [39, 40]. Compared with young adults, seniors experience more

barriers to online shopping due to risks and habits [41]. Similar to gender, age is likely to play a moderating role.

H8a: Age moderates the relationship between Personal Innovativeness and Behavioral Intention.

H8b: Age moderates the relationship between Perceived Risk and Behavioral Intention.

Computer technologies require users to have certain knowledge and skills, and their education levels make a difference in adoption and usage behaviors [42]. How well a user is educated is associated with the person's perception of a system in terms of its usability [43]. Also, education level is found to be negatively correlated with user anxiety in computer use [44]. For an innovative technology like mobile payment, therefore, education may interact with personality traits related to innovativeness and risk-averseness.

H9a: Education moderates the relationship between Personal Innovativeness and Behavioral Intention.

H9b: Education moderates the relationship between Perceived Risk and Behavioral Intention.

Whether mobile payment is for online shopping or face-to-face purchase (e.g., restaurant and taxi), it is the last step to complete the transaction. Due to other constraints, a person's intention to use mobile payment may not always be converted into actual usage. Among them, the individual's previous mobile payment experience and income level cannot be ignored [45]. On the technological side, the previous experience with an innovation can influence an individual's perceived ease-of-use, which affects usage volume and frequency [43]. On the socioeconomic side, personal income is closely related to purchasing power and risk tolerance associated with online transactions [46]. In a study of mobile wallet, for instance, income is found positively associated with an individual's acceptance and use of the technology [10]. All else being equal (especially intention), users at the different experience and income levels are likely to use the mobile payment to different extents.

H10: Experience moderates the relationship between Behavioral Intention and Usage Behavior.

H11: Income moderates the relationship between Behavioral Intention and Usage Behavior.

Finally, mobile payment platforms are based on specific currency systems, and people's usage behavior is likely to vary from one country to another. In particular, national culture concerns the fundamental values and shared beliefs among people in a country [47]. There are six cultural dimensions along which people's behavior may vary: (a) Power Distance: acceptance to unequal power distribution in society; (b) Individualism versus Collectivism: tendency integrate into strong cohesive groups; (c) Masculinity versus Femininity: preference between male-associated qualities (e.g., assertiveness and material success) and female-associated ones (e.g., modesty and quality of life); (d) Uncertainty Avoidance: fear of unknown situations; (e) Long-term Orientation: persistence and thrift leading to future rewards; (f) Indulgence: tendency to seek happiness [48]. Among them, some are closely related to the extrinsic factors and personal characteristics pertaining to mobile payment. In particular, the dimension of Individualism versus Collectivism concerns social influence and the dimension of Uncertainty Avoidance concerns risk aversion. Thus, the hypothesized relationships as mentioned above may vary significantly across different cultures.

4. Methodology

4.1. Research Design. The target population comprises mobile payment users in multiple countries that have relatively high population penetration of smartphone technology yet very different cultures. The two countries that lead the trend of smartphone diffusion are the USA and China [49]. Whereas mobile payment in the USA had an early start, China is catching up quickly, and the total annual transaction exceeded 5 trillion US dollars (50 times that of the USA and more than Japan's GDP) in 2016 [50]. Their cultures are distinct, as shown in Figure 3. In particular, the USA is high in individualism whereas China is high in collectivism. This suggests that the effects of social influence on mobile payment user behavior vary across two countries. Also, the USA is noticeably higher than China in uncertain avoidance, which makes a difference in perceived risk associated with the mobile payment.

This study tests the invariance in the hypothesized relationships with the observations collected from different cultures. If a large proportion of the relationships vary significantly across the samples, there is supporting evidence of the cultural influences on mobile payment adoption. Thus, this study conducted a survey with working professionals in both China and the USA, most of whom own smartphones and are more likely to use mobile payment than students and retirees. Invitations to the online survey were sent to full-time workers in two countries based on snowball sampling. Initial contacts were gathered from three profession training programs in China and USA. The participants were encouraged to send the invitation to their friends and relatives who might have used mobile payment.

4.2. Subjects. Altogether, there were 162 valid responses in the China sample and 136 in the USA sample, leading to a total sample size of 298. The sample size is sufficient for statistical analyses used in this study, mainly factor analysis for measurement validation and partial least squares for model estimation [51]. As shown in Table 1, the participants in the two samples had somewhat different profiles. Whereas more than one-third participants in the USA sample had used mobile payment for three years or more, one-fourth had limited experience (i.e., <6 months). Meanwhile, very few in the China sample were new to the technology but more than 80% were quite experienced with more than one-year history. On average, the China sample was younger, and the USA sample had higher levels of education. Gender and income distribution were relatively balanced between the

two samples (e.g., around two-thirds in both had low or medium low income).

Potential nonresponse bias was assessed by comparing early and late responses [52]. In both samples, the first 45 and last 45 responses had insignificant differences on the means of any variables based on *t* tests. The invariance suggested no serious threat of nonresponse bias.

4.3. Measurement. The Appendix section lists all the measurement items used in the questionnaire. All psychometric scales were adapted from previous studies. Items measuring Perceived Usefulness and Perceived Ease-of-Use were adapted from Davis [9]. Behavioral Intention and Subject Norm measures were adapted from Fassnacht and Köse [53] and Schierz et al. [54]. Personal Innovativeness and Perceived Risk were measured with items adapted from Yang et al. [16]. Other more objective variables, such as mobile payment frequency, scope, and amount, were measured with self-developed items. The score of mobile payment scope was calculated as the count of total options (e.g., dining and bill pay) checked.

As most of the questionnaire items were psychometric measures, their potential common method bias was assessed following Harman's one-factor test [55, 56]. Exploratory factor analysis (EFA) results revealed that 40.11% common variance was captured by the first principal component (less than half), whereas all the major components (eigenvalue >1) explained 69.23% (more than two-third). Confirmatory factor analysis (CFA) results as reported in Table 2 compared method-only, trait-only, and trait/method models. The goodness-of-fit indices of the method-only model were much worse than those of trait-only model, which were even slightly better than those of trait-and-method model. Together, the EFA and CFA results indicated that common method bias was not a serious issue.

5. Results

The descriptive statistics in Table 3 show the response patterns of all variables, the possible range of each is between one and five. In the overall sample, the average responses of psychological variables related to technology acceptance were higher than the midpoint of three, but those of actual mobile payment usage variables were lower. The distinct response patterns support the use of relatively objective Usage Behavior measures that not only mitigates common method bias but also gauges the gap between psychological behavior and overt behavior. Cross-country comparison shows that the USA sample exhibited more positive responses on psychological constructs (except for Perceived Risk), but the China sample showed relatively active mobile payment usage.

The results in Table 4 validate the reflective psychological constructs in terms of convergent validity, discriminant validity, and nomological validity. Convergent validity was supported as all the coefficient alpha and composite reliability (CR) values were above 0.7, and average variance extracted (AVE) values were above 0.5. Discriminant

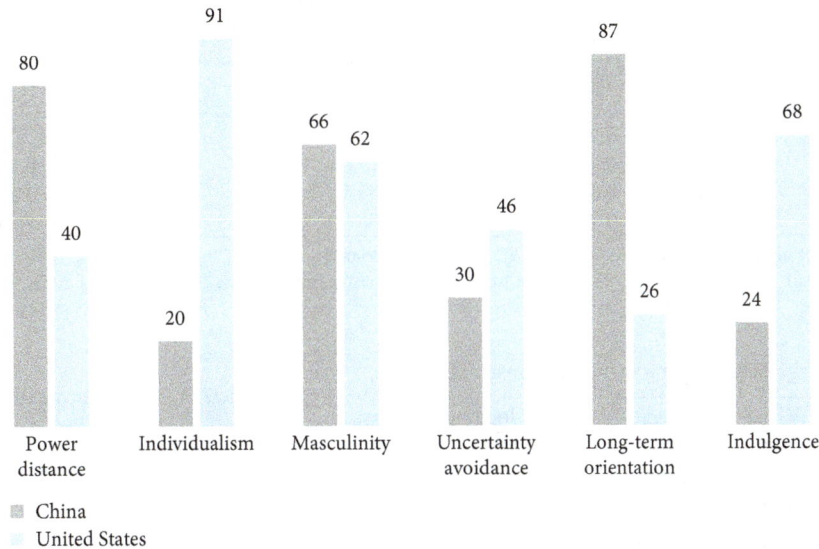

Figure 3: Cultural comparison between China and the USA (source: Hofstede [48]).

Table 1: Profiles of participants.

Characteristics	China ($n = 162$)	USA ($n = 136$)
Gender		
Male	86 (53.1%)	61 (44.9%)
Female	76 (46.9%)	75 (55.1%)
Age		
25 or younger	125 (77.2%)	32 (23.5%)
26–40	27 (16.7%)	81 (59.6%)
Older than 40	10 (6.2%)	23 (16.9%)
Education		
High school	15 (9.3%)	1 (0.7%)
Associate	27 (16.7%)	1 (0.7%)
Bachelor	91 (56.2%)	98 (72.1%)
Master	16 (9.9%)	32 (23.5%)
Doctoral	13 (8.0%)	4 (2.9%)
Income		
Low	63 (38.9%)	78 (57.4%)
Medium low	46 (28.4%)	12 (8.8%)
Medium	27 (16.7%)	13 (9.6%)
Medium high	9 (5.6%)	4 (2.9%)
High	17 (10.5%)	29 (21.3%)
Experience		
0–6 months	4 (2.5%)	33 (24.3%)
6–12 months	20 (12.3%)	15 (11.0%)
1-2 years	50 (30.9%)	17 (12.5%)
2-3 years	38 (23.5%)	21 (15.4%)
>3 years	50 (30.9%)	50 (36.8%)

Table 2: Common method bias assessment with CFA.

Model	χ^2	df	χ^2/df	RMSEA	CFI	NFI
Method-only (1-factor)	1243.32	135	9.21	0.166	0.648	0.624
Trait-only (6-factor)	231.96	120	1.93	0.056	0.964	0.930
Trait-and-method (7-factor)	243.85	114	2.14	0.062	0.959	0.926

Note. RMSEA: root mean square error of approximation; CFI: comparative fit index; NFI: normed fit index.

Table 3: Descriptive statistics and sample comparison.

Variable	Overall	China	USA	Difference
Perceived Ease-of-Use	4.52 (0.68)	4.28 (0.68)	4.80 (0.56)	−0.53
Personal Innovativeness	4.16 (0.81)	3.91 (0.79)	4.46 (0.74)	−0.55
Perceived Usefulness	4.61 (0.62)	4.48 (0.60)	4.76 (0.60)	−0.28
Perceived Risk	3.80 (1.15)	4.00 (1.02)	3.57 (1.24)	0.43
Subjective Norm	4.07 (0.78)	4.11 (0.70)	4.02 (0.86)	0.09[ns]
Behavioral Intention	4.30 (0.84)	4.07 (0.79)	4.59 (0.83)	−0.52
Frequency	2.50 (0.68)	2.74 (0.53)	2.21 (0.72)	0.53
Amount	3.57 (1.34)	3.83 (1.21)	3.27 (1.44)	0.56
Scope	2.02 (1.03)	1.95 (0.85)	2.10 (1.21)	−0.15[ns]

Note. Standard deviations in the parentheses beside the means; [ns]not significant at the 0.1 level, all other differences were significant at the 0.01 level.

validity was supported as the square roots of AVE were all greater than the correlation coefficients. As expected, all variables were positively correlated with each other, except for perceived risk. Thus, nomological validity was also supported.

The validation of Usage Behavior as a formative construct has different requirements. Instead of being consistent with each other, formative indicators are supposed to be somewhat distinct and have nontrivial contributions to the construct in question. As shown in Table 5, all the variance inflation factors (VIF) were well below 5, indicating nonsalient multicollinearity among formative indicators. The relationship between each indicator and the construct was significant as indicated by multiple regression weight, and

TABLE 4: Measurement validation for reflective constructs.

Construct	α	CR	AVE	V1	V2	V3	V4	V5	V6
V1: Perceived Ease-of-Use	0.84	0.90	0.76	**0.87**					
V2: Personal Innovativeness	0.82	0.89	0.73	0.49	**0.86**				
V3: Perceived Usefulness	0.85	0.91	0.77	0.66	0.55	**0.88**			
V4: Perceived Risk	0.89	0.93	0.82	−0.15	−0.16	−0.17	**0.91**		
V5: Subjective Norm	0.75	0.85	0.66	0.39	0.20	0.39	−0.03^{ns}	**0.81**	
V6: Behavioral Intention	0.90	0.94	0.83	0.66	0.55	0.75	−0.20	0.50	**0.91**

Note. α: Cronbach's coefficient alpha; CR: composite reliability; AVE: average variance extracted; ^{ns}not significant at 0.05 level, all other correlation coefficients were significant at 0.01 level. The bold on the diagonal of correlation matrix indicates the squared root of AVE.

TABLE 5: Validation of usage behavior as a formative construct.

Formative indicator	VIF	Weight	Outer loading
Frequency	1.317	0.738	0.925
Amount	1.332	0.246	0.672
Scope	1.098	0.286	0.533

Note. VIF: variance inflation factor. All weights and outer loadings were significant at 0.01 level.

TABLE 6: Coefficients of determination (R^2) of endogenous variables.

Endogenous variable	Overall	China	USA
Perceived Ease-of-Use	0.234	0.134	0.206
Perceived Risk	0.001	0.000	0.007
Perceived Usefulness	0.464	0.447	0.456
Behavioral Intention	0.678	0.612	0.824
Usage Behavior	0.442	0.262	0.618

outer loading (i.e., simple regression weight) was significant, suggesting none removable [51].

To test the hypothesized relationships that involve both reflective and formative constructs, partial least squares (PLS) structural equation modeling is appropriate [51]. Table 6 reports the endogenous variables' coefficients of determination (R^2) for the overall sample as well as two country samples. In the overall sample, more than two-thirds of variance was explained for Behavioral Intention, less than half for Perceived Usefulness and Usage Behavior, around one-fourth for perceived ease-of-use, and almost none for Perceived Risk. This is somewhat consistent with the number of predictors that each construct has. In particular, the majority of variation in Behavioral Intention was accounted for, suggesting that most important predictors are included. Across the USA and China samples, the coefficients of determination from the former were more or less higher than those from the latter, especially in the case of Usage Behavior. Thus the gap between mobile payment intention and actual usage seems wider for people in China than those in the USA.

Table 7 reports the standardized estimates of each path coefficient obtained from overall and split samples. In the overall sample, all the hypothesized linear relationships (i.e., H1–H6) were significant except for that between Subjective Norm and Perceived Risk. Meanwhile, three out of eight moderating effects turned out to be significant, including the two from Experience and Income to the relationship between Behavioral Intention and Usage Behavior. In either country sample, however, only the moderator Education did not yield any significant effects. Thus, there is more or less supporting evidence for each research hypothesis, except for those related to Education. In addition, a multigroup analysis (MGA) was conducted to examine cross-culture differences in path coefficient estimates. The observed significance level of each difference was obtained with the permutation method of MGA based on the two-tailed test [51]. About half of the relationships were

found to be quite different across the USA and China samples. Thus, culture did make noticeable differences in hypothesized relationships.

In particular, Perceived Usefulness and Perceived Ease-of-Use had stronger effects on Behavioral Intention in the USA sample than in the China sample. Personal innovativeness' effect on Behavioral Intention, however, was the other way around. Perceived Risk, as another user characteristic, had more negative effect on Perceived Usefulness in the USA sample than in the China sample. Social influence (i.e., Subjective Norm) on Behavioral Intention was stronger in China than in the USA, yet its effect on Perceived Risk switched in strength between two samples. Gender interacted with Personal Innovativeness in China but with Perceived Risk in the USA in their effects on Behavioral Intention. Age is the opposite: it interacted with Perceived Risk in China but with Personal Innovativeness in the USA. Finally, Experience and Income played more negative moderating roles on the relationship between Behavioral Intention and Usage Behavior in the China sample than in the USA sample. Their direct impacts on Usage Behavior were also more positive in the USA sample.

Figure 4 illustrates the salient moderating effects in each country sample. Their f-square values indicate the effect sizes of moderation [57]. All were well above 0.009, the average moderating effect found in a meta-analysis [58]. When the age increased, the effect of perceived risk on behavior intention got less negative in the China sample and that of personal innovativeness got more positive in the USA sample. Compared with males (Gender = 0), females (Gender = 1) saw a more positive relationship between Personal Innovativeness and Behavior Intention in the China sample, but more negative effect of Perceived Risk in the USA sample. For both countries, more Experience in mobile payment meant more active Usage Behavior, yet the effect of behavioral intention diminished due to habitual use

TABLE 7: Standardized PLS estimates.

Path	Overall	China	USA	Diff.
H1: Perceived Usefulness → Behavioral Intention	0.422***	0.312***	0.615***	−0.303**
H2a: Perceived Ease-of-Use → Behavioral Intention	0.202***	0.144*	0.208***	−0.064
H2b: Perceived Ease-of-Use → Perceived Usefulness	0.588***	0.580***	0.553***	0.027
H3a: Personal Innovativeness → Behavioral Intention	0.154***	0.156***	0.073	0.083
H3b: Personal Innovativeness → Perceived Ease-of-Use	0.484***	0.366***	0.454***	−0.088
H4a: Perceived Risk → Behavioral Intention	−0.059**	−0.074**	−0.037*	−0.037
H4b: Perceived Risk → Perceived Usefulness	−0.079**	−0.017	−0.148***	0.132**
H5a: Subjective Norm → Behavioral Intention	0.222***	0.386***	0.106***	0.280***
H5b: Subjective Norm → Perceived Usefulness	0.156***	0.158***	0.156***	0.002
H5c: Subjective Norm → Perceived Risk	−0.030	−0.002	−0.082*	0.081
H6: Behavioral Intention → Usage Behavior	0.112**	0.202**	0.274***	−0.072
H7a: Gender × Innovativeness → Behavioral Intention	−0.043	0.133**	0.031	0.101
H7b: Gender × Perceived Risk → Behavioral Intention	0.000	0.045	−0.079**	0.124**
H8a: Age × Innovativeness → Behavioral Intention	0.098**	0.069	0.138**	−0.069
H8b: Age × Perceived Risk → Behavioral Intention	0.012	0.084*	−0.021	0.105*
H9a: Education × Innovativeness → Behavioral Intention	0.013	0.054	−0.008	0.061
H9b: Education × Perceived Risk → Behavioral Intention	−0.039	0.000	0.037	−0.037
H10: Experience × Intention → Usage Behavior	−0.107***	−0.147**	−0.110	−0.037
H11: Income × Intention → Usage Behavior	0.100**	−0.147*	0.080*	−0.227**
Gender → Behavioral Intention	−0.046	−0.050	−0.076**	0.026
Age → Behavioral Intention	0.016	−0.025	0.008	−0.033
Education → Behavioral Intention	0.013	−0.018	0.013	−0.031
Income → Usage Behavior	0.072	−0.074	0.144**	−0.218**
Experience → Usage Behavior	0.590***	0.344***	0.612***	−0.268**

Note. *significant at 0.1 level; **significant at 0.05 level; ***significant at 0.01 level.

[15]. Income played different moderating roles across two countries: people with higher income were more likely to convert Behavioral Intention to Usage Behavior in the USA sample, but it was the opposite in the China sample.

6. Discussions

The findings yield some important theoretical and practical implications. First of all, they support the conceptualization of mobile payment adoption as a system that involves the interactions among behavioral processes, personal characteristics, and extrinsic factors. Three layers of personal characteristics and extrinsic factors affect two stages of behavioral processes in terms of technology acceptance and actual usage in different ways. The internal-layer elements at the bottom of extrinsic factors and personal characteristics in Figure 1, including past behavior and socioeconomic status, are found to mainly moderate the relationship between technology acceptance and actual usage. The midlayer elements in the middle, including personality traits and social influence, have direct impacts on technology acceptance. The external-layer elements at the top, including demographics and culture, mainly make differences in the linear relationships involved in technology acceptance and other moderating relationships.

The multistage and multilayer conceptualization and modeling yield a deeper understanding of mobile payment adoption. Compared with extant research on mobile payment adoption, this study helps bridge the gap between the psychological behavior of technology acceptance and the overt behavior of actual usage with additional variables associated with both. Due to their different natures, personal

characteristics and extrinsic factors exert influences on technology acceptance and actual usage through direct, mediating, and moderating routes. In addition, the comparison between the samples from China and the USA suggests that their cultures make differences in the strengths of many relationships. Responding to the call for more meaningful and generalizable research on mobile payment adoption [5], this study contributes to the mobile payment literature by deepening the understanding of adoption stages and expanding the scope of explanatory variables at the same time.

Some specific findings may be interesting to researchers and practitioners. For instance, the effect of Subjective Norm on Behavioral Intention was significantly stronger in the China sample than that in the USA sample. The two countries are very different along the relevant cultural dimension of Individualism versus Collectivism. Compared with American people, Chinese people are more likely to form strong and coherent groups and influence each other. On the other hand, the negative effect of Perceived Risk on Perceived Usefulness was stronger in the USA sample than that in the China sample. Along the relevant dimension of Uncertainty Avoidance, correspondingly, the finding suggests that American users worry more about the potential security and privacy breaches from using mobile payment than Chinese users.

In addition to the moderation of linear relationships, Culture makes a difference in how Income moderates the relationship between Behavioral Intention and Usage Behavior. In the USA sample, higher income is conducive to the conversion from technology acceptance to actual usage, but it is the opposite in the China sample. In China, credit

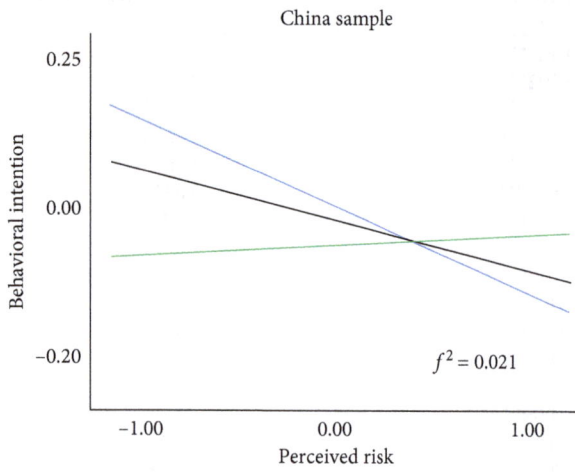

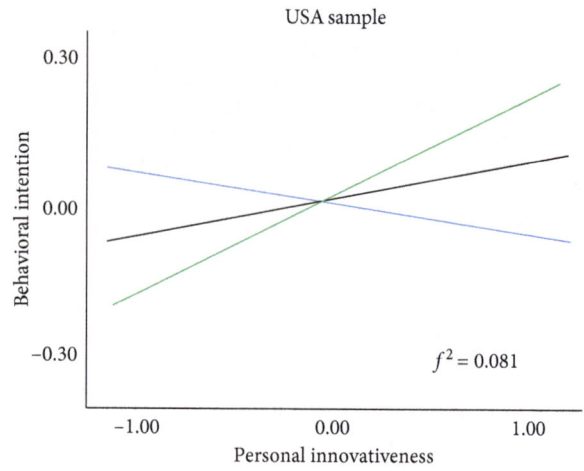

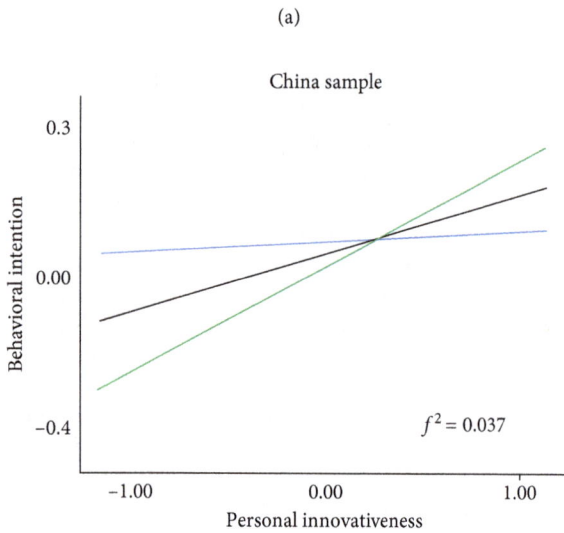

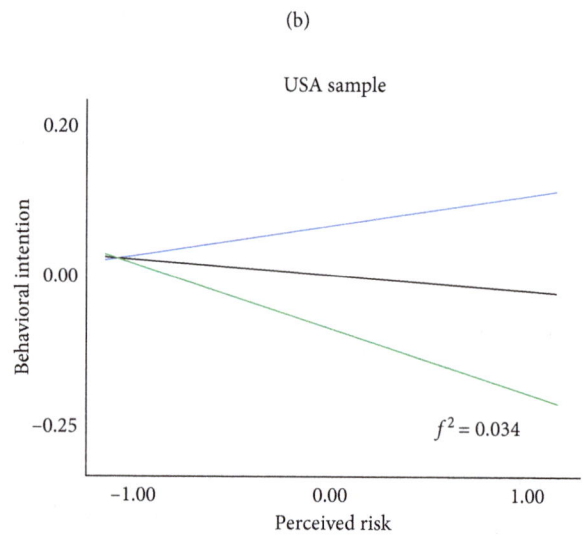

(a)

(b)

(c)

(d)

FIGURE 4: Continued.

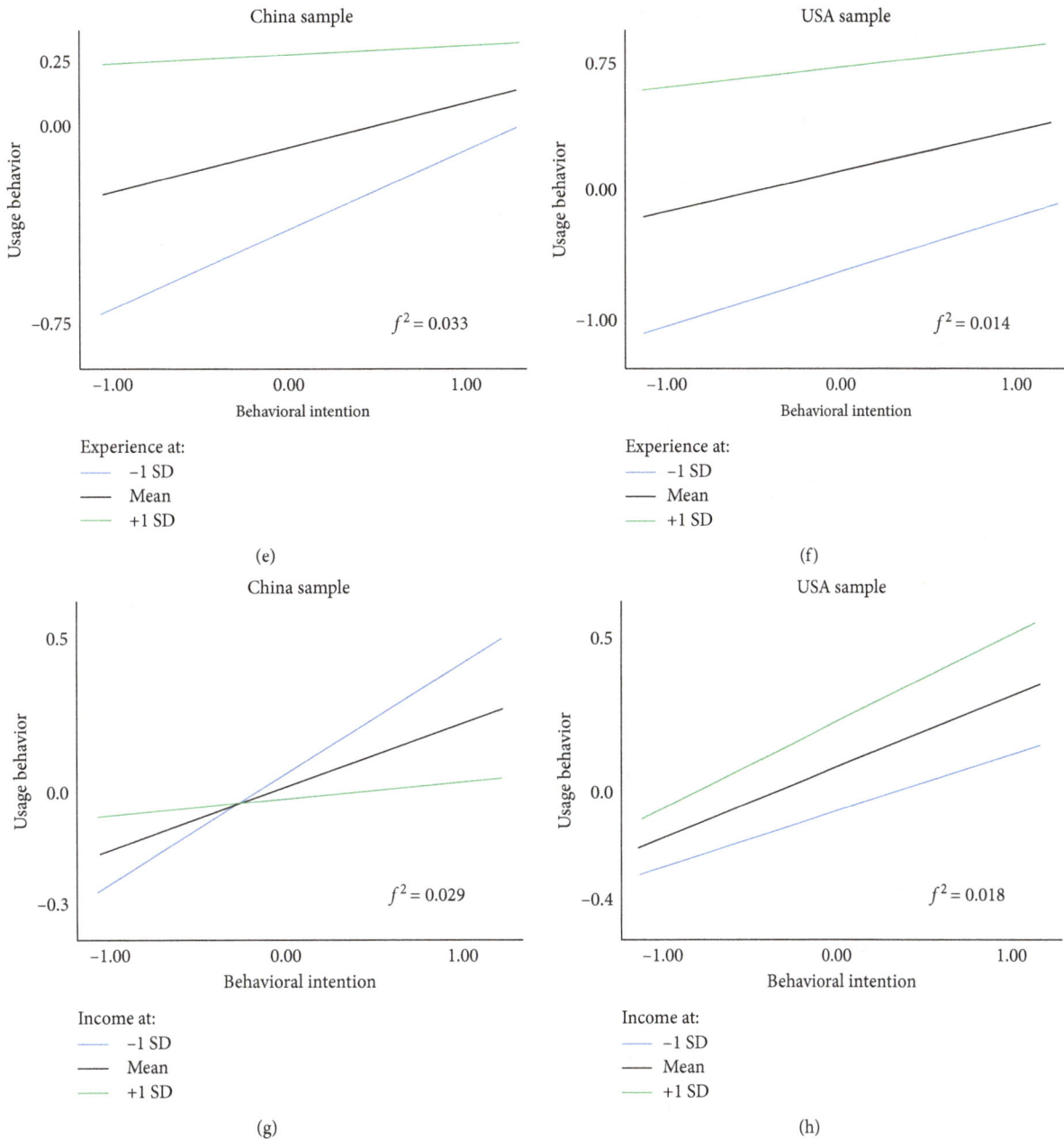

FIGURE 4: Salient moderation effects.

card transactions are still rare and the primary method of payment is still cash. Mobile payment provides a viable means to shop online, dine in restaurants, and pay for services (e.g., taxi). Online stores usually offer lower prices than brick-and-mortar stores. To encourage the use of mobile payment (e.g., so as to keep track of customers), vendors often offer additional discounts. For people of relatively low income, the saving from mobile payment constitutes a major incentive. Yet mobile payment is riskier than cash transactions. For people with relatively high income, they are more concerned about the potential loss associated with security and privacy breaches than monetary saving. In the USA, however, the differences in prices and risk levels between mobile payment and other methods are

not that obvious, and income plays a positive moderating role as expected.

Insignificant results also deserve a close look. Especially, Education does not moderate the effects of Personal Innovativeness and Perceived Risk on Behavioral Intention in either country. In theory, education may provide potential mobile payment users more background knowledge about it and facilitate adoption decision-making [59]. Yet there is no supporting evidence from the observations. One explanation is that mobile payment is an innovation that brings people not only benefits but also risks. When people are well-educated, they also become aware of its cons as well as pros, canceling out the positive effect of Education. This explanation is more applicable to developing countries like China where mobile payment is still relatively new to most people. In developed countries like USA, however, mobile payment is no longer a cutting-edge innovation, and it is possible that people at all education levels are familiar with it.

The findings provide some helpful clues on the best practices to promote mobile payment adoption. There are two possible routes: one to enhance technology acceptance and another to materialize actual usage. At an early stage of mobile payment development, it is worth the effort to help people accept the technology first. It is more effective to target potential users of relatively high personal innovativeness and low risk aversion, who will then influence others through word-of-mouth. After mobile payment gained a certain level of popularity, the main challenge is how to convert technology acceptance financial transaction-making. This highlights the importance of investigating actual usage in addition to behavioral intention at the current stage of mobile payment diffusion. As the findings indicate, the strategy needs to be based on case-by-case analyses considering cultural factors, business environment, income levels, and so on.

7. Conclusion

This study examines how personal characteristics and extrinsic factors influence the behavioral processes of mobile payment adoption in terms of technology acceptance and actual usage. Based on the understanding of embedding relationships, it proposes a research model that hypothesizes their direct and moderating effects on user behavior. The survey observations collected from the USA and China provide supporting evidence to most of the research hypotheses and reveal some interesting cross-country differences. The findings suggest that social influence and personality traits have direct impacts on technology acceptance, whereas demographics, past behavior, socioeconomic status, and culture play different moderating roles.

This study has limitations that point to the directions of future research. A major limitation of this study is due to the fact that the observations were collected from only two countries. China and the USA are selected because they have large populations of smartphone users and are distinct in culture. Yet they cannot represent other countries and regions, which limits the generalizability of findings. In addition, culture is used as a grouping variable in this study, but its different dimensions may have different effects on mobile payment adoption. Future studies may collect data from more countries and include specific cultural dimensions in analyses. This will not only enhance the generalizability of findings but also reveal the specific roles that different cultural dimensions play.

Appendix

Survey Questionnaire Items

Gender

☐Male ☐Female

Age

☐<25 ☐25–30 ☐31–35 ☐36–40 ☐41–45 ☐46–50 ☐>50

Education

Highest degree owned:
☐High School ☐Associate ☐Bachelor ☐Master's ☐Doctoral

Income

Monthly Income:
☐<5,000 ☐5,000–5,999 ☐6,000–6,999 ☐7,000–7,999 ☐8,000–8,999 ☐9,000–9,999 ☐>10,000

Experience

How long have you been using mobile payment?
☐0–6 months ☐6–12 months ☐1–2 years ☐2-3 years ☐>3 years

Scope

You use mobile payment for (please check all that apply):
☐Restaurant dining ☐Bill pay ☐Digital products ☐Traditional products/services ☐Others (please specify): _____

Frequency

How many mobile payments did you make during the last month?
☐None ☐1–3 times ☐4–7 times ☐8–10 times ☐>10 times

Amount

The largest amount of mobile payment that you are willing to make:
☐<50 ☐50–100 ☐100–300 ☐300–500 ☐>500

Perceived Usefulness

Using mobile payment makes my life more convenient.
Compared with other methods, mobile payment has many advantages.
To me, mobile payment is useful.

Perceived Ease-Of-Use

It is easy for me to become skillful at using mobile payment.
The steps to follow for mobile payment are clear to me.
In general, I find mobile payment easy to use.

Behavioral Intention

I intend to use mobile payment.
I will use mobile payment if there is a chance.
I will recommend mobile payment to my friends/relatives/colleagues.

Personal Innovativeness

I am ready to try out new ideas and innovations.

Among my peers, I am usually the first to explore new things.

I like to experiment with new technologies and services.

Perceived Risk

I would not feel totally safe providing personal privacy information during mobile payment.

I am worried to use mobile payment because someone else may be able to access my account.

I would not feel secure sending sensitive information through mobile payment.

Subjective Norm

Many of my friends/relatives/colleagues are using mobile payment.

My friends/relatives/colleagues think that I should use mobile payment.

People around me believe it is a good idea to use mobile payment.

References

[1] F. Liébana-Cabanillas, J. Sánchez-Fernández, and F. Muñoz-Leiva, "The moderating effect of experience in the adoption of mobile payment tools in virtual social networks: the m-Payment Acceptance Model in Virtual Social Networks (MPAM-VSN)," *International Journal of Information Management*, vol. 34, no. 2, pp. 151–166, 2014.

[2] D. Shrier, G. Canale, and A. Pentland, "Mobile money & payments: Technology trends," in *MIT Connection Science's Series on Financial Technology*, MIT, Cambridge, MA, USA, 2016.

[3] C. Sanford and H. Oh, "The role of user resistance in the adoption of a mobile data service," *Cyberpsychology, Behavior, and Social Networking*, vol. 13, no. 6, pp. 663–672, 2010.

[4] M. Barkhordari, Z. Nourollah, H. Mashayekhi, Y. Mashayekhi, and M. S. Ahangar, "Factors influencing adoption of e-payment systems: an empirical study on Iranian customers," *Information Systems and e-Business Management*, vol. 15, no. 1, pp. 89–116, 2017.

[5] T. Dahlberg, J. Guo, and J. Ondrus, "A critical review of mobile payment research," *Electronic Commerce Research and Applications*, vol. 14, no. 5, pp. 265–284, 2015.

[6] E. L. Slade, M. D. Williams, and Y. K. Dwivedi, "Mobile payment adoption: classification and review of the extant literature," *Marketing Review*, vol. 13, no. 2, pp. 167–190, 2013.

[7] A. Zhanga, X. Yue, and Y. Kong, "Exploring culture factors affecting the adoption of mobile payment," in *Proceedings of Tenth International Conference on Mobile Business, (ICMB 2011)*, pp. 263–267, Como, Italy, June 2011.

[8] R. Stichweh, "Systems theory," in *International Encyclopedia of Political Science*, Sage Publications, New York, NY, USA, 2011.

[9] F. D. Davis, "Perceived usefulness, perceived ease of use, and user acceptance of information technology," *MIS Quarterly*, vol. 13, no. 3, pp. 319–340, 1989.

[10] D.-H. Shin, "Towards an understanding of the consumer acceptance of mobile wallet," *Computers in Human Behavior*, vol. 25, no. 6, pp. 1343–1354, 2009.

[11] V. Venkatesh, M. G. Morris, G. B. Davis, and F. D. Davis, "User acceptance of information technology: toward a unified view," *MIS Quarterly*, vol. 27, no. 3, pp. 425–478, 2003.

[12] J. Lu, J. E. Yao, and C.-S. Yu, "Personal innovativeness, social influences and adoption of wireless Internet services via mobile technology," *Journal of Strategic Information Systems*, vol. 14, no. 3, pp. 245–268, 2005.

[13] G. Filbeck, P. Hatfield, and P. Horvath, "Risk aversion and personality type," *Journal of Behavioral Finance*, vol. 6, no. 4, pp. 170–180, 2005.

[14] R. Thakur and M. Srivastava, "Adoption readiness, personal innovativeness, perceived risk and usage intention across customer groups for mobile payment services in India," *Internet Research*, vol. 24, no. 3, pp. 369–392, 2014.

[15] V. Venkatesh, J. Y. Thong, and X. Xu, "Consumer acceptance and use of information technology: extending the unified theory of acceptance and use of technology," *MIS Quarterly*, vol. 36, no. 1, pp. 157–178, 2012.

[16] S. Yang, Y. Lu, S. Gupta, Y. Cao, and R. Zhang, "Mobile payment services adoption across time: an empirical study of the effects of behavioral beliefs, social influences, and personal traits," *Computers in Human Behavior*, vol. 28, no. 1, pp. 129–142, 2012.

[17] C. Kim, M. Mirusmonov, and I. Lee, "An empirical examination of factors influencing the intention to use mobile payment," *Computers in Human Behavior*, vol. 26, no. 3, pp. 310–322, 2010.

[18] T. Oliveira, M. Faria, M. A. Thomas, and A. Popovič, "Extending the understanding of mobile banking adoption: when UTAUT meets TTF and ITM," *International Journal of Information Management*, vol. 34, no. 5, pp. 689–703, 2014.

[19] M. K. Chang, W. Cheung, and V. S. Lai, "Literature derived reference models for the adoption of online shopping," *Information and Management*, vol. 42, no. 4, pp. 543–559, 2005.

[20] M. Y. Yi, K. D. Fiedler, and J. S. Park, "Understanding the role of individual innovativeness in the acceptance of IT-based innovations: Comparative analyses of models and measures," *Decision Sciences*, vol. 37, no. 3, pp. 393–426, 2006.

[21] Z. Kalinic and V. Marinkovic, "Determinants of users' intention to adopt m-commerce: an empirical analysis," *Information Systems and e-Business Management*, vol. 14, no. 2, pp. 367–387, 2016.

[22] B. Tariq, "Exploring factors influencing the adoption of mobile commerce," *Journal of Internet Banking and Commerce*, vol. 12, no. 3, pp. 32–42, 2007.

[23] Y. A. Au and R. J. Kauffman, "What do you know? Rational expectations in information technology adoption and investment," *Journal of Management Information Systems*, vol. 20, no. 2, pp. 49–76, 2003.

[24] N. Donthu and A. Garcia, "The internet shopper," *Journal of Advertising Research*, vol. 39, no. 3, pp. 52–58, 1999.

[25] M. S. Featherman and P. A. Pavlou, "Predicting e-services adoption: a perceived risk facets perspective," *International Journal of Human-Computer Studies*, vol. 59, no. 4, pp. 451–474, 2003.

[26] IResearch, *IResearh Wisdom: Date of Chinese Netizens Use Cell Phone Banking in 2009*, IResearch, Beijing, China, 2009.

[27] N. Mallat, "Exploring consumer adoption of mobile payments—a qualitative study," *Journal of Strategic Information Systems*, vol. 16, no. 4, pp. 413–432, 2007.

[28] V. Venkatesh and F. D. Davis, "A theoretical extension of the technology acceptance model: four longitudinal field studies," *Management Science*, vol. 46, no. 2, pp. 186–204, 2000.

[29] E. Karahanna, D. W. Straub, and N. L. Chervany, "Information technology adoption across time: a cross-sectional comparison of pre-adoption and post-adoption beliefs," *MIS Quarterly*, vol. 23, no. 2, pp. 183–213, 1999.

[30] V. Venkatesh, C. Speier, and M. G. Morris, "User acceptance enablers in individual decision making about technology: toward an integrated model," *Decision Sciences*, vol. 33, no. 2, pp. 297–316, 2002.

[31] B. E. Whitley, "Gender differences in computer-related attitudes and behavior: a meta-analysis," *Computers in Human Behavior*, vol. 13, no. 1, pp. 1–22, 1997.

[32] S. L. Bem, "The BSRI and gender schema theory: a reply to Spence and Helmreich," *Psychological Review*, vol. 88, no. 4, pp. 369–371, 1981.

[33] V. Venkatesh and M. G. Morris, "Why don't men ever stop to ask for directions? Gender, social influence, and their role in technology acceptance and usage behavior," *MIS Quarterly*, vol. 24, no. 1, pp. 115–139, 2000.

[34] H. Nysveen, P. E. Pedersen, and H. Thorbjørnsen, "Explaining intention to use mobile chat services: moderating effects of gender," *Journal of Consumer Marketing*, vol. 22, no. 5, pp. 247–256, 2005.

[35] V. Venkatesh, M. G. Morris, and P. L. Ackerman, "A longitudinal field investigation of gender differences in individual technology adoption decision-making processes," *Organizational Behavior and Human Decision Processes*, vol. 83, no. 1, pp. 33–60, 2000.

[36] C. Van Slyke, C. L. Comunale, and F. Belanger, "Gender differences in perceptions of web-based shopping," *Communications of the ACM*, vol. 45, no. 8, pp. 82–86, 2002.

[37] H. E. Riquelme and R. E. Rios, "The moderating effect of gender in the adoption of mobile banking," *International Journal of Bank Marketing*, vol. 28, no. 5, pp. 328–341, 2010.

[38] Y. S. Wang, M. C. Wu, and H. Y. Wang, "Investigating the determinants and age and gender differences in the acceptance of mobile learning," *British Journal of Educational Technology*, vol. 40, no. 1, pp. 92–118, 2009.

[39] P. Sorce, V. Perotti, and S. Widrick, "Attitude and age differences in online buying," *International Journal of Retail and Distribution Management*, vol. 33, no. 2, pp. 122–132, 2005.

[40] K. A. Passyn, M. Diriker, and R. B. Settle, "Images of online versus store shopping: have the attitudes of men and women, young and old really changed?," *Journal of Business and Economics Research*, vol. 9, no. 1, pp. 99–110, 2011.

[41] J.-W. Lian and D. C. Yen, "Online shopping drivers and barriers for older adults: age and gender differences," *Computers in Human Behavior*, vol. 37, pp. 133–143, 2014.

[42] R. W. Zmud, "Individual differences and MIS success: a review of the empirical literature," *Management Science*, vol. 25, no. 10, pp. 966–979, 1979.

[43] A. Burton-Jones and G. S. Hubona, "The mediation of external variables in the technology acceptance model," *Information and Management*, vol. 43, no. 6, pp. 706–717, 2006.

[44] M. Igbaria and S. Parasuraman, "A path analytic study of individual characteristics, computer anxiety and attitudes toward microcomputers," *Journal of Management*, vol. 15, no. 3, pp. 373–388, 1989.

[45] F. Liébana-Cabanillas, F. Muñoz-Leiva, J. Sánchez-Fernández, and M. I. Viedma-del Jesús, "The moderating effect of user experience on satisfaction with electronic banking: empirical evidence from the Spanish case," *Information Systems and e-Business Management*, vol. 14, no. 1, pp. 141–165, 2016.

[46] B. Hernández, J. Jiménez, and M. José Martín, "Age, gender and income: do they really moderate online shopping behaviour?," *Online Information Review*, vol. 35, no. 1, pp. 113–133, 2011.

[47] V. Taras, J. Rowney, and P. Steel, "Half a century of measuring culture: review of approaches, challenges, and limitations based on the analysis of 121 instruments for quantifying culture," *Journal of International Management*, vol. 15, no. 4, pp. 357–373, 2009.

[48] G. Hofstede, *Culture's Consequences: Comparing Values, Behaviors, Institutions and Organizations Across Nations*, Sage Publications, Thousand Oaks, CA, USA, 2003.

[49] M. Meeker, "Internet trends 2015-code conference," *GLOKALde*, vol. 1, no. 3, 2015.

[50] Yicai, *Total Volume of Chinese Users' Mobile Payments Exceeds Japan's GDP, Say Japanese Media Reports*, Yicai Global, Shanghai, China, 2017.

[51] J. F. Hair, G. T. M. Hult, C. Ringle, and M. Sarstedt, *A Primer on Partial Least Squares Structural Equation Modeling (PLS-SEM)*, Sage Publications, Thousand Oaks, CA, USA, 2014.

[52] J. S. Armstrong and T. S. Overton, "Estimating nonresponse bias in mail surveys," *Journal of Marketing Research*, vol. 14, no. 3, pp. 396–402, 1977.

[53] M. Fassnacht and I. Köse, "Consequences of web-based service quality: uncovering a multi-faceted chain of effects," *Journal of Interactive Marketing*, vol. 21, no. 3, pp. 35–54, 2007.

[54] P. G. Schierz, O. Schilke, and B. W. Wirtz, "Understanding consumer acceptance of mobile payment services: an empirical analysis," *Electronic Commerce Research and Applications*, vol. 9, no. 3, pp. 209–216, 2010.

[55] P. M. Podsakoff, S. B. MacKenzie, and N. P. Podsakoff, "Sources of method bias in social science research and recommendations on how to control it," *Annual Review of Psychology*, vol. 63, no. 1, pp. 539–569, 2012.

[56] H. A. Richardson, M. J. Simmering, and M. C. Sturman, "A tale of three perspectives: examining post hoc statistical techniques for detection and correction of common method variance," *Organizational Research Methods*, vol. 12, no. 4, pp. 762–800, 2009.

[57] L. S. Aiken, S. G. West, and R. R. Reno, *Multiple Regression: Testing and Interpreting Interactions*, Sage Publications, Thousand Oaks, CA, USA, 1991.

[58] H. Aguinis, J. C. Beaty, R. J. Boik, and C. A. Pierce, "Effect size and power in assessing moderating effects of categorical variables using multiple regression: a 30-year review," *Journal of Applied Psychology*, vol. 90, no. 1, pp. 94–107, 2005.

[59] H.-F. Lin, "An empirical investigation of mobile banking adoption: the effect of innovation attributes and knowledge-based trust," *International Journal of Information Management*, vol. 31, no. 3, pp. 252–260, 2011.

6

A Framework for Exploiting Internet of Things for Context-Aware Trust-Based Personalized Services

Abayomi Otebolaku ⓘ **and Gyu Myoung Lee** ⓘ

Department of Computer Science, Faculty of Engineering and Technology, Liverpool John Moores University, Liverpool, UK

Correspondence should be addressed to Abayomi Otebolaku; a.m.otebolaku@ljmu.ac.uk

Academic Editor: Floriano Scioscia

In the last years, we have witnessed the introduction of the Internet of Things (IoT) as an integral part of the Internet with billions of interconnected and addressable everyday objects. On one hand, these objects generate a massive volume of data that can be exploited to gain useful insights into our day-to-day needs. On the other hand, context-aware recommender systems (CARSs) are intelligent systems that assist users to make service consumption choices that satisfy their preferences based on their contextual situations. However, one of the key challenges facing the development and deployment of CARSs is the lack of functionality for providing dynamic and reliable context information required by the recommendation decision process. Thus, data obtained from IoT objects and other sources can be exploited to build CARSs that satisfy users' preferences, improve quality of experience, and boost recommendation accuracy. This article describes various components of a conceptual IoT-based framework for context-aware personalized recommendations. The framework addresses the weakness whereby CARSs rely on static and limited contexts from user's mobile phone by providing additional components for reliable and dynamic context information, using IoT context sources. The core of the framework consists of a context classification and reasoning management and a dynamic user profile model, incorporating trust to improve the accuracy of context-aware personalized recommendations. Experimental evaluations show that incorporating context and trust into personalized recommendation process can improve accuracy.

1. Introduction

The Internet of Things is an emerging global Internet-based information infrastructure, which is now positioned as the de facto platform for ubiquitous sensing and personalized service delivery. It promises a new information infrastructure in which all objects around us are connected to the Internet, possessing the capability to communicate with one another with minimal conscious interventions [1–3]. IoT allows people and things to connect at anytime, and anywhere, with anything and anyone, ideally using any network to facilitate global exchange and delivery of intelligent and relevant services [2]. It is the new Internet where things and humans become addressable and readable counterparts [4]. The IoT infrastructure consists of heterogeneous physical and virtual objects that can cooperate on social interactions, where each entity is capable of producing or consuming intelligent services [5]. With this revolutionary and innovative development, it is now

possible for our everyday objects to understand our needs: what we want or prefer and where and when we need them. However, for practical deployment of intelligent applications on this infrastructure, one of the associated problems that need to be addressed is how to find relevant services. A recent and an excellent proposal is context-aware recommendation system that exploits contextual information obtained from devices to learn the user preferences to provide services of interest [6]. CARSs are an extension of traditional recommendation systems, which are generally categorized into three main types: the collaborative filtering (CF) [7], the content-based filtering (CBF) [8, 9], and the hybrid recommendation (HR) systems [10]. On one hand, the traditional CF systems rely on the similarity (or the so-called correlation) between each pair of service consumers, who have consumed or rated the same items, to predict preferences of the target user. If the system predicts the preferences of the target user accurately, then it can suggest relevant and interesting services that the target user has not yet seen or consumed. On the other hand, the CBF systems use information about the user's consumption history to provide new and relevant services that the user has not yet seen or consumed. The hybrid recommender system is a combination of both CF and CBF, where the strengths of both are harnessed to address their peculiar weaknesses. The most important step therefore in the recommendation process, depending on the recommendation algorithm, is either the determination of the target user's neighbors and the aggregation of the preference information of each neighbor to generate a predicted preference or the determination of services that are similar to the ones the target user has consumed in the past. Note that, in the case of collaborative recommendation, every neighbor must have rated some services in order to participate in the preference prediction process. This means that if none of the neighbors has rated the service, the rating prediction cannot be computed. This problem is referred to as cold start problem [11]. Similarly, one key problem of CBF is overspecialization. An overspecialized CBF system always tends to suggest items or services that are similar to those consumed in the past by the user. However, these problems have been addressed by various excellent proposals [6]. And of late, context-aware recommendation techniques have been proposed and extensively explored with success [6, 12, 13]. The traditional recommendation systems consider user preferences and assume that these preferences do not change as users move from one location to another engaging in various activities. However, in addition to user preferences, CARSs use contextual information such as location, activities, environment situations, traffic information, device characteristics, and network conditions to provide relevant recommendations according to the user preferences in those contextual situations. Despite their widely reported success, CARSs are limited in the type of contextual information they can use and the knowledge they can infer from such context information. In addition, even though CARSs have been explored to understand diverse preferences of users to suggest relevant services [14], nevertheless, apart from the user's mobile phones, some existing CARSs have not taken into consideration other context sources such as everyday objects with which users interact in the IoT environment. Therefore, we argue that existing CARSs lack the adequate capability to provide dynamic and flexible context information required for making dynamic and intelligent recommendation decisions. Thus, the IoT is a novel computing paradigm for developing a new ubiquitous and pervasive network of addressable interconnected heterogeneous everyday objects [14] that can provide dynamic sources of context information. Since IoT objects interact with the environment and other objects, for example, object-to-object and object-to-human interactions sharing vast volume of data, value-added services based on these data can be delivered to users, providing impressive user experience. The IoT possesses the potentials to take context awareness to the next level considering that massive and dynamic data coming from diverse IoT objects can be exploited, via available software interfaces, to build more dynamic and intelligent CARSs [4].

An IoT-based infrastructure is a potential platform to address CARS context problem because it provides an important infrastructure to collect information from various objects that users interact with. This context information can then be exploited to address many challenges of context-aware personalized recommendation systems. There are two main advantages of IoT as a solution to CARS problem. Firstly, IoT provides ubiquitous sensing functionality allowing better understanding of object and human environments. This understanding can be used to gain useful insights into user preferences in a way traditional recommendation approaches, including CARS, have not been able to achieve. Secondly, as illustrated in Figure 1, IoT also allows, for example, collaborative recommendation algorithms, to exploit social characteristics of IoT objects to learn about those objects or humans with similar interests and preferences. However, collecting data, such as context data, from IoT devices poses a new challenge in terms of security and reliability. Thus, trust management is emerging as a powerful tool that can be used to address reliability of not only context data itself but also that of context providers, context information consumers, users, and services [5, 15–18].

The aim of this article is fourfold as illustrated by the high-level functional architecture of the proposed system in Figure 2:

(1) It proposes a conceptual context-aware framework that can collect, analyze, and infer high-level context information from IoT objects. The high-level context is a complex contextual information obtained by combining more than one atomic context using cognitive reasoning or machine intelligence; for example, from "*activity = walking*" and "*location = gym*," a more complex context such as "*walking to the gym*" or "*walking in the gym*" can be inferred.

(2) It provides suggestions to users based on their current and historical context information, taking into account the contextual preferences of the users.

(3) It proposes an extended contextual profile model for entities to manage their service consumption preferences.

(4) It presents how trust can be incorporated into context-aware personalized recommendations.

Thus, in Section 3, the article presents a conceptual framework addressing these fourfold issues. The framework can collect, analyze, and infer context information from mobile sources such as IoT devices. The analyzed context information can be exploited to gain useful insights into the preferences of an entity in an IoT environment in terms of what services they consume or can provide in diverse contextual situations thus satisfying the first aim of the proposed framework. To realize the second aim, the framework provides the components for recommendation processes that can use the inferred contextual information to provide personalized and relevant suggestions. Thirdly, a contextual preference model capable of capturing entity/user preferences, contexts in which such preferences have been expressed, and the level of confidence in such contextual preferences, measured

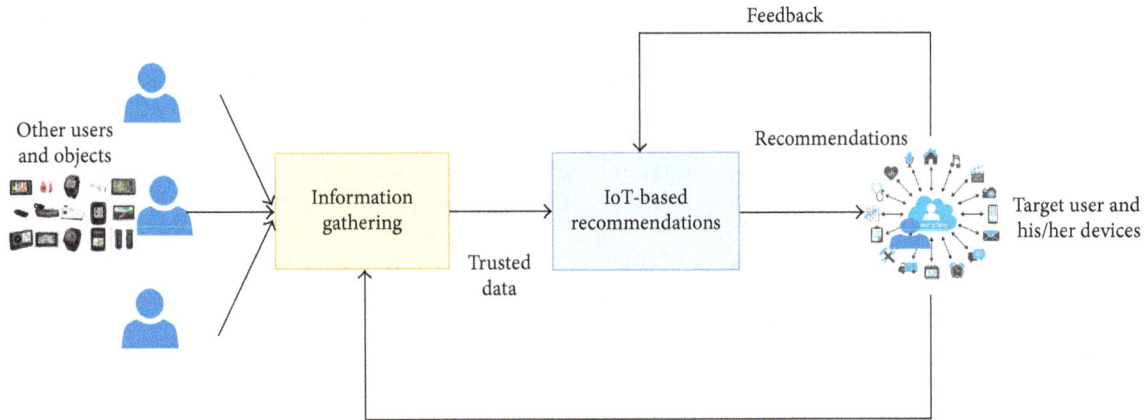

FIGURE 1: A high-level view of IoT-based recommendation system.

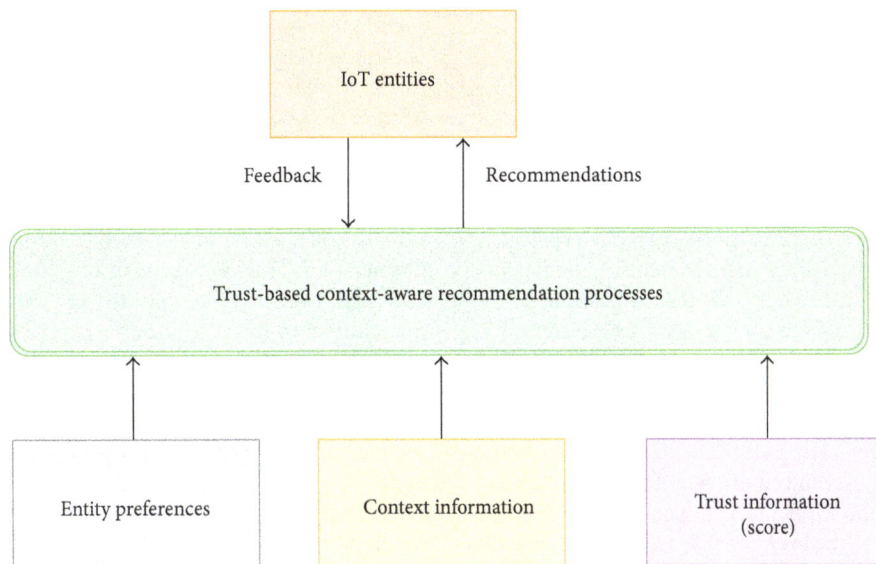

FIGURE 2: A high-level architecture of the proposed trust-based context-aware recommender systems.

through trust evaluation, is also presented. Finally, we discuss how to incorporate contexts and trust information to improve the relevance of recommendations. The contribution of the current article is a conceptual framework for the provisioning of relevant service recommendations in IoT-based infrastructure.

In Section 2, we discuss background and related work. Section 3 presents the proposed system and its components. In Section 4, we present the experimental setup, evaluations, and results. In Section 5, we discuss the results and provide some insights about our findings. In Section 6, we conclude the article and discuss the future direction of our work.

2. Background and Related Work

2.1. Recommender Systems: Traditional Approaches. Considering the overwhelming volume of information generated by billions of devices today, the information overload problem cannot be overemphasized [19]. To address this so-called "*information overload problem*," in which information consumers spend invaluable time to find relevant services

due to the humongous volume of myriad available alternatives, personalized recommendation techniques have been extensively explored to deliver such relevant services to users according to their personal interests [6, 13, 20–23]. The traditional recommendation systems take information about items and users and process this information to suggest items of interest to target users. In [24], Resnick and Varian define recommender systems as systems that provide recommendations to people by aggregating and directing them to the appropriate recipients.

Burke [10] define recommendation system as "*any system that produces individualized recommendations as output or has the effect of guiding the user in a personalized way to interesting and useful objects in a large space of possible options.*" This definition is much broader than the one given by Resnick and Varian [24], in which they essentially define a recommendation system as a collaborative system. In [12], Adomavicius and Tuzhilin provide a formal and widely cited definition of RSs, covering all types of traditional RSs; thus, "*... Let C be the set of all users and let S be the set of all possible items that can be recommended. Let u be a utility* *function that measures the usefulness of item s to user c, that is, u: $C \times S \rightarrow R$, where R is a totally ordered set (e.g., nonnegative integers or real numbers within a certain range). Then, for each user $c \in C$, we want to choose such item $s' \in S$ that maximizes the user's utility."* Ricci et al. [13] also define RSs *"as software tools and techniques, providing suggestions for items to be of use to a user. The suggestions relate to various decision-making processes such as what items to buy, what music to listen to or what online news to read."* They have also been exploited to help movie (e.g., Netflix) and music (e.g., Last.fm) lovers make informed decisions on the best and interesting movies or music they prefer to watch or listen to, respectively. Large and leading online industrial service providers such as Yahoo, IMDB, YouTube, and Facebook have deployed RSs as part of their services.

Nevertheless, conventional recommendation systems as defined above have focused on suggesting services to users, based on the information about the service items the user has consumed in the past, designated as content-based filtering (CBF) [9], or using information about those items that other like-minded users have consumed, designated as collaborative-based filtering (CF) [12]. Sometimes, the combination of these techniques, which is designated as hybrid recommendation, is used to push relevant services to users [10]. The core task of a traditional RS is to predict the subjective rating a user would give to an item he/she has not yet seen. Normally, to realize this task, they exploit either the ratings given by friends of the user to services consumed in the past to predict his/her preference for an item he/she has not yet consumed as shown in Figure 3 or they explore the content of the candidate items and the preferences stored in the profile of the target user. Essentially, a recommendation system takes as inputs user information and the description of items and processes the inputs to produce as outputs items likely to be of interest to the user. Additionally, to improve the relevance of future recommendations, a recommendation system learns the user's *"likeness"* (via feedback information) of the recommended items (Figure 1).

Even though recommendation techniques have been developed to provide users with relevant services from very large corpus of information items, however, their main objective is the explicit service suggestion that is relevant to the user preferences; without considering that, such preferences are dynamic and change according to the user's contexts [11, 20, 25]. In addition, the recommendation is made when the user explicitly requests for the RS assistance, and the system does not expect that the user's preferences would change with contexts such as time, activities, and locations. However, users generally prefer certain items in certain contexts. For example, a user who prefers horror and adventure movies would not want to watch horror movies while with his/her two-year-old daughter. Thus, Adomavicius et al. [21] introduced context dimensions into the recommendation systems by extending the traditional 2D (user × item) recommendations to multidimensional recommendation systems (user × item × context), now known as context-aware recommendation systems.

Since the introduction of contexts into recommendation systems, research in context-aware personalized recommendations has explored contexts extensively for its potentials to improve the relevance of recommendations and thus improve the user's quality of experience and recommendation accuracy [6, 11, 21]. Nevertheless, CARSs lack the capability to consider context information from various sensing objects in the mobile environments. IoT devices provide new dimension and capability that can be exploited by CARSs for flexible and dynamic access to various context information characterizing not only the user and his/her consumption environment but also the objects in such an environment.

2.2. Context-Aware IoT-Based Recommendation Frameworks. Software frameworks are designed to provide some additional layers of abstraction suitable for specific types of applications [26]. In the IoT environment, there is a limited research work exploiting and integrating IoT and recommendation systems to provide relevant services in this newly emerging ubiquitous environment [4, 14, 23, 27]. Generally, in the design and development of frameworks for CARSs, diverse contextual information has been used, for example, location, time, weather information, environment conditions such as temperature, illumination, and noise level, device characteristics, and network conditions. This information is usually collected from the mobile devices, especially smartphones [20]. However, relying only on context information obtained from mobile phones to provide service recommendations in the ubiquitous environment limits the ability of the system to truly learn the preferences of users as their contextual situations and environment conditions change. Often, the system cannot obtain context information when mobile users do not carry their mobile phones especially at night or when the device has run out of power. Useful contextual information might not be readily captured because in ubiquitous environments, context is dynamic and user preferences change according to the changing contextual situations. Thus, it is important to capture context information from other objects in the user's consumption environments. This gap can be bridged by exploring the potentials of IoT-based context sensing [2]. It allows dynamic, flexible, and seamless sensing of context data whether the user is with his/her device or not. In this section, we review some existing recommendation systems based on IoT context sensing.

In the literature, only limited work is available providing comprehensive frameworks for personalized service recommendations in IoT. The COMPOSE project [28] proposed an innovative framework for building smart and context-aware mobile applications. It utilizes the cloud computing infrastructure and IoT technologies to provide seamless integration of smart objects and external services for delivering scalable resources for data and application management in the IoT environment. Essentially, the framework consists of tools that enable end-to-end development and deployment of context awareness by providing functionality for collecting contextual information on smart devices, functionality for communication with external resources, and an infrastructure for hosting data storage and processing.

In [23], the authors propose a context-aware recommender system framework with a novel temporal interaction

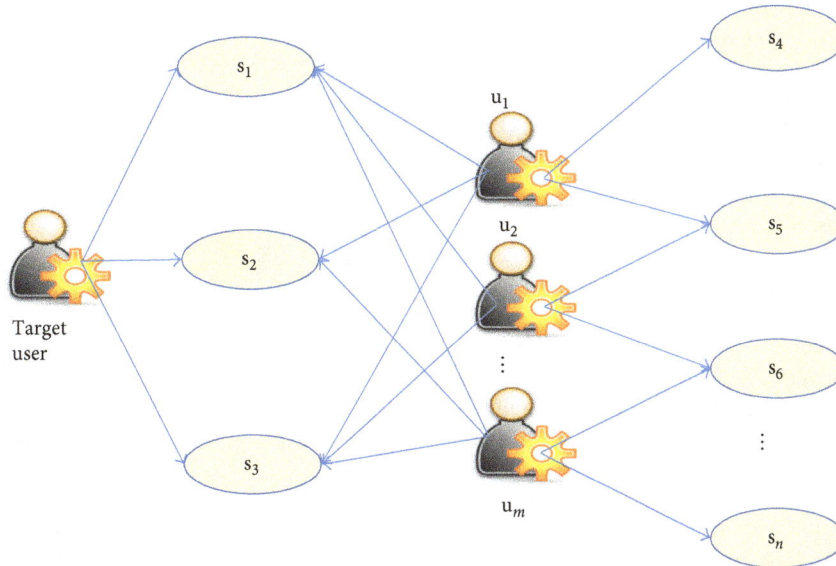

Figure 3: Users 1, 2, and 3 are similar to the target user considering their consumption history (s_1–s_3). Based on this knowledge, CF will recommend a set of items from s_4 to s_n (that target user has not yet seen) to the target user.

scheme for IoT-based interactive digital signage, which can be deployed in urban spaces to engage anonymous viewers. The framework was developed using the hybrid recommendation algorithm, combining both content- and collaborative-based recommendation techniques. In the framework, context information is provided by IoT sensing and context data pre-processing modules. What is not clear from this work is how to manage contextual preferences of these "anonymous viewers" to provide personalized recommendations according to their preferences and contexts.

Among other proposals that attempt to address the concern in [23] is the one presented in [14] by Mashal et al. They proposed a recommender system for suggesting services, which are relevant to users' preferences according to the contextual information obtained from IoT objects. They developed a weighted undirected tripartite graph-based service recommendation algorithm for delivering contextual personalized services. Another interesting and an excellent work with similar direction is the one presented in [4] by Hussein et al. They have proposed a social Internet of Things (SIoT) based intelligent recommendation framework that was developed on top of the SIoT context infrastructure, which uses cognitive reasoning mechanisms on context elements obtained from SIoT sensing objects. They developed ontologies for inferring user or object contextual situations. They emphasized and developed a task navigator algorithm that matches situational goals against the quotidian task ontology to determine which tasks match the situational goals and which smart services can fulfill such tasks. Similar to this proposal, the main goal of our proposal in this article is to provide a generic framework for personalized service recommendations, adapted to the target users according to their contextual situations and environment conditions obtained from IoT devices.

In [27], Saleem et al. proposed a concept for exploiting the SIoT for service recommendations among various IoT applications. The authors highlight how the SIoT can be used to provide recommendations by presenting an interesting sample application scenario. Whilst being an excellent proposal that intends to exploit opportunities provided by the thing-to-thing and thing-to-human social relationships of the Internet of things, it is not clear how they will exploit IoT and CARS to provide context-aware personalized recommendations. In [22], the authors proposed a novel algorithm for providing venue recommendations that include information sourced from IoT, Web services, and applications, as well as social networks (SNs) in the context of IoT. The proposed algorithm, according to the authors, also considers qualitative attributes and semantic information of the venues (e.g., price and atmosphere), the profile and habits of the user for whom the recommendation is generated, and the opinions of the user's influencers.

These solutions are some of the excellent proposals using IoT-based contextual information to develop intelligent and adaptive service recommendations. These proposals show that recommendation systems are effective tools for solving information overload problems, especially in the newly emerging IoT infrastructure. However, judging from the analysis above, some key issues still need to be addressed. Apart from some areas pointed out above, these systems lack the capability to improve future recommendations to users based on their experience of the current consumption. It means they lack the means to collect feedbacks from users and use these feedbacks to better understand the user preferences. Thus, the conceptual framework developed in this article provides a bridge between those proposals that do not provide means to personalize and improve service recommendations by developing a reliable contextual profile model able to learn user preferences from their ever-changing contextual situations, environment conditions, and feedbacks. The framework has been designed not only with the flexibility to allow implementation of new recommendation algorithms but also to accommodate the extension of the traditional recommendation algorithms by using IoT-based context sensing. It also proposes the incorporation of a trust management component to enforce the reliability of context information sources as well as those users who consume services provided by the system. Thus, instead of having the CARSs with (*user × items × contexts*) *dimensions*, we now have *user × items × items × trust* dimensions.

3. Proposed System Framework

3.1. Overview of the Proposed System Architecture. In this section, we present an overview of the proposed conceptual framework for context-aware personalized service delivery in IoT. The predicted deployment of billions of devices and their global access require layered architecture for the management of information from the physical devices to the cloud computing [1–3]. IoT will take the concept of Ambient Intelligence to the next level as it supports the concrete realization of building a digital ecosystem that is fully aware of the environments without conscious human intervention [14, 26, 29]. To realize such technological development, some key features must be present: (1) the IoT-based systems should be adaptive, and they should change in response to contextual situations. (2) They should be dynamic by anticipating preferences of entities or users. (3) They should be context aware to recognize the environment, objects, or entity's situational contexts. (4) The IoT-based systems should support tailoring of services to the users' contextual preferences, and (5) the sources of information and information itself must be reliable. Thus, we propose trust as a means for improving the reliability of personalized recommendations by using feedback from users to compute trust after consuming services in specific contexts.

Firstly, we give a high-level overview of a typical IoT platform, such as the one being developed by the Wise-IoT project consortium [30]. Wise-IoT project is an EU H2020 funded project, being executed by leading European research and academic institutes and their counterparts in South Korea. Wise-IoT aims to provide a worldwide interoperable IoT that utilizes a large variety of different IoT systems and combines them with contextualized information from various data sources. Such an IoT platform can be viewed broadly as a three-layered architecture as illustrated in Figure 4. Thus, the first layer of the architecture is the sensor or physical layer where many heterogeneous IoT devices are deployed. These devices can communicate with each other, using standardized connectivity such as Wi-Fi, 3GPP, 3G/4G, Bluetooth, ZigBee, and LoRaWAN (supporting low-power connectivity) based on such standards as OneM2M. The OneM2M standard is a global IoT standard that addresses communication issues for a common IoT service layer and interoperability on the IoT connectivity layer [31]. This sensor layer is also responsible for the IoT management.

The second layer is the contextual information management layer and can be considered as the cloud layer responsible for the acquisition, aggregation, and processing of context data. Furthermore, low-level context data from the physical layer are preprocessed for cognitive inference

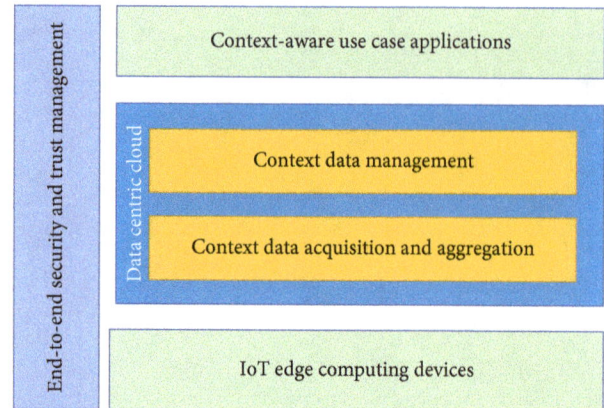

FIGURE 4: IoT context abstraction layers.

purposes using various techniques such as data mining or machine-learning algorithms as well as semantic web models for context reasoning, consisting of a context knowledge base and query and inference engines [32, 33]. One of the current and popular implementations of this layer is the FIWARE context broker, which is based on the NGSI standard [17]. FIWARE provides the capabilities to obtain IoT context data in addition to context information obtained directly from the user's mobile devices. Nevertheless, presently, these context brokers such as the one developed in the FIWARE project, do not provide the capability for context classification and knowledge processing and logical inference [31]. The last and the upper most layer is the application layer where context information is exploited to deliver intelligent services. Different kinds of applications can be deployed at this layer, for example, car parking or route recommendation applications.

Similarly, the framework proposed in this article is a three-layer architecture as illustrated in Figure 5. The three layers align with the two layers of Figure 4. The first layer in Figure 5 consists of those components that provide the functionality for dynamic context sensing, context recognition, context information distribution, cognitive context reasoning, and context information model. In the second layer, we have the context-aware trust-based personalization-related components. Among these components are the trust (evaluation) manager, which is responsible for computing trustworthiness of an entity of the platform [16]. We also have a contextual profile model, recommendation manager, feedback manager, and contextual profile manager. The contextual profile model is responsible for learning the entity's preferences in contexts of service consumption. The recommendation manager is responsible for using the contextual preferences to filter candidate services and suggest the most relevant ones to the users. The recommendation manager manages the selection of recommendation techniques or algorithms to use during the filtering process. The feedback manager is responsible for using implicit or explicit feedback obtained from users after consuming recommended services. The IoT context information access and processing (CIAP) layer comprises various components for context sensing, recognition, cognitive context reasoning, and context broker for context distribution.

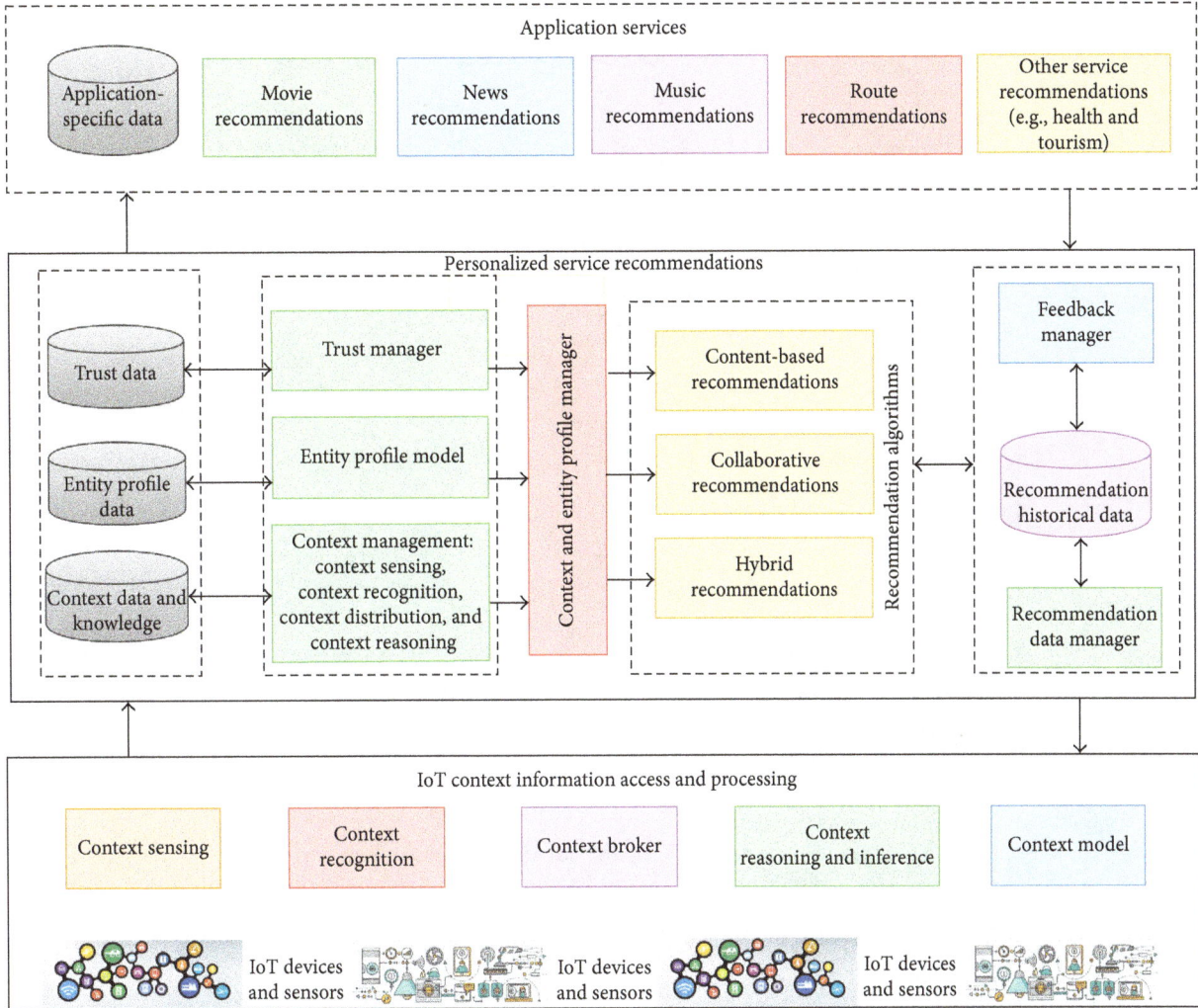

FIGURE 5: Components of the proposed trust-based context-aware recommendation system.

3.2. Context Information Access and Processing. This section introduces the context awareness component (as illustrated in Figure 6) that we have designed, for deployment on the IoT platform to process context data obtained from IoT context brokers and other IoT platforms such as the FIWARE broker [31]. Each of these components will be discussed in the next sections. The framework consists of various processing modules to be incorporated into our trust-based platform for intelligent service delivery in the IoT environment [16]. The context framework has been designed to provide the capability for high-level context classification, using machine learning and ontologies for context classification, semantic processing, and reasoning. In [34], we have described in detail context sensing, context classification, and reasoning using the neural network algorithm and various statistical features to identify the context in IoT. In this section, we only provide various steps for context classification and reasoning. However, we extend the concepts developed in that paper by providing here the cognitive model of context information and inference using ontologies, reasoners, and Semantic Web Rule Language (SWRL) [35]. Figure 7 illustrates the relationships between IoT

sensing, cognitive context management, and intelligent IoT-based applications. Figure 8 illustrates an example of how the context framework can infer high-level contexts (watching TV) from raw data accelerometer, GPS, and so on directly obtained from sensors.

3.2.1. Context Sensing and Preprocessing. The ability of a system to identify contextual situations and to make informed decisions is one of the most important functions of any context-aware system. However, sensors emit data that are in low-level formats, which are not suitable for decision-making by mobile applications [36]. Context recognition process collects raw data from sensors and transforms them into information that can be used to build intelligent applications. To provide an accurate context information about service consumers, the proposed framework uses the context recognition process to identify contextual information such as user activities, from smartphone-embedded and IoT sensors (obtained in the form of Web Services APIs from IoT platforms). The framework's sensing module gathers events from these sensors such as accelerometer, gyroscope,

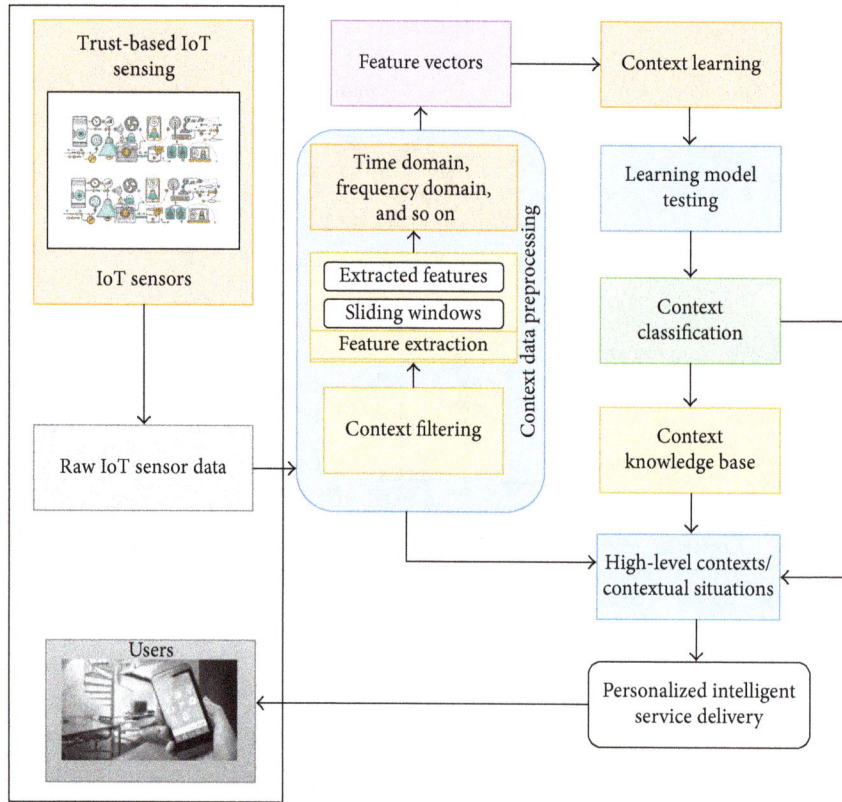

FIGURE 6: Context recognition processes of the proposed framework.

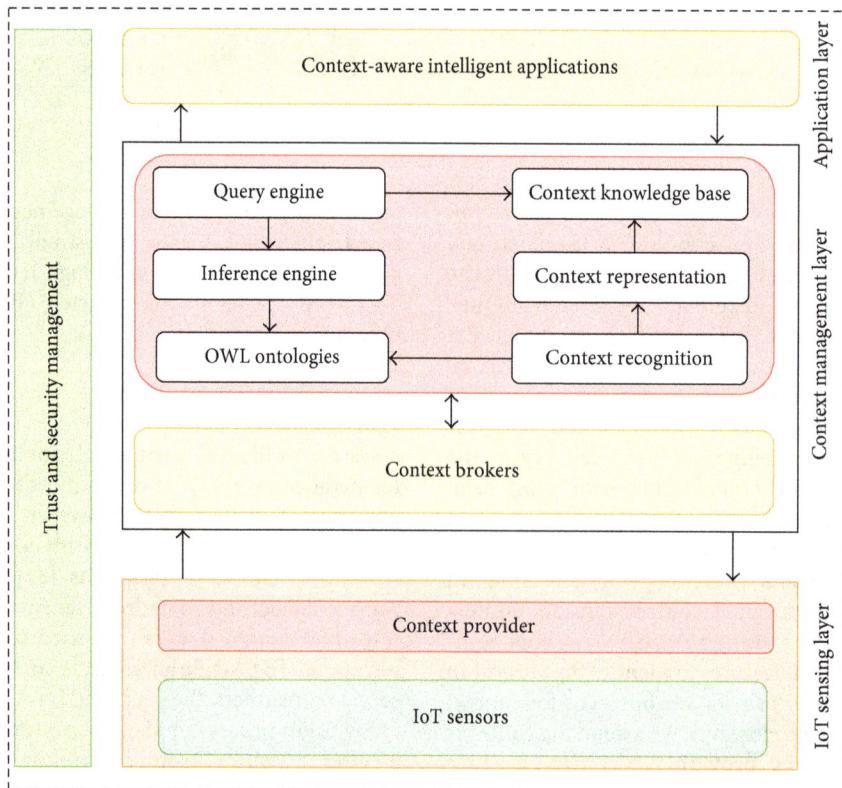

FIGURE 7: High-level architecture of the proposed cognitive context awareness, an extension of Figure 6.

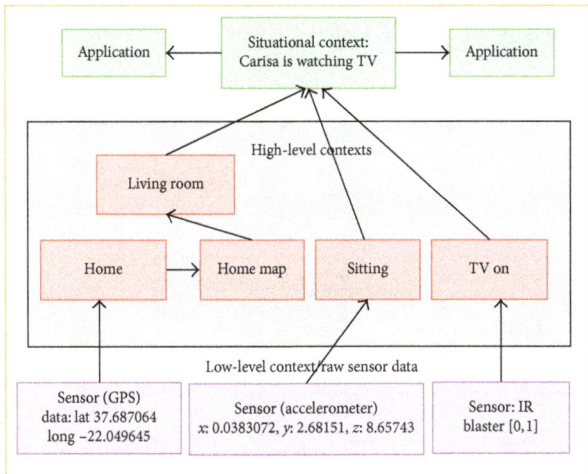

FIGURE 8: Situational context from high-level context and low-level context data.

rotation vector, orientation, proximity, microphone, light sensors, and other sensing devices, preprocesses the data, and then uses classification algorithm(s) to derive a more meaningful context information. It collects sample data from each sensor in predetermined time intervals overlapping the next predetermined time space. In the data-preprocessing phase, the framework removes the data outliers [34, 36]. This is achieved by eliminating samples from the beginning and ending of each example to reduce the influence of noise in the streams of sensor data.

3.2.2. Feature Extraction.
The user's dynamic contexts, such as activities, occur in relatively long period, in seconds or minutes, compared to the sampling rate of the sensors. This sampling rate usually does not provide sufficient data to describe user activities. Therefore, activities are usually identified on time window basis rather than sampling rate basis [37]. Comparing a series of time windows to identify activities is almost impossible even if the signals being compared come from the same user performing the same activity context [13]. The feature extraction process addresses this problem by filtering relevant data and obtaining quantitative data from each time window. Many approaches have been explored in literature such as statistical and structural properties of the sensor signals. Structural features such as Fourier transform are quite complex and require more computational resources [37]. This may not be ideal for an energy-starved smart device. On the other hand, statistical features are simple and require less computational resources. Thus, this process uses simple labeled statistical features (e.g., range, maximum, minimum, mean, and standard deviation), which have been validated in our previous work to be very effective to discriminate between time windows [34, 36]. These features are extracted into feature vectors that are then used in the next process.

3.2.3. Context Classification.
After extracting the time window features from the raw sensor signals, without deriving the context knowledge from them, the example features are limited in how they can be used. The context

classification uses supervised machine learning algorithms such as the neural network (NN) to derive high-level context from the statistical features. Details of the modeling and evaluations can be found in [34]. These models are integrated into the recognition service to obtain independent future activities and contexts of the users.

3.2.4. High-Level Context/Contextual Situations: Context Inference.
The context information obtained from context recognition and classification processes is not able to provide cognitive meaning of complex context information, in addition to not being able to recognize complex context information. Thus, it is important to develop a cognitive model based on semantic web technologies that can infer higher-level contextual situation from these individual contexts as illustrated in Figure 7. For example, it is important to know what a user is presently doing, when, and where he/she is doing it. Nevertheless, without relating this information to provide a more high-level contexts, this information will not be useful to the application. This is one of the weaknesses of some of the existing work. The proposed context framework thus uses the knowledge-based model on top of the classification process to relate different atomic context information to obtain contextual information at a higher semantic level. For example, having known that a user sitting at home is in the living room, if we know that the TV is on, then it is important to relate this information and conclude that the user is watching TV. For example, having known that a user located at home is sitting in the living room, if we know that her TV is on, then it is important to relate this information and conclude that the user is watching TV. The user is watching TV, obtained from sitting, home, living room, TV, and so on, is inferred using the knowledge-based cognitive process. The details of this process can be found in [35, 36]. It can also determine complex context such as a user is "jogging at the sport arena." This kind of information is crucial to offering rich quality of experience to users.

(1) Ontology Concepts. In this section, we introduce the design of the ontology model in the proposed framework for the inference of higher or complex contextual information from the atomic context classified by the context recognition model. To infer higher-level or complex contextual situation from the recognized and classified contexts from the previous section, we have developed semantic web-based ontology reasoning for context in the IoT environment. As illustrated in Figure 9, we have provided some important concepts from which the ontology was designed and developed. To infer high-level contexts from those concepts, context reasoners and context rules have been used. The conceptual context framework incorporates nine generic interrelated concepts, which were defined for providing context information for the context-aware media recommendation domain. Eight of the core concepts are based on our previous context representation model in [35] but now extended to incorporate trust as an additional concept that relates with other core concepts as a property.

The ontology is being developed using OWL DL with adequate provision for adaptation decisions, relying on

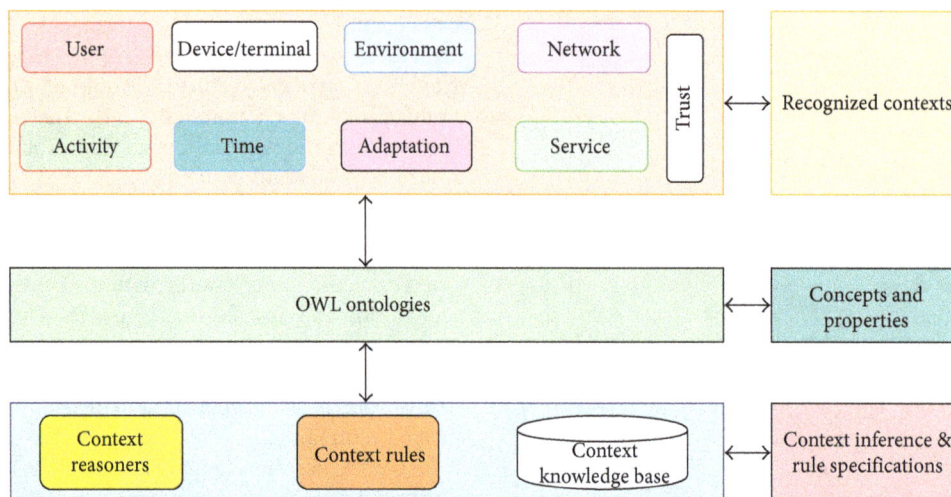

FIGURE 9: The framework inference model.

various rule-based mechanisms, which support an automatic knowledge inference from contextual information [38, 39]. The cognitive model has been designed as a two-layered model consisting of lower and upper layers representing upper and lower domain ontologies as illustrated in Figure 10. The first layer is the primary or upper domain ontology with 9 core concepts: *User, Device, Environment, Service, Activity, Time, Trust, Adaptation, and Network*. The concepts in this domain are generic and are valid for many application scenarios. The second layer incorporates specific domain ontologies, integrating reasoning and rules for inferring contexts specific to each domain. The specific or lower domain ontology is designed based on the combination of user specific activities and locations and other contextual situations. For example, the user *home domain activity* and the related contexts determine contextual information that characterizes the user's activities when at home. Figure 11 illustrates the relationships between semantic concepts, the building blocks of the proposed cognitive context reasoning model, ontology, and the SWRL-based context model. We define the primary concepts of the context model as follows:

User: the central focus of any context-aware recommendation system is the user. In these systems, information such as location, activities, and preferences is very crucial to building a system that adapts to the user's needs. This contextual information influences the user's preferences for service consumption, and therefore, when recommending service items to the user, it is important that this information be incorporated into the recommendation process.

Time: the time concept is primarily used to capture the "when question" of the other concepts. Time duration, start time, end time, and so on are some of the elements of this concept. These elements of the time concept are used to infer, for instance, high-level time information such as weekdays or weekends which is subsequently used in addition to location information to infer a user's contextual situation including her activity contexts.

Device: the characteristics and capabilities of IoT devices can have significant impact on the personalized

recommendation process. The device characteristics can be used to adapt recommendations to the user's device.

For example, the device battery level can be used to determine the appropriate format to present recommended items. In fact, it may present the item such as streaming video in a format that requires less power such as in text form. If the battery level is low, the recommendation system could decrease the brightness or lower the spatiotemporal resolution of the content in case of an audiovisual content. In addition, this concept also describes the capabilities of the devices.

Environment: another important concept of the ontology is the natural environment. This concept influences the kind of items users would prefer to consume depending on the environmental condition or situational contexts. It provides information about the environment in which a user interacts with the recommendation system or his device. There are two groups of fundamental concepts related to the environment concepts: location, including logical (e.g., city and street) and physical (e.g., GPS coordinates), and environment conditions (e.g., noise level, illumination, and weather information). For example, information about the current weather can be used to provide more relevant media recommendations to users, that is, the category and genre of contents preferred by the users can be influenced by weather information.

Activity: user activities can be used in addition to the other context information by the recommendation process, given that they have a strong impact on the preferences and needs of mobile users, as previously stated. As explained in the previous sections, to address this problem, low-level data obtained from device-embedded sensors are channeled to a context recognition model running on the device, which recognizes high-level activity being performed by the user. To relate this activity information, as illustrated in Figure 12, with other contextual information, an activity concept was designed as part of the ontology.

Network: the network concept describes the conditions and characteristics of the network connection of the user

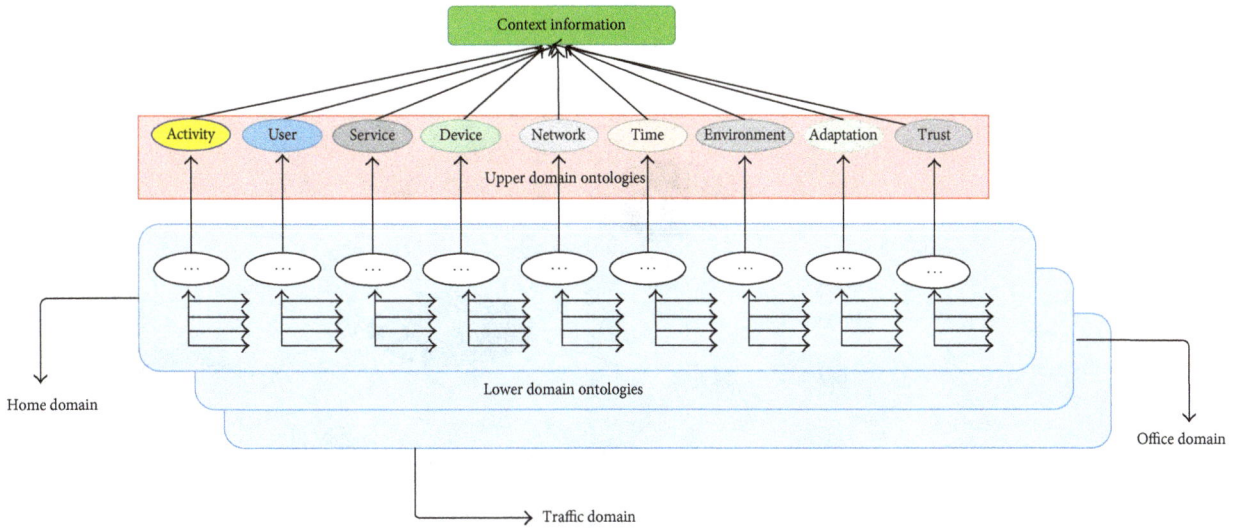

FIGURE 10: Cognitive upper and lower domain context ontology model.

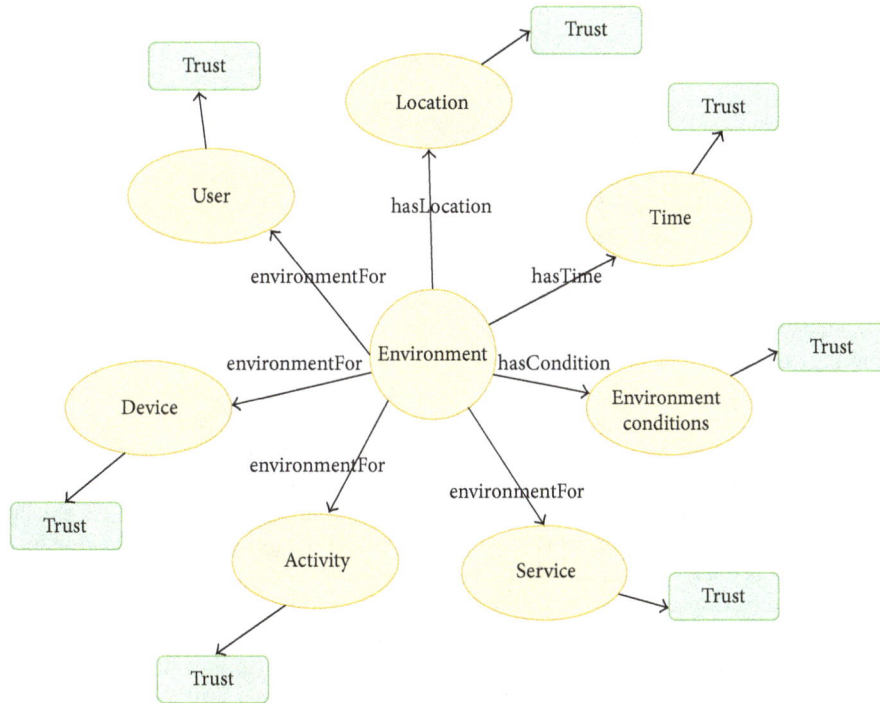

FIGURE 11: Relationship between the framework ontology concepts and trust.

device. It describes such aspects as the network maximum bandwidth, network minimum guaranteed bandwidth, error rate, network delay formation, and currently available network bandwidth.

These characteristics can be used either to enhance a minimum level of quality of media consumption or even to help to decide between different versions of the same media item.

Service: information about candidate services, for example, multimedia content such as streaming movies, to recommend is also an important concept. The most important features of the class are the metadata and the presentation format of the service. From this concept, let us take a streaming movie as an example, from which information about the genre, titles, and so on can be obtained. This information could be used in the recommendation process to filter important characteristics of the audiovisual content, such as video-audio formats and so on.

Adaptation: the adaptation concept describes the presentation of the service items based on the device characteristics, environment conditions, network conditions, and so on [20, 35]. The adaptation concept relates to other concepts such as the environment, device, network, time, and so on. The concept can be used to define rules to determine when and how a service should be delivered. For example, if

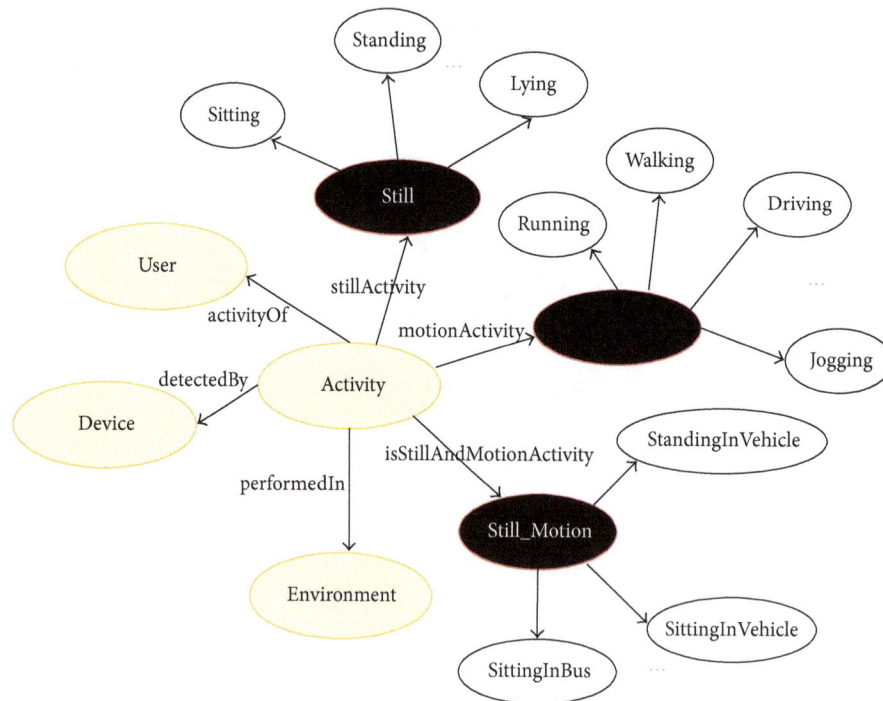

FIGURE 12: Activity concept [40].

the network condition such as bandwidth is very poor while delivering a streaming movie in HD, the adaptation decision could decide to deliver the content in formats that consume less computational resources.

Trust: trust has been used as a measure of confidence that an entity will behave in an expected manner despite the lack of the capability to monitor or control the environment in which it operates [5]. Trust is an important concept as we enforce confidence in items being suggested to the target users. In our framework, trust is used to enforce confidence of the target users in the recommendations that are provided in the given contexts. In the semantic model, trust is used as a property of other concepts (Figure 11), whereby every concept has an associated trust property and it is measured based on the feedback obtained from users after consuming services in specific contextual situation.

(2) Cognitive context reasoning using SWRL. The knowledge-based or cognitive context model incorporates reasoning mechanisms based on two methods. The first method is the inherent ontology reasoning mechanism that is responsible for checking class/concept consistency and implied relationships. This method uses the inference engine such as Pellet or Jena to provide functionality for checking the consistency of ontologies, computing the classification hierarchy, explaining inferences, and answering queries [41].

The second method is based on the Semantic Web Rule Language (SWRL) [42, 43]. Because OWL does not provide the mechanisms for expressing all relations between concepts in the ontology model, OWL has been extended with SWRL to allow inferring new knowledge from multiple facts or conditions at the same time by providing mechanisms for expressing complex relations. For example, OWL cannot express the relation between a child and married parents

because it does not have the mechanism to express the relation between individuals with which an individual has relations. For instance, OWL cannot relate that a user, say Ana, is the child of the married parents James and Comfort. Nevertheless, with SWRL, this can be expressed in the following way: $user(?u)^{\wedge}hasParent(?u,?f)^{\wedge}hasSpouse(?f,?m) \rightarrow childOfMarriedParents(?u)$. This specification adds rules to OWL ontologies, while also providing an extra layer of expressivity and the capability to infer additional information from an OWL knowledge base. Basically, SWRL consists of antecedent and consequent parts, which in turn consist of positive connections of atoms. Informally, SWRL can provide high-level reasoning in the following form: *if all atoms in the antecedent are true, then all the consequents of the rule must be true or vice versa.* Besides, SWRL comes with built-in unary predicates for describing relations between classes and data types, binary predicates for data properties, and some *n*-ary predicates such as arithmetic operators to compute desired behaviors. For example, it is easier using SWRL to infer if Ana, in the previous example, is an adult or not using these built-in predicates as follows: $user(?u)^{\wedge}hasAge(?u,?age)^{\wedge}swrl: greaterThan(?age,18) \rightarrow adult(?u)$. Likewise, by combining time and location data, the type of the device, and whether the user is accompanied or not, as well as user activity, the system will be able to infer the situational contexts of the user, for example, *Ana is sitting at home.*

In listings 1–3 of Figure 13, we provide some examples of specific domain rules supported by the context ontology model. Listing 1 is an example that infers that the user is indoor, sitting at home, and on a weekend. In listing 2, the user is in outdoor in the office walking, whereas in listing 3, the user is indoor, sitting in the office. Using this contextual information, personalization systems can be designed to

Listing 1
Context:hasUserId (?user, ?userId) ∧context:atHome (?user, ?location) ∧ context:weekEndAtLocation(?location, ?weekend) ∧
Context:locationTimeIs (?location, ?time) ∧context:isMorning(?location, ?time) ∧ context:inIndoorLocation (?user, "true") ∧
Context:hasUserActivity (?derived, ?activity) → context:hasHomeLocationActivity (?user, ?sitting)

Listing 2
Context:hasUserId (?user, ?userId) ∧ context:atOffice (?user, ?location) ∧ context:weekDayAtLocation (?location, ?weekday) ∧
Context:locationTimeIs (?location, ?time) ∧ context:isEvening(?location, ?time) ∧context:inOutdoorLocation (?user, "true") ∧
Context:hasUserActivity (?derived, ?activity) → context:hasOfficeLocationActivity (?user, ?walking)

Listing 3
Context:hasUserId (?user, ?userId) ∧context:atOffice (?user, ?location) ∧ context:weekDayAtLocation (?location, ?weekday) ∧
Context:locationTimeIs (?location, ?time) ∧ context:isEvening(?location, ?time) ∧context:inOutdoorLocation (?user, "true") ∧
Context:hasUserActivity (?derived, ?activity) → context:hasOfficeLocationActivity (?user, ?sitting)

FIGURE 13: Listings 1, 2, and 3: example cognitive rules for inferring higher-level contextual information using SWRL.

recommend relevant service items that suit these specific contextual situations and preferences of the user [40].

3.3. Trust Evaluation.

The *Trust Evaluation* aggregates trust-related information and feedback obtained from users whenever they consume services in specific contexts to derive an evaluation for the trustworthiness of context data. Such evaluation for the trustworthiness supports decision-making of users. The trust evaluation is done using the following key modules as illustrated in Figure 14:

(i) *Trust Data Collection and Preprocessing*: the data collection implements the trust data collection agents using the available interfaces or APIs provided by the IoT platform data access layer such as the FIWARE [44] context broker or other REST interfaces, to collect or gather trust-related data. The data could be opinions of entities (users) as feedbacks on services consumed and so on.

(ii) *Trust Feature Computation*: the feature computation is responsible for computing/extracting features from the collected trust-related data. It also preprocesses the collected data to eliminate erroneous or repetitive and other irrelevant data for trust indicators' evaluation.

(iii) *Trust Data Store*: this module is responsible for storing trust-related data and the history of interactions between entities of the IoT platform. Additionally, this historical data store stores trust scores for the entities of the platform. Any trust score consumers, for example, recommendation manager, can interact with the trust data store via a web service interface, or via the Wise-IoT data access component such as the FIWARE broker to request such data.

(iv) *Trust Indicator Computation*: trust indicator computation implements one of the functionalities of the trust evaluation, realized based on the REK model [16]. It is responsible for calculating trust indicators such as experience, reputation, and knowledge.

(v) *Trust Score Evaluator*: this implements the REK model, the main module responsible for evaluating and generating trust scores for IoT entities of the framework. The trust evaluator computes the trust score or index from the combination of trust indicators, for example, experience, reputation, and knowledge.

In summary, the trust evaluation module subscribes to the IoT context broker. Whenever a user submits a feedback about the service he/she has just consumed, the feedback manager sends the feedback information to the trust component (TC), using the provided API. Details of mechanisms for trust score computation can be found in [16].

3.4. Contextual Entity Profiling.

We define a user as any entity that has the capability to provide and consume services be it data or any other information service. In this section, we present a profile/preference model capable of using the user's contextual situation, preferences, and other related information to personalize and deliver relevant services.

3.4.1. Contextual User Preference Definition.

The user profile describes preferences of a user in the form of a summary, normally based on the history of the user's actions such as consumption or provisioning of services [9]. The contextual user profile service (CUPS), based on our previous work in [11, 20], summarizes the user's service consumptions into a limited set of categories. Categories can be characterized by one or more services, and such services can be characterized by many properties or attributes. Several services can be associated with one or more service categories. In addition, the profiling model incorporates contextual dimension, associating one or more inferred context with each *category-service-property* relationship, with each context having an associated trustworthiness score. Thus, we have *category-service-property-context-trust concept*, presenting each user profile as a five-level tree, as shown in Figure 15, and the root of the tree represents the entity or user's optional information such as entity ids, timestamp, and so on. The first level of nodes corresponds to the service category; the second level represents the services associated with the specific service category; the third level contains the attributes of a given category-service. This level provides the service item's context, characterizing at a finer detail, the consumed service, and thus the preferences of the user. A limited set of properties is used for each service to obtain a good compromise between sufficient degree of characterization of service (hence, sufficient ability to make distinctions) and reasonable dimensions of the

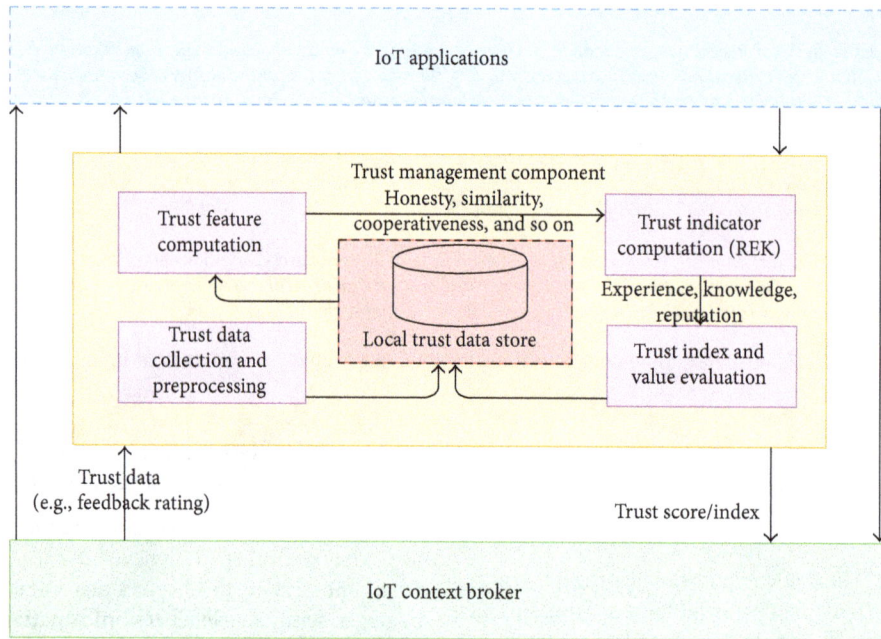

FIGURE 14: Trust computation component.

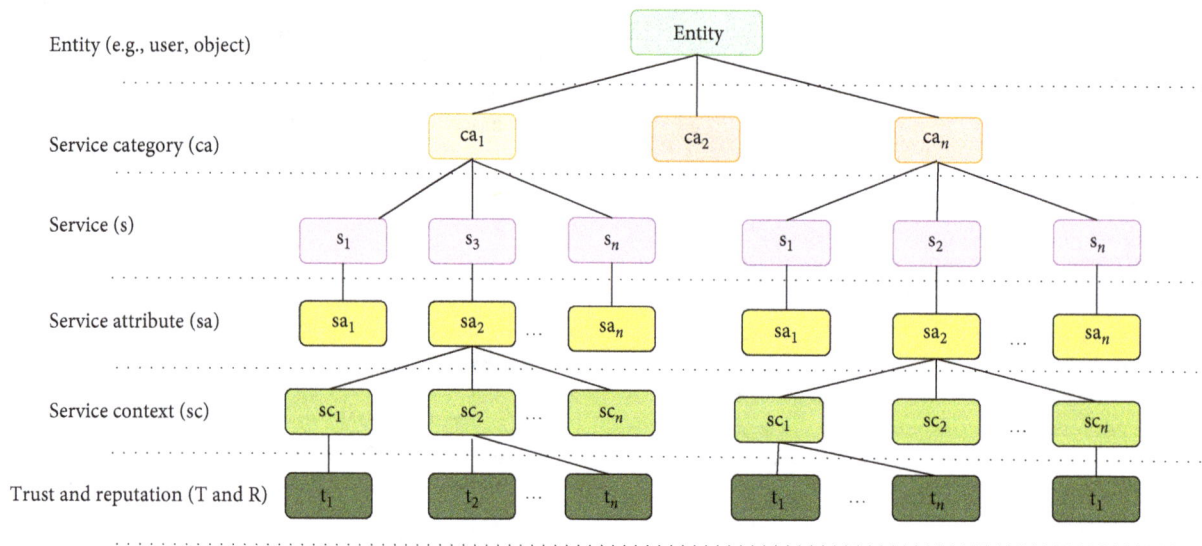

FIGURE 15: The preference model, showing the structure and relationships between entities in the profile model.

user profile. The leaf nodes provide information about the contexts (where the user preferences have been expressed) and associated trust score representing trustworthiness of the context information. Leaf nodes have four numerical fields (type, trust score, intensity, and *lifetime*), whereas all other nodes have only the type field. In the leaves, the types represent the type of context. The introduced concepts in the user profile, the intensity, trust score (as an extension), and lifetime, track the user's contextual consumption history, allowing incorporation of the trustworthiness of the contextual situation wherein the users consume services.

Using these weighted parameters, the system can determine the services that are important and relevant to the target user based on his/her contextual preferences with some measure of confidence. The *intensity* provides information about the number of times the user has consumed items of that *category-service-attribute* in that specific context. The intensity (the dynamic preference of the user) of those elements belonging to the item's content tis calculated by summing up the products (weight × lifetime) of all their occurrences. The intensity value of the retained elements at this level is obtained by visiting their child nodes in a breadth-first traversal. The same applies to the retained elements of the category level. The intensity of the elements that belong to the genre level is the largest value of their children. Accordingly, these values are obtained by performing a depth-first traversal. This way, the user profile can handle any category of services, for example, services providing multimedia items such as movie, news, and music to name a few.

The intensity (the dynamic preference of the user) is computed by summing up the product (weight × lifetime × trust score), given in (4), of all their occurrences.

$$P_{ri} = \sum_{i=1}^{n} (\text{weight} \times \text{lifetime} \times \text{trust}) \quad (1)$$

3.4.2. Entity Profile Definition. Let E be the set of N things or entities, and entities can be humans (users), objects, or devices. Each entity $E = \{e_1, e_2, \ldots, e_n\}$ is defined by m set of optional demographic attributes $A = \{A_1, A_2, A_3, \ldots, A_m\}$ and a set of preferences $\Pr(u_i) = \{pr_1, pr_2, pr_3, \ldots, pr_{ir}\}$, where each preference $\Pr_{im}$ of user a is further defined by a set of attributes $SA = \{sa_1, sa_2, sa_3, \ldots, sa_p\}$. Furthermore, we have a set of high-level context information $SC = \{sc_1, sc_2, sc_3, \ldots, sc_n\}$ associated with each preference $\Pr_i$.

In the above definitions, context SC has a very complex structure, which reflects the nature of the difficulties in accurately representing context information. However, in this paper, it is assumed that the context information is defined by a structure that allows atomic contexts to be related in a way that allows high-level contextual information to be inferred, that is, a three-layer structure of inferring from low-level sensor data, a high-level atomic context information, which can be related to another context information to arrive at a higher-level semantic context information. For instance, from the above definition, an entity e_i where $e_i \in E = \{e_1, e_2, e_3, \ldots, e_n\}$ is defined by a set of demographic information $A = \{\text{entityId, Name, etc.}\}$, by a set of preferences $\Pr = \{pr_1, pr_2, pr_3, \ldots, pr_n\}$. Each service is classified into a category $Ca = \{cr_1, cr_2, cr_3, \ldots, cr_n\}$. Each category is characterized by a set of services $S = \{s_1, s_2, s_3, \ldots, s_n\}$, and each service is characterized by a set of attributes $Sa = \{sa_1, sa_2, sa_3, \ldots, sa_n\}$. In addition, each context sc_i defined above is associated with the set $Tr = \{tr_1, \ldots, tr_n\}$ of trust. The trust value defines the confidence of the user in services recommended to his/her in that context. The model of the user preference can then be built in a general form as follows:

$Y = f(X_1, X_2, X_3, \ldots, X_n)$, where $X_1, X_2, X_3, \ldots, X_n$ represent the profile information such as the set of service attributes, A, and preferences, $\Pr$, characterized by the sets $S, Ca,$ and SC with associated trust values Tr. Y is the dependent variable (representing user preference) to be predicted, and function f is the predictive function as provided in Section 3.4.3. For example, we can predict the services that user u_i will like to consume using this model. This model can be defined such that we can have context-based user preferences, context-aware trust-based user preferences, and user preferences without context and trust information as defined in (2)–(4), respectively.

$$Y_{ct} = f_{ct}(X_1C_1T_1, X_2C_2T_2, \ldots, X_nC_nT_n), \quad (2)$$

$$Y_c = f_c(X_1C_1, X_2C_2, X_3C_3, \ldots, X_nC_n), \quad (3)$$

$$Y = f(X_1, X_2, X_3, \ldots, X_n). \quad (4)$$

These models can be used to build the contextual user profile and can also be switched to build the traditional user profile.

The parameters $c_{i=1}, \ldots, c_{i=n}$ represent the contextual information $t_{i=1}, \ldots, t_{i=n}$; $f_c, f_{ct},$ and f are the functions for retrieving users' contextual preferences, user context-aware trust-based preferences, and ordinary user preferences, respectively. As (2)–(4) illustrate, the profile model can be used to learn the user profile in three modes. It works in the contextual mode and in the noncontextual or traditional mode to determine the user preferences with and without using contextual information, respectively. The traditional mode (non-contextual mode) is executed in situations where the user's contextual information is not available or when it is difficult or impossible to acquire. In this mode, an entire profile considering the user consumption history (if any) is used. Alternatively, it could use the consumption history of users who have similar preferences to the current user to learn the target user's preference. The default mode is the contextual mode in which the user's contextual information is used to learn the user's preferences. The contextual mode allows us to incorporate trust information of the context information, and thus, the ability to evaluate the preferences is not only based on contexts but also based on the trust information of the contexts.

3.4.3. Feedback Mechanism. To learn the user's service consumption preferences based on the entity profile model in the last section, the proposed system adopts the relevance feedback method, which we have used in our previous solution [11]. Relevance feedback is a technique used to obtain a user's opinion on recommendations. It can speed up the user preference learning process and improve the quality of recommendations [42]. There are two methods in literature used for learning user preferences based on user feedback, namely, implicit and explicit relevance feedback [9]. In implicit user feedback, the system observes the content consumptions by the user and records information about the consumed content. In our case, such information includes category, genre, property of the item, and most importantly the context in which users consume this item. The user feedback process, without asking for any information from the user, assigns a relevance value to the consumed content using the content metadata information as well as the context of consumption. The explicit user feedback involves asking the user to provide ratings or some form of evaluation of the relevance of the consumed content, using metrics such as like or dislike of the recommended items. In mobile environments, this approach can be obtrusive as it distracts and consequently bores users, discouraging them from using the recommendation system, which consistently requires them to provide rating information every time they consume content [12]. For this reason, we adopted an implicit user feedback method in combination with a simpler form of explicit user feedback. Our approach neither asks users for explicit rating information nor measures how long a user has spent during the consumption of any service, for example. Measuring time spent on recommended items as argued by Bjelica [42] may lead to wrong conclusions. Thus, rather than measuring time spent when consuming content, we used a combination of the contextual user profile learning model and the context in which users respond to recommendation either by clicking or not (e.g., the device's screen) to learn the user's feedback on the recommended items. It then extracts implicit information about such items to learn whether users like or dislike such items by assigning what we call relevance values to the extracted information, such as category of the content and its genre, taking into consideration the contextual situation of the user at that given time. The system automatically assigns these relevance values in two numeric formats as illustrated in (5)–(7) consisting of *weight w* and *lifetime γ* parameters. Equation (5) represents positive feedback, whereas (6) represents negative feedback. Equation (7) is used as a decay function, which updates the preferences as the elapses.

An intensity w_{ij} represents the relevance of a service with the preference pr_i belonging to the contextual situation C that entity e_i consumes in context C_i. Whenever the context awareness model detects that the entity is in a context C_i, for a continuous period of time $[0, T]$, the reasoner finds the preference of service $s_1, s_2, s_3, \ldots, s_n$, whose contextual situations match the present contextual situation of the entity; then, weights $w_i c_i \in [0, 1]$ are associated with this service preference, which at time T are updated as follows:

$$w_i c_i = w_{i-1} c_{i-1} + \gamma (\alpha - w_{i-1} c_{i-1}), \quad i = 1, 2, 3, \ldots, n. \quad (5)$$

Then, for those services $b_1, b_2, b_3, \ldots, b_n$ with associated weights $w_1 b_1, w_2 b_2, w_3 b_3, \ldots, w_n b_n$ not consumed by the user, these weights are updated as follows:

$$w_i c_i = w_{i-1} c_{i-1} - \gamma (\alpha - w_{i-1} c_{i-1}), \quad i = 1, 2, 3, \ldots, n, \quad (6)$$

where γ is a learning parameter whose value is obtained by

$$\gamma = 1 - \left(\frac{t}{45}\right)^5. \quad (7)$$

$\alpha \in [0, 1]$ and its value is set to 1 in (5) and 0 in (6).

Equation (7) was derived so that certain attributes in the user profile that represent items that have not been seen for a long period (note that we have considered the long period to be 30 days) will eventually have no impact on the preference evaluation. However, unlike our previous solution in [11], where the weight w_i provides information on the number of times the user has consumed items in specific contexts, here w_i represents the trust tr_i of context of consumption computed from feedback information obtained from users. The lifetime parameter, γ, provides an indication of the time elapsed since the last consumption occurred. Its value is set to 1 when the user consumes a service that belongs to a *category-service-attribute* of Figure 14 and periodically decrements it if the consumed service does not belong to such category-service-attribute. Smaller values indicate that the user has lost interest in that type of service, regardless of the value of its weight or contexts. We also use this value to determine if a trustworthiness can be considered valid or not.

For example, in practice, it allows to give more importance to contexts of items consumed recently and less to those consumed long time ago. The factor α has a value in the range $[0, 1]$, assigning less or more importance to new category-service-attribute. For newly created category-service-attribute, α assumes the value of 1; otherwise its value, determined based on experiments, is set at 0.5. The value of -1 is used to indicate that the user has rejected or not preferred a service item with those properties.

The *lifetime* parameter as explained earlier is set to 1 for matching items in the user profile, by assigning the value 0 to the factor t. The factor t represents the number of days elapsed since the last time the user has consumed an item with the characteristics described by the profile. With (7), the relative importance of the *category-service-property* of an item consumed by a user remains above 0.9 in the first 30 days after it has been visited, rapidly decreasing to zero after that period (nonnegative values are automatically converted to zero). For all other *category-service-attribute* in the user profile, the update of the *lifetime* parameter is performed by linearly increasing the value of t. Thus, category-service-attribute of a service that has been consumed before but has not been seen or consumed for a long period will have either low or no impact on the user preference evaluation. Also, note that the trust score influences the relevance of a given service.

3.5. Incorporating Trust as an Extension of the Context-Aware Recommendation Process. In this section, we demonstrate how trust can be incorporated into context-aware collaborative recommendations. The process of incorporating trust into the recommendation system, especially context-aware collaborative recommendations, is in three key important steps, namely, building entity profile, generating entity trust, and predicting service preferences for the target user.

In recommendation systems, especially collaborative recommender systems, opinions of the neighbors (or so-called friends) of a target user are used to suggest interesting items. Thus, it is required that such neighbors whose opinions are used to compute suggestions are those neighbors that are trusted by the target user. In a social network environment, people are usually influenced by the opinions of friends with whom they have established trust relationships. Trust-based recommendations will enhance the acceptability of recommendations provided by the system when it uses information from trusted sources. Therefore, incorporating trust into recommendation processes as shown in Figure 6 will not only improve the accuracy of the recommendations but also has the potentials to enhance the user experience. Service consumers will only receive recommendations from those who are in their trust networks. With trust, it is also possible to address the challenges of the traditional recommender systems [15]. However, there is no clear explanation on how it can be used in CARSs. We proposed to incorporate trust into CARSs with three main components as we have earlier illustrated in Figure 2, namely, trust-related information or trust scores, context information, and recommendation mechanisms (including contextual preferences) as analyzed in the previous sections.

3.5.1. Building Entity Profile. Generally, this process involves associating user preferences with contexts in which such preferences have been expressed. As discussed in Section 3.3, every time a user consumes a service, information such as the context of consumption, relevance of that service in such a context, which is given as rating provided by the user after consuming the service or through an implicit means, and the computed trust score for the given context is captured in the user profile. Also, the user profile is updated using a relevance feedback mechanism explained in Section 3.4.3 Thus, the context, the preference, the relevance feedback, and other information about the consumed service will make up the entity or user profile.

3.5.2. Trust Score Computation. The second step, which involves how the trust score is computed for any entity such as the context of a user, is beyond the scope of the current paper. However, Jayasinghe et al. have provided computational models for trust computation in [16], interested readers can check the reference for details. However, the trust score for a given context of consumption is computed using the implicit or explicit rating information obtained from the relevance feedback process, and this score is associated with the context, service, and preference of the service for a given user in every consumption session.

3.5.3. Generating Predicted Preference. In this section, we illustrate how trust can be used to generate predicted preference for a given item or service for a target user.

In a typical collaborative recommendation system, the goal is to predict rating (preference) of the target users based on the similarity of other users with similar taste. This concept has been extended in context-aware collaborative recommendation where the goal is to first determine users who prefer certain items in a similar context and then exploit this information to predict the preferences of the target user in such contexts [11]. Thus, in a traditional collaborative recommendation, typically the Pearson Correlation Coefficient (PCC) is used to find the degree of linear relationships between two entities. In this article, we modified the PCC to incorporate contexts. This means that the degree of linear relationships between two entities takes their contexts into account. Thus, the similarity between an entity v and its neighbor u in context c can be defined by the PCC equation as follows:

$$w_{v,u,c} = \frac{\sum_{i=1}^{m}\left(P_{v,i,x} - \overline{P_v}\right)\left(P_{u,i,c} - \overline{P_u}\right)}{\sigma_v \sigma_u}. \qquad (8)$$

Using this equation, the CARSs can combine the similarity between the relative preferences of users v and u for the same item in context c for all the items they have both consumed and preferred in that context or other similar contexts. Usually, it generates one of the two values: +1 or −1. +1 means that users v and u have similar taste, and −1 means that they have dissimilar taste.

After obtaining the similarity between v and u, CARSs process the profiles of all friends of v, that is, those users who have similar taste as the target user by computing the average of their previous preferences in the rating form, using the PCC value generated above as a weight [25]. The predicted preference for the target user for the item i can then be computed using

$$\text{pr}_{v,i,c} = \overline{P_v} + j\sum_{u=i}^{n}\left(p_{u,i} - \overline{P_u}\right)w_{v,u}. \qquad (9)$$

To introduce trust into the context-aware collaborative model, each contextual preference is associated with at least a context and a trust score as explained in Section 3.3. Now, we define a contextual trust-based preference as follows:

$$P_{u,i,c,t} = j\sum_{x \in C}\sum_{i=1}^{m}\left(p_{u,i,c}\right)\text{sim}_t\left(c, x\right)T_c, \qquad (10)$$

where $\text{sim}_t\left(c, x\right)$ is the similarity between the target user's current context and context of his/her friends. T_c represents the trust scores for the contexts of those friends. The inner sum loops over each context dimension and the inner sum loops over all context trust scores in that dimension to obtain their trust scores. Interested readers can see [11] where we presented a model for computing similarity between two contexts.

We now substitute $P_{u,i,c,t}$ in (8) to have

$$\text{pr}_{v,i} = \overline{P_v} + j\sum_{u=i}^{n}\left(P_{u,i,c,t} - \overline{p_u}\right)w_{v,u}. \qquad (11)$$

This model combines all weighted preferences with respect to similarity between the target user's context, contexts of his/her friends, and their corresponding trust scores to provide an overall context-aware trust-based predicted preference for the target user in the current context. In Section 4, we would be evaluating the impact of context and trust on the relevance of recommendations.

4. Experimental Evaluation and Results

In this section, we provide our initial experimental evaluations of the proposed conceptual framework based on the context-aware preference models. But first, we define the metrics used in the evaluations and then we provide details of the evaluations.

4.1. Evaluation Metrics

4.1.1. Precision. Precision can be broadly defined as the proportions of the top n recommended items that are relevant. It is the number of relevant items selected from recommended items to the number of items that are relevant in the recommended set. It represents the probability that a selected recommended item is relevant according to the preferences of the user.

In context-aware recommendation systems, precision represents the probability that the selected items among those recommended in the current context are relevant in such context as defined in (12):

$$P = \frac{N_{rc}}{N_c},\tag{12}$$

where N_{rc} is the number of items from the recommendation list selected by the users as relevant in the context of use. N_c is the number of relevant items.

4.1.2. Recall.

Recall is the ratio of the recommended media items relevant and preferred by the users in the current context (for contextual recommendations) to the total number of relevant items in the recommendation set as defined in (13). In other words, it measures the proportions of all relevant items included in the top n ranked recommended items.

$$R = \frac{N_{rc}}{N_r},\tag{13}$$

where N_r is the number of selected relevant items in the recommendation set.

4.1.3. F-score.

F-score is the harmonic mean or weighted average of precision and recall. It is a measure of the accuracy of the recommendation system in consideration of relevant and nonrelevant items included in the recommendation set. In practice, precision and recall are inversely proportional and are both dependent on the number of recommended items returned to the user. When more items are returned, recall tends to increase whilst precision decreases and vice versa. Thus, evaluating the detailed accuracy of the performance of a recommendation system becomes difficult using only precisions and recalls individually. However, both metrics can be combined using the standard F-score metric to evaluate the quality of a recommendation system to address this conflict as shown in

$$F\text{-score} = \frac{2PR}{P + R}.\tag{14}$$

An example of how to compute these metrics is given in Figure 16, and the precision, recall, and F-score of item number 5 in the ordered recommendation list are obtained as follows.

From the recommendation list, which contains ten items, only seven are relevant.

$N_r = 7$, $N_{rc} = 4$, $N_c = 5$, $P = 4/5 = 0.8$, $R = 4/7 = 0.57$, and F-score $= ((2 \times 0.8 \times 0.57)/(0.8 + 0.57)) = 0.67$.

4.1.4. Average Precision (AP@K).

Average precision (AP@K) is the mean precision score obtained for each relevant item at top k recommendations in the test cases for every test profile computed as follows:

$$AP@K = \frac{1}{m} \sum_{i=1}^{m} \frac{1}{k_i},\tag{15}$$

where m is the number of relevant items and $1/k_i = 0$, if item i in the set is not relevant.

4.2. Evaluation Data.

Currently, no available open data suit and address the methods in this article. Thus, to evaluate the feasibility of the proposed framework, we conducted preliminary evaluations by using our existing anonymous survey data solicited from online users capturing movie consumption preferences of various users on mobile devices [11, 20, 40]. With these data, we created 200 unique user profiles, each with 19 different entries at the attribute level (the entry in the category level was the same for all users—movie-streaming service). High-level contexts, as presented in Section 3.2, were associated to these entries. Examples of such high-level contexts are provided in Table 1. The anonymous users were asked to associate these terms with location, time, and activity in which they consume any content with such terms. Using these data, we instantiated those contextual profiles representing contextual preferences. For the candidate movie items, online movie databases were explored. Thus, specific movie data were crawled from two popular online movie databases, namely, the IMDB and The Movie Database (TMDB). The crawled candidate content metadata for 5000 movie records from The Movie Database (themoviedb.org) further enhanced with additional metadata retrieved from the IMDB.

This metadata set consists of unique movie genres representing the service attributes in the preference model, and each record contains, on the average, 3 different genre labels. Genres were further characterized by language, cast, country, duration, and release date. These terms thus constitute the media item's attributes in the profile model. Given that users were anonymous and thus were not available to provide feedback, we devised an alternative approach to allow marking of recommended items as *relevant* or as *irrelevant* in an experimental context-aware movie streaming recommendation application we have developed [11]. This allowed to simulate the acceptance or rejection of those items by the user (Figure 17). Hence, an item is selected as relevant if two-thirds of the terms that appear in the item record (but not less than 2 terms) also appeared in the user profile with a weight larger than the average weight of all terms in the user profile. Specifically, a recommended movie item with a metadata record presenting 3 terms is marked as relevant if 2 out of these 3 terms appear in the user profile with weights larger than the average weight of all terms. Otherwise, it is marked as irrelevant. We adopted this approach because we observed that the number of terms contained in the item metadata profile or record along the *attribute-context* concept influenced the classification of candidate items.

4.3. User Profile-Centered Evaluation.

To understand the impact of context and trust on quality of recommendations according to the user preferences either with or without context and trust, we examine the three preference models. In this section, we evaluate the three preference models defined in Section 3.4.3. These models represent those preferences and service consumption that are characterized by context information, trust, and traditional user preferences that are not characterized by these additional properties.

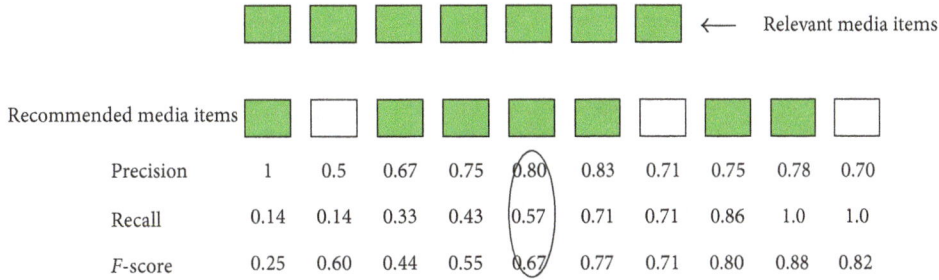

FIGURE 16: Computing precision, recall, and *F*-score for the recommendation system [40].

TABLE 1: Sample contextual information.

User	High-level contexts
User 1	*{(DayOfWeek: Sunday), (TimeOfDay: Evening), (Activity: Sitting), (Location: Cinema), (Illumination: Bright), (Noise Level: Normal)}*
	. . .
User 2	*{(DayOfWeek: Friday), (TimeOfDay: Evening), (Activity: Sitting), (Location: Home), (Illumination: Bright), (Noise Level: Normal)}*
	. . .
User 3	*{(DayOfWeek: Monday), (TimeOfDay: Evening), (Activity: Jogging), (Location: Sport Complex), (Illumination: Bright), (Noise Level: Normal)}*
	. . .
User 4	*{(DayOfWeek: Monday), (TimeOfDay: Evening), (Activity: Sitting), (Location: Office), (Illumination: Bright), (Noise Level: Normal)}*
	. . .
User 5 . . .	*{(DayOfWeek: Friday), (TimeOfDay: Evening), (Activity: Sitting), (Location: Home), (Illumination: Bright), (Noise Level: Normal)}*
	. . .

4.3.1. Traditional Preference Evaluation. In the first experiment, the user preferences were computed without considering the consumption contexts. Of course, this also means that we do not consider whether consumption context is trustworthy or not. The user preferences have been retrieved with no consideration for these additional characterizations by context and trust. Thus, recommendations were generated based only on the preference values. After analyzing the recommendations based on the content of user profiles and movie items, we computed the *F*-score for each recommendation accordingly. Note that, in all experiments, 5 rounds (R1-R5) of recommendations were generated. Also, note that $n@j[j = 5, 10, 15, 20, 25]$ represents the number of recommendations generated.

The rationale was to observe if the model could find more relevant recommendations from the corpus of candidate' items as we increased the number of items in the recommendation set. As explained in Section 4.2, each recommendation set is generated, and its *F*-score is computed. Figure 18 shows the *F*-score as a measure of the accuracy of recommendations. In this figure, as the number of recommended items increased, the *F*-score also increased, but from $n = 20$ and 25, the *F*-score begins to fall. However, the best accuracy was obtained between $n = 10$ and 15,

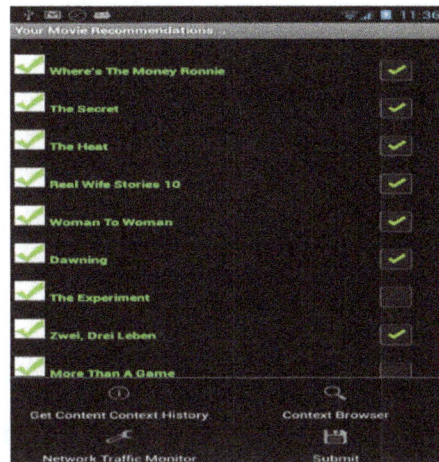

FIGURE 17: Item relevance/irrelevance simulation interface on the mobile device.

getting up to 70%. The possible reason for this is that as the number of items in each set of recommendation increases, the relevant items left in the corpus also decrease, which means that it is difficult for the system to obtain relevant items as this number increases.

4.3.2. Contextual User Preference Evaluation. The second evaluation involves generating the user preferences with specific contexts, for example, movies in the category-service-attribute-context. In this case, the records of services (streaming movies in this instance) consumed by users in specific contexts being processed have been recommended according to the contextual situations of the users. In practice, this translates into setting average weights for preferences assigned to services and properties in specific contexts, whereas others are assigned very low weights or even zero by the system since the user might not prefer such a service in that specific contextual situation. This means that if a user consumes a service/item in each context, such preferences receive bigger weights than those services consumed where contexts of consumption were not considered. This preference model does not consider the trust scores for the context entities in the evaluation process. We call this scenario contextual profiling. The *F*-scores were computed as in the last experiments for the contextual profiles. Figure 19 illustrates the *F*-scores obtained for the context-based preference model. We observed an improvement in the *F*-scores

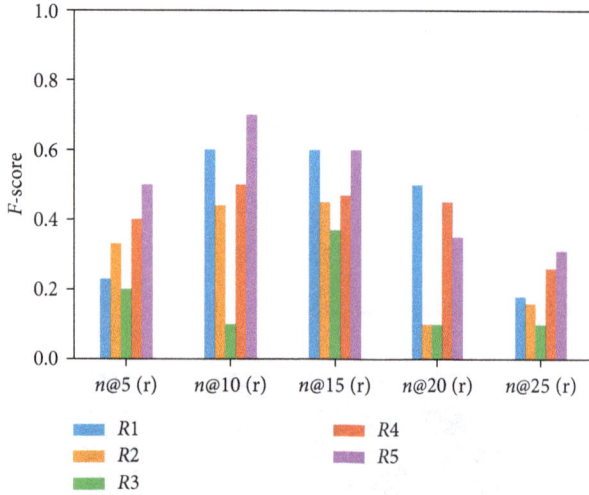

FIGURE 18: *F*-score for noncontextual preferences.

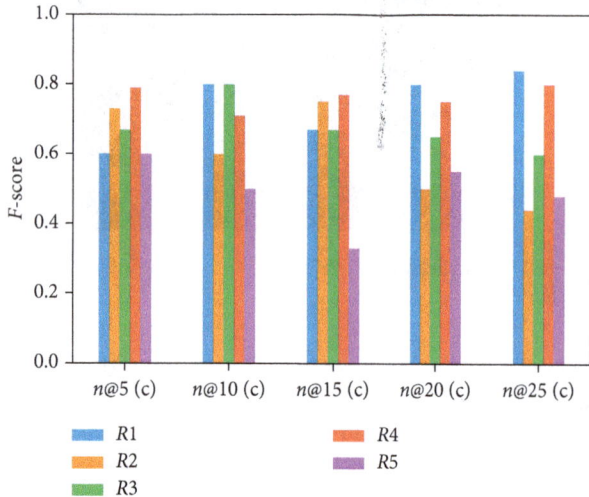

FIGURE 19: Comparison of *F*-score for contextual preferences.

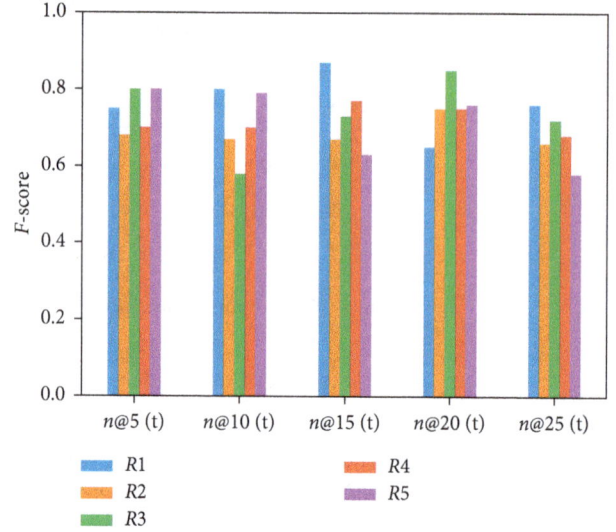

FIGURE 20: Comparison of *F*-scores for preference with trust-based contextual preferences.

computed for each round of recommendations compared to the traditional preference model in the last section. The improvement can be explained as being the result of the capability of the system for using contexts to filter the candidate items according to the contextual preferences of the users. We also observed that the system maintains better *F*-score as the number of items in the recommendation set increases, sometimes up to 0.8.

4.3.3. Context-Aware Trust-Based User Preference Evaluation. In this section, we provide an experiment conducted to study the impact of context and trust on recommendation accuracy using *F*-score as an evaluation metric as we did in the previous sections. In the current evaluation, however, we evaluate the possible impacts of trust on recommendation quality. Other possible evaluations could be to determine the impact of using untrustworthy profiles to compute the preferences of the target users. We would leave that for the next phase of our project. As explained in Section 3.3, we extended and developed a dynamic context-based

preference model with the capability to incorporate trustworthiness of contexts in which preferences are expressed by any consumption entity (users). For each preference expressed, there exist associated contexts and trust scores. Trust can be used in the recommendation filtering process for eliminating or excluding preferences of any profile considered untrustworthy based on its associated trust score. In fact, the profile containing such profile can be entirely excluded from the recommendation process. The ground truth is that the possible consequence of this would be an improvement in the accuracy of recommendations.

Thus, the goal of the current evaluation is to study the impact of trust on CARSs. In the recommendation process, preferences whose contexts' trust scores are below the threshold, which we set at 0.5 [0, 1] being the range of the trust values, are excluded from the recommendation process. Using this mechanism, recommendations were generated as done in the previous experiments. *F*-score for each round of recommendations is computed, and the obtained results in comparison to the other preference models are as shown in Figure 20. The figure shows that using context and trust-related information in the recommendation filtering process can improve the recommendation quality. For example, average *F*-scores at $N = 5$ for the 3 models are 0.68, 0.25, and 0.74, respectively, showing a progressive improvement in the recommendation accuracy.

In addition to evaluating the proposed solution using *F*-score, we have also compared the performances of traditional, context-, and trust-based recommendations by computing the average precision as illustrated in (15) of Section 4.1. The results obtained are illustrated in Figure 21. The result follows similar trends in the other evaluations presented above.

4.3.4. Comparison of Processing Time for Traditional, Context-, and Trust-Based Recommendations. One of the key

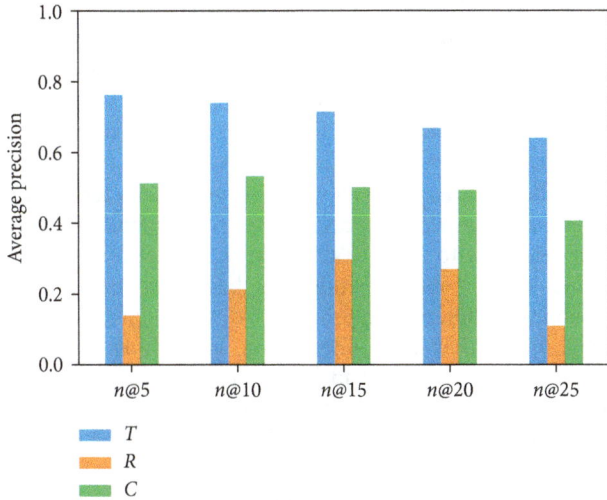

FIGURE 21: Comparison of average precision for traditional recommendation (*R*), context-based recommendation (*C*), and trust-based recommendation (*T*).

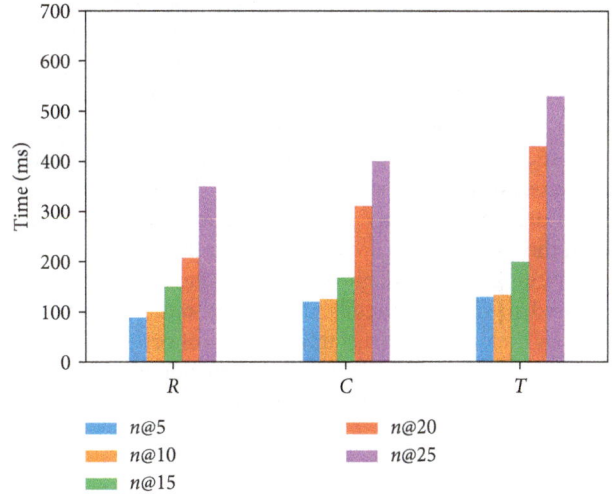

FIGURE 22: Comparison of time taken by traditional recommendation (*R*), context-based recommendation (*C*), and trust-based recommendation (*T*) to generate ranked recommendations.

performance metrics we have considered for measuring the quality of experience and thus of recommendation is the time taken to execute the recommendation execute and generate a set of ranked recommendations. In this regard, we computed the time taken to obtain recommendations with context and trust and without context and trust. The experimental system contains Intel® Core™ i7-3610 QM CPU @ 2.30 GHz, with 8 GB RAM, running 64-bit Windows 10 Pro. Like the previous experiments, recommendations were generated and ranked with 5, 10, 15, 20, and 25. Figure 22 shows the $n@j[j = 5, 10, 15, 20, 25]$ and the processing time. Generally, the computational time increases with the number of items in the recommendation sets. However, we observed that computation time for the trust-based process is higher compared to the context-based and the traditional processes. This is due to the additional computation required to determine the trust-related information from the user profile information.

5. Discussion

In this article, our goal is to present a conceptual framework including an architecture for context-aware personalization using cognitive contexts obtained from diverse sources including IoT devices and present how trust can be incorporated into the recommendation filtering process. The framework comprises components for realizing its specific functionality. From context sensing, context classification and inference to contextual entity preference modeling and recommendation mechanisms utilizing contexts, preferences, and trust to filter candidate services during the recommendation process. The work presented in this article is the initial stage of our work, which is based on developing a framework for context-aware trust-based personalized service delivery in IoT using mechanisms incorporating the elements of trust, context, and recommendation to personalize service delivery. The work, especially the trust

management component, is part of a European-Korean H2020 research project on interoperable global IoT [30].

The context awareness component of the framework was designed to utilize context sensing and classification mechanisms using machine learning algorithms, whereas the cognitive inference of higher-level or complex contextual situations of entities can be performed using semantic web technologies: ontological model, based on OWL, reasoning, and inference. The goal is to combine contexts from different sources including IoT objects to characterize the contextual situations of any entity in the ubiquitous environment.

To predict the preferences of the entity, a component has been proposed with the capability to learn such preferences in the contexts where they have been expressed. In many existing personalized systems, contexts of consumption have not been considered when filtering candidate items. And even those that utilize contexts of consumption do so with a limited set of static contexts, which are obtained directly from the users or from their mobile phones. The developed contextual preference model includes mechanisms to obtain feedback from users to adapt and improve the future personalization process. In addition, such feedback information can be used to compute and incorporate trustworthiness in the personalization process. The component can utilize the contextual preference model to filter and personalize candidate services using recommendation mechanisms. The goal is to use the contextual preferences and trust score associated with the consumption to filter such services. With this capability, we aim to improve the relevance of recommendations and improve user experience as our initial evaluation has confirmed.

To understand how contexts and trust can influence user preferences and consequently the quality of provided recommendations, we have performed some preliminary experiments to analyze the impacts of context and trust on recommendations. We evaluated the traditional preference model, which does not consider the contexts of consumption as well as the trustworthiness. We also evaluated the contextual preference model. Finally, we evaluated the impact of trust on context-aware personalized recommendations as presented in the last section. The initial results show that traditional preference model performed poorly in terms of recommendation accuracy measured by F-score as a statistical test of the accuracy of recommendations. Further evaluation of the impact of contexts on the user preferences shows that more relevant recommendations can be provided when the consumption context is used in the filtering process. This result aligns with previous work where trust has been explored in traditional collaborative recommendations [15]. Further, incorporating trust into contexts of consumption can further improve the accuracy of recommendations. However, this comes with an additional computational cost in terms of processing time.

6. Conclusion

In this article, we have proposed a conceptual framework for exploiting the IoT's context awareness for predicting the user's preferences for personalized services. The article has discussed and elaborated the components of the proposed context-aware framework addressing some important functionality. First, the design addresses fourfold requirements, namely, (1) it proposes a context-aware framework that can collect, analyze, and infer high-level context information from IoT objects. (2) It can provide suggestions to users based on the context information, taking into account the contextual preferences of the users. (3) It proposes a context-based entity profile model for managing preferences. (4) It also proposes the incorporation of trust into context-aware personalized recommendations. Secondly, an integrated layered architecture of the system was proposed, and each component of the architecture has been elaborated based on how it realizes the system's functionality. One of the salient features of the framework is the contextual profile model with the capability to incorporate trustworthiness of contextual information. The proposed conceptual framework thus provides the following important features supporting context-aware personalization: context sensing, recognition, cognitive reasoning, and inference and modeling supports that provide a generic context awareness framework for mobile service personalization to address the problem of static contextualization of existing approaches. Contextual entity profiling supports the automatic learning of the entity's service consumptions in contexts, thereby allowing dynamic determination of its contextual preferences.

To validate our proposal, we conducted preliminary context-aware preference-centered evaluations of the profile models to establish the importance of contextual user preferences for filtering candidate service items in recommendation processes. We have also evaluated the importance of trustworthiness for context-aware personalization during the recommendation process. We have presented and discussed some initial results of the experimental validation of the proposed system. The results obtained emphasize the importance of these parameters for improving the quality of recommendations.

One key advantage of the proposed system is its unique ability that allows incorporating various context-aware recommendation algorithms, which can be implemented as components of the framework to deliver different kinds of recommendations.

Although the system shows some promise in terms of its potentials to improve the accuracy of recommendations and their relevance to the user's contextual preferences, we have evaluated only recommendations based on an extended collaborative filtering approach. We would like to experiment with other recommendation algorithms, especially content-based or hybrid-based recommendations incorporating context and trust. In addition, trust management mechanisms have been used in some existing works to provide security, reliability, and dependability of entities among others. However, we have not used trust in the current work to enforce security, but it has been used as a means to improving the quality of recommendations. Our future goal is to use trust as a means to ensuring that data or information from malicious entities is not allowed in the context-aware personalized service recommendation process. Finally, we would be implementing the framework for real-world deployment in the future and then perform extensive user studies with real users.

Acknowledgments

This work was supported by the EU-funded Horizon 2020 Wise-IoT Project (The EC Grant Agreement no. 723156, Award ID Worldwide Interoperability for Semantics IoT) and Institute for Information & Communications Technology Promotion (IITP) Grant funded by the Korean Government (MSIT) (2015-0-00533, Award ID Development of TII (Trusted Information Infrastructure) S/W Framework for Realizing Trustworthy IoT Eco-System).

References

[1] B. Guo, D. Zhang, Z. Yu, Y. Liang, Z. Wang, and X. Zhou, "From the internet of things to embedded intelligence," *World Wide Web*, vol. 16, no. 4, pp. 399–420, 2013.

[2] C. Perera, A. Zaslavsky, P. Christen, and D. Georgakopoulos, "Context aware computing for the internet of things: a survey," *IEEE Communications Surveys & Tutorials*, vol. 16, no. 1, pp. 414–454, 2013.

[3] P. Guillemin and P. Friess, *Internet of Things Strategic Research Roadmap*, Tech. Rep., The Cluster of European Research Projects, European Commission, Brussels, Belgium, August 2017, http://www.internet-of-things-research.eu/pdf/IoT_Cluster_Strategic_Research_Agenda_2009.pdf.

[4] D. Hussein, S. Park, and N. Crespi, "A cognitive context-aware approach for adaptive services provisioning in social internet of things," in *Proceedings of the IEEE International Conference on Consumer Electronics*, Taiwan, June 2015.

[5] T. W. Um, G. M. Lee, and J. K. Choi, "Strengthening trust in the future social-cyber-physical infrastructure: an ITU-T perspective," *IEEE Communications Magazine*, vol. 54, no. 9, pp. 36–42, 2016.

[6] Q. Liu, H. Ma, E. Chen, and H. Xiong, "A survey of context-aware mobile recommendations," *International Journal of Information Technology and Decision Making*, vol. 12, no. 1, pp. 139–172, 2013.

[7] B. M. Sarwar, G. Karypis, J. Konstan et al., "Item-based collaborative filtering recommendation algorithms," in *Proceedings of the 10th International World Wide Web Conference*, pp. 285–295, Hong Kong, May 2001.

[8] J. Son and S. B. Kim, "Content-based filtering for recommendation systems using multiattribute networks," *Expert Systems with Applications*, vol. 89, pp. 404–412, 2017.

[9] M. J. Pazzani and D. Billsus, "Content-based recommendation systems," in *The Adaptive Web*, vol. 4321, pp. 325–341, Springer-Verlag, Berlin, Heidelberg, Germany, 2007.

[10] R. Burke, "Hybrid Web recommender systems," in *The Adaptive Web: Methods and Strategies of Web Personalization*, LNCS, P. Brusilovsky, A. Kobsa, and W. Nejdl, Eds., vol. 4321, pp. 377–408, Springer, Heidelberg, Germany, 2007.

[11] A. M. Otebolaku and M. T. Andrade, "Context-aware personalization using neighborhood based context similarity," *Wireless Personal Communications*, vol. 94, no. 3, pp. 1595–1618, 2017.

[12] G. Adomavicius and A. Tuzhilin, "Towards the next generation of recommender systems: a survey of the state-of-the-art and possible extensions," *IEEE Transactions on Knowledge and Data Engineering*, vol. 17, no. 6, pp. 734–749, 2005.

[13] F. Ricci, L. Roback, and B. Shapira, "Introduction to recommender systems," in *Recommender Systems Handbook*, F. Ricci, L. Rokach, B. Shapira, and P. B. Kantor, Eds., Springer Science+Business Media LLC, New York, NY, USA, 2011.

[14] I. Mashal, T. Y. Chung, and O. Alsaryrah, "Toward service recommendation in internet of things," in *Proceedings of the Ubiquitous and Future Networks (ICUFN), 2015 Seventh International Conference*, pp. 328–331, Sapporo, Japan, July 2015.

[15] M. A. Abbasi, J. Tang, and H. Liu, "Trust-aware recommender systems," in *Machine Learning Book on Computational Trust*, Chapman & Hall/CRC Press, Boca Raton, FL, USA, 2014.

[16] U. Jayasinghe, A. M. Otebolaku, T. Um, and G. M. Lee, "Data centric trust evaluation and prediction framework for IoT," in *Proceedings of the IEEE/ITU Kaleidoscope: Challenges for a Data-Driven Society*, Nanjing, China, September 2017.

[17] H.-S. Choi, N.-S. Kim, and W.-S. Rhee, "Trusted cross-domain metadata model for context aware services," in *Proceedings of the Ninth International Conference on Ubiquitous and Future Networks (ICUFN)*, pp. 338–343, Milan, Italy, July 2017.

[18] Z. Yan, P. Zhang, and A. V. Vasilakos, "A survey on trust management for internet of things," *Journal of Network and Computer Applications*, vol. 42, pp. 120–134, 2014.

[19] Y. Koren, R. Bell, and C. Volinsky, "Matrix factorization techniques for recommender systems," *Computer*, vol. 42, no. 8, pp. 30–37, 2009.

[20] A. M. Otebolaku and M. T. Andrade, "Context-aware media recommendations for smart devices," *Journal of Ambient Intelligence and Humanized Computing*, vol. 6, no. 1, pp. 13–36, 2015.

[21] G. Adomavicius, A. Tuzhilin, R. Sankaranarayanan, and S. Sen, "Incorporating contextual information in recommender systems using a multidimensional approach," *ACM Transactions on Information Systems*, vol. 23, no. 1, pp. 103–145, 2005.

[22] D. Margaris and C. Vassilakis, "Exploiting Internet of Things information to enhance venues' recommendation accuracy," *Jounal of Service Oriented Computing and Applications*, vol. 11, no. 4, pp. 393–409, 2017.

[23] M. Tu, Y. K. Chang, and Y.-T. Chen, "A context-aware recommender system framework for IoT based interactive digital signage in urban space," in *Proceedings of the Second International Conference on IoT in Urban Space (Urb-IoT'16)*, Tokyo, Japan, May 2016.

[24] P. Resnick and H. R. Varian, "Recommender systems," *Communications of the ACM*, vol. 40, no. 3, pp. 56–58, 1997.

[25] A. Chen, "Context-aware collaborative filtering system: predicting the user's preference in the ubiquitous environment," in *Proceedings of the International Workshop on Location-and Context-Awareness*, Oberpfaffenhofen, Germany, May 2005.

[26] A. Badii, M. Crouch, and C. Lallah, "A context-awareness framework for intelligent networked embedded systems," in *Proceedings of the Advances in Human-Oriented and Personalized Mechanisms Technologies and Services (CENTRIC) 2010 Third International Conference*, pp. 105–110, Nice, France, August 2010.

[27] Y. Saleem, N. Crespi, M. H. Rehmani, R. Copeland, D. Hussein, and E. Bertin, "Exploitation of social IoT for recommendation services," in *Proceedings of the 2016 IEEE 3rd World Forum on Internet of Things (WF-IoT)*, Reston, VA, USA, pp. 359–364, December 2016.

[28] C. Doukas and F. Antonelli, "COMPOSE: building smart & context-aware mobile applications utilizing IoT technologies," in *Proceedings of the 5th IEEE Global Information Infrastructure & Networking Services*, Trento, Italy, 2013.

[29] L. Atzori, A. Iera, and G. Morabito, "The internet of things: a survey," *Computer Networks*, vol. 54, no. 15, pp. 2787–2805, 2010.

[30] Wise-IoT, 2017, http://wise-iot.eu/en/home/.

[31] E. Kovacs, M. Bauer, J. Kim, J. Yun, F. Le Gall, and M. Zhao, "Standards-based worldwide semantic interoperability for IoT," *IEEE Communications Magazine*, vol. 54, no. 12, pp. 40–46, 2016.

[32] E. Mingozzi, G. Tanganelli, C. Vallati, B. Martnez, I. Mendia, and M. Gonzlez-Rodrguez, "Semantic-based context modeling for quality of service support in IoT platforms," in *Proceedings of the 2016 IEEE 17th International Symposium on A World of Wireless, Mobile and Multimedia Networks (WoWMoM)*, pp. 1–6, Coimbra, Portugal, June 2016.

[33] X. Amatrian, A. Jaimes, N. Oliver, and J. M. Puriol, "Data mining methods for recommender systems," in *Recommender Systems Handbook*, F. Ricci, L. Rokach, B. Shapira, and P. B. Kantor, Eds., pp. 39–71, Springer, Berlin, Germany, 2011.

[34] A. M. Otebolaku and G. M. Lee, "Towards context classification and reasoning in IoT," in *Proceedings of the 14th International Conference on Telecommunications (ConTEL)*, pp. 147–154, Zagreb, Croatia, June 2017.

[35] A. M. Otebolaku and M. T. Andrade, "Context representation for context aware mobile multimedia recommendation," in *Proceedings of the 15th IASTED International Conference on Internet and Multimedia Systems and Applications*, Washington, DC, USA, May 2011.

[36] A. M. Otebolaku and M. T. Andrade, "User context recognition using smartphone sensors and classification models," *Journal of Network and Computer Applications*, vol. 66, pp. 33–51, 2016.

[37] D. Figo, P. C. Diniz, D. R. Ferreira, and J. M. P. Cardoso, "Preprocessing techniques for context recognition from accelerometer data," *Personal and Ubiquitous Computing*, vol. 14, no. 7, pp. 645–662, 2010.

[38] C.-H. Liu, K. L. Chang, J. Y. Jason, and S. C. Hung, "Ontology-based context representation and reasoning using OWL and SWRL," in *Proceedings of the 8th Annual Communication Networks and Services Research Conference*, Montreal, QC, Canada, May 2010.

[39] D. Riboni and C. Bettini, "OWL 2 modeling and reasoning with complex human activities," *Pervasive Mobile Computing*, vol. 7, no. 3, pp. 379–395, 2011.

[40] A. M. Otebolaku, "Context-aware personalization for mobile multimedia," Ph.D. thesis, Faculty of Engineering, University of Porto, Porto, Portugal, 2015, https://repositorio-aberto.up.pt/bitstream/10216/78954/2/35037.pdf.

[41] Pellet, 2017, http://github.com/stardog-union/pellet.

[42] M. Bjelica, "Unobtrusive relevance feedback for personalized TV program guides," *IEEE Transactions on Consumer Electronics*, vol. 57, no. 2, pp. 658–663, 2011.

[43] X. H. Wang, D. Q. Zhang, T. Gu, and H. K. Pung, "Ontology based context modeling and reasoning using OWL," in *Proceedings of the Second IEEE Annual Conference Pervasive Computing and Communications Workshops*, pp. 18–22, Orlando, FL, USA, March 2004.

[44] FIWARE NGSI, 2017, https://forge.fiware.org/plugins/mediawiki/wiki/fiware/index.php/Monitoring_Open_RESTful_API_Specification.

7

A Platform for e-Health Control and Location Services for Wandering Patients

Samantha Yasivee Carrizales-Villagómez ⓘ,[1] **Marco Aurelio Nuño-Maganda** ⓘ,[1] **and Javier Rubio-Loyola** ⓘ[2]

[1]*Universidad Politécnica de Victoria, Av. Nuevas Tecnologías, Parque Científico y Tecnológico Tecnotam, Km. 5.5 Carretera a Soto la Marina, Ciudad Victoria, TAMPS, Mexico*
[2]*Centro de Investigación y de Estudios Avanzados del Instituto Politécnico Nacional, Av. Nuevas Tecnologías, Parque Científico y Tecnológico Tecnotam, Km. 5.5 Carretera a Soto la Marina, Ciudad Victoria, TAMPS, Mexico*

Correspondence should be addressed to Samantha Yasivee Carrizales-Villagómez; carrizalesvillagomez.samantha@gmail.com

Academic Editor: Alessandro Bazzi

Wandering patients frequently have diseases that demand continuous health control, such as taking pills at specific times, constant blood pressure and heart rate monitoring, temperature and stress level checkups, and so on. These could be jeopardized by their wandering behavior. Mobile applications that focus on health care have received special interest from medical specialists. These applications have been widely accepted, due to the availability of smart devices that include sensors. However, sensor-based applications are highly energy demanding and as such, they can be unaffordable in mobile e-health control due to battery constraints. This paper presents the design and implementation of a platform aimed at providing support in e-health control and provision of location services for wandering patients through real-time medical and mobility information analysis. The platform includes a configurable mobile application for heart rate and stress level monitoring based on Bluetooth Low Energy technology (BLE), and a web service for monitoring and control of the wandering patients. Due to battery limitations of smart devices with sensors, the mobile application includes energy-efficient handling and transmission policies to make more efficient the transmission of medical information from the sensor-based smart device to the web service. In turn, the web service provides e-health control services for patients and caregivers. Through the platform functionality, caregivers (and patients) can receive notifications and suggestions in response to emergency, contingency situations, or deviations from health and mobility patterns of the wandering patients. This paper describes a platform that conceals continuous monitoring with energy-efficient applications in favor of e-health control of wandering patients.

1. Introduction

Dementia refers to the loss of cognitive functioning. People suffering from dementia are affected in their ability to think, remember, and reason. The most common causes of dementia are Alzheimer's disease, vascular dementia, or multi-infarct dementia with Lewy bodies and frontotemporal dementia [1]. Although vascular dementia accounts for 20% to 30% of all cases of dementia, Alzheimer's disease is the most common cause covering 50% to 75% of cases. It destroys brain cells and nerves, affecting the transmitters that send messages to the brain, particularly those responsible for storing memories. As mentioned in [2], age is the most important risk factor for dementia. It is expected that by 2030, 60 countries will have more than 2 million elderly persons over 65 years. This will have a great impact on public health in several countries because dementia is the leading cause of disability in elderly people, accounting for 11.9% of years lived with disability from noncommunicable diseases. Symptoms of Alzheimer's disease are the sudden loss of memory, confusion in everyday activities, adoption of disturbing behavior such as getting up in the middle of the night or wandering errantly and getting lost [1]. People with dementia are at constant risk of wandering away, being exposed to physical and emotional damage and even death. In addition, wandering patients, mostly elderly people, frequently have other diseases that demand additional health control processes (taking pills, blood pressure and temperature checkups, heart rate and stress level checkups, etc.), which could be jeopardized by their wandering behavior. In this regard, there is a need for e-health monitoring platforms that can provide support to caregivers who take care of patients with wandering behavior.

At present, a large number of technology companies worldwide have already announced innovations of portable electronic devices that facilitate the accomplishment of certain tasks. An example is mobile health (*mHealth*). The term *mHealth* refers to health care using mobile technologies [3]. That is, with the integration of sensors, *mHealth* applications allow for the monitoring of chronic diseases through vital signs sensing, raising the mobile technology as a new era in health care. Moreover, there are factors that influence this transformation, such as aging in the population and the requirement to have fast and available access to monitoring and health care. It is here that mobile devices act as a fundamental axis, simplifying monitoring and delivering critical information at the time users require it.

The improvements made in the creation and design of sensors that measure vital parameters in the human body represent a huge change for the conventional health-care systems, allowing the creation of patient-centered [4] systems and applications, transforming the current health system by reducing costs, anticipating chronic episodes, and improving the patients' quality of life. This is the case of the research [5], where the authors designed and implemented a remote monitoring medical system for the prediction and management of heart failure, predicting the risk of heart failure as a function of changes in pressure arterial, and body weight of patients who are in a noninvasive environment. Integration of several functions into a monitoring application has also been achieved, such as a visualization interface of data relating to monitored parameters, graphically presenting the patient's ECG [6]. Vital signs are indicators of the physiological state of vital organs such as the brain, heart, lungs, and so on. Therefore, systems that perform the monitoring of one or more vital signs (heart rate, respiratory rate, and blood pressure among others) [7] through sensors represent a huge change for conventional health-care systems. The improvements made in the design of intelligent sensors allow for improving the quality and efficiency of *mHealth*, offering advantages over conventional health care such as ease of use, lower risks of infection, reduced user discomfort, and greater mobility [8].

Recent efforts have proposed conceptual frameworks like the PAVISALE framework proposed in reference [9], which is proposed to exploit the monitoring of biological signals by means of novel sensors (invasive and noninvasive), Cloud-based storage and analysis, and the use of big data techniques to analyze massive amount of biological data. Nevertheless, unlike the work presented in this paper, the PAVISALE framework lacks implementation details; it has not been demonstrated in real patients, and its feasibility to address real-life scenarios has not been demonstrated.

This paper presents the design and implementation of a configurable platform aimed at providing support in e-health control and provision of location services for wandering patients through real-time medical and mobility information analysis. The platform is a configurable mobile application and a system for heart rate and stress level monitoring (CMA-HR-SL). The CMS-HR-SL platform consists of a mobile application and a web service for monitoring and control of wandering patients. The mobile application allows integrating to a smartphone, an external sensor or Smart-Band through Bluetooth Low Energy (BLE) technology. Due to battery limitations of smart devices, the mobile application includes energy-efficient handling and transmission policies to make more efficient the transmission of medical information from the smartphone. The mobile application extends the smartphone's capabilities to sensing and transferring medical information through communications networks (Wi-Fi and cellular network) to an external storage server (database (DB)) [10] for further analysis. The medical and location-based information is stored temporally in a Cloud-based service, from where, eventually, a web service retrieves and analyses such information to follow up the evolution of the state of wandering patients. The mobile application processes the heart rate (HR) and physical activity (PA) data in order to calculate the stress level (SL), which is continuously monitored. The mobile application and the web service allow taking precaution or reactive actions based on abnormal heart rate readings, abnormal SL, and abnormal patient locations.

The rest of this paper is structured as follows: Section 2 describes the structure of our configurable mobile application and system for heart rate and stress level monitoring (CMA-HR-SL); Section 3 describes the components of the mobile application, namely, the proposed mechanisms for heart rate (HR) monitoring, GPS sensing, and the procedure to detect anomalies in the smartphone side of the platform; Section 4 describes the procedure proposed to send the medical information from the smartphone to a Cloud-based server to store relevant information about the medical state of the patients; Section 5 describes the components of our platform system aimed at providing assistance to the patient in the event of abnormal heart rate readings, abnormal SL, and abnormal patient locations; finally, Section 6 presents the experimental scenarios and results of the evaluation of our platform; and finally Section 7 concludes the paper.

2. Overview of the Configurable Mobile Application and System for Heart Rate and Stress Level Monitoring (CMA-HR-SL)

The Configurable Mobile Application and System for Heart Rate and Stress Level Monitoring (CMA-HR-SL) manages the information in three phases: (1) a mobile middleware monitors vital signs, GPS, and stress level making use of a smartphone with external sensors; (2) the medical and location information is transmitted to a Cloud-based storage system, which works as an interface between the mobile application and the web service; and (3) the information is

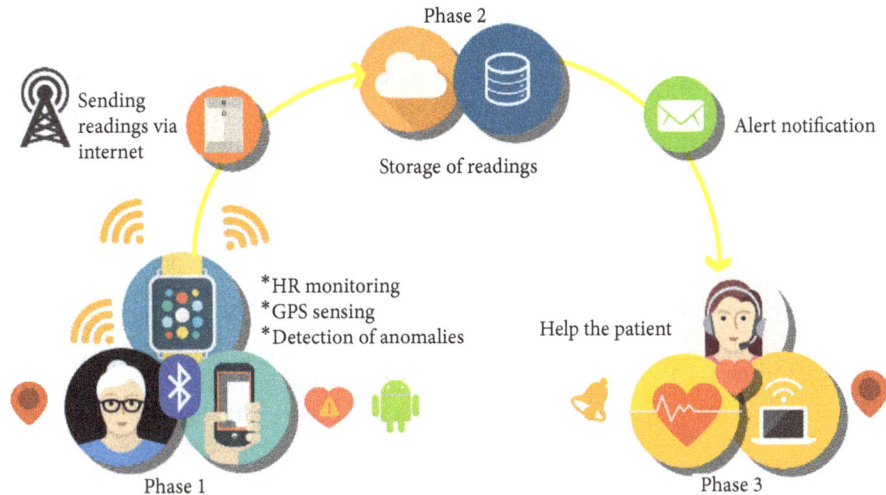

FIGURE 1: Overview of the CMA-HR-SL platform.

appropriately managed and analyzed in the web service to assist the patient when abnormal medical states and location behavior are detected. Figure 1 illustrates the interactions among the proposed phases and an overview of the CMA-HR-SL platform.

In Phase 1, the CMA-HR-SL platform manages the connections between the sensor and the smartphone, collects the data of each sensor, configures the sensing periodicity of the sensor, and decides when it is necessary to transmit the collected data of vital signs and mobility information to the Cloud-based storage. In Phase 2, the CMA-HR-SL implements functions for storing the data in the Cloud, which allows accessing the information when it is required. The Cloud-based service is in charge of sending alerts due to abnormal state of the wandering patients, to the mechanisms implemented in Phase 3. The Phase 2 allows the CMA-HR-SL platform to manage the information of wandering patients easily, with scalable mechanisms, concentrate the information of several patients in a single logical entity (the Cloud), and, most importantly, secure the information of the wandering patients for further assistance when it is required. In Phase 3, the CMA-HR-SL platform gives support to caregivers who register on the web service, allowing them to manage personal information from wandering patients, as well as review and follow up biomedical readings, alert statuses, and abnormal patient locations so as to assist whenever it is needed.

3. Phase 1: A Middleware for Monitoring Vital Signs, GPS Position, and Stress Levels of Wandering Patients

Despite the advances in the production of wearables by hardware vendors, there is a need for platform systems that incorporate the sensing of vital signs, such as the blood pressure, heart rate and stress levels, and location services that can provide an API or real-time access to the collected data for proactive or reactive assistance of patients. Such systems are more relevant when taking care of patients with cognitive disability or wandering behavior. This section presents the mobile application of the CMA-HR-SL platform to address the aforementioned critical need.

This section describes a middleware that allows monitoring the vital signs with external devices by using the Bluetooth Low Energy (BLE) standard to connect the external devices to Android-based smartphones. This approach allows us to (i) take decisions based on the sensed data; (ii) secure the sensed data in an external, accessible, and scalable Cloud-based storage system for further analysis; and (iii) parameterize the way of storing and sensing data to save energy. The mobile device selected for this type of architecture is a smartphone that has built-in sensors such as GPS, gyroscope, proximity, accelerometer, barometer, and so on. The configurable mobile application is based on an architecture that allows the management of the data acquired from the monitoring of specified vital signs (heart rate and stress state) through a BLE communication and energy consumption based on a reading and GPS transmission of the patient. Because smartphones have access to the Internet through a wireless network or access point, the application is able to send biomedical and location (GPS) data from the wandering patient to a virtual cloud storage server from practically everywhere.

The mobile application of the CMA-HR-SL platform is composed of four components: (1) a controller for each sensor connected to the smart device, (2) a sensing scheduler for vital signs and GPS information, (3) a set of energy-efficient handling and transmission policies, and (4) an abnormality and stress level detector. These four components will enable the sensing of multiple vital signs and GPS information through several sensors, orchestrating the periodicity of sensing of each vital sign, managing the storage process of the collected data in the Cloud service accordingly, and triggering warning alerts to caregivers based on detected abnormalities in the patients. These four components are briefly described hereafter.

3.1. A Specific Controller for Each Connected Sensor. Every time a new sensor is added to the middleware, a new controller is instantiated. This controller creates a session with the sensor in order to manage the connections as well as the data that will be sensed. This component also controls the sensing periodicity. According to its configuration, the controller sends information to the sensing scheduler (described below) to keep updated the vital signs that are monitored.

3.2. A Sensing Scheduler for Vital Signs and GPS Information. The scheduler component manages the addition of new sensors by creating new controllers. The scheduler orchestrates all the controllers with their connected sensors and receives the data related to the vital signs. The scheduler also prepares the information for the abnormality and stress level detector (described later). Every sensor datum is locally stored in this scheduler until the energy-efficient handling and transmission policies indicate to the scheduler that the data must be sent to the cloud storage service.

For the case of a heart rate (HR) sensor, the sensor provides HR data through BLE wireless communication to the mobile application. At the same time, the GPS sensor provides the location of the patient. With the collected data, the application model will be able to classify the stress level (SL) according to the identification of an abnormal HR episode (AHRE) with the help of the abnormality and stress level detector and send this collected data to the private cloud whenever the transmission policy indicates to the scheduler to do it.

3.3. Energy-Efficient Handling and Transmission Policies. In order to find trade-offs between energy efficiency and transmission requirements of the patients' information, we proposed three energy-efficient handling and transmission policies. The policies give caregivers the flexibility to select the criteria that better fits the patients' needs. The policies are (a) storing by quantity, in which a certain number of data packages are collected, and they are transmitted to the cloud storage server; (b) storing by time, in which a certain number of data packages are collected for a specific time period, and they are transmitted to the cloud storage server; and (c) storing by abnormality detection, every time a collected data package is evaluated as abnormal, this data package is sent to the cloud storage.

The data packages transmitted to the Cloud service are integrated by the following sets of sensor-based information: a unique identification (ID), GPS position where the medical data was collected, time stamp (TS), heart rate (HR), stress level (SL), and physical activity (PA). This information is made available to the web service (described later) from which caregivers can analyze and take proactive or reactive decisions.

3.4. Abnormality and Stress Level Detector. The abnormality and stress level collector use the collected vital signs data to identify abnormal heart rate episodes (AHREs).

AHREs are identified when the HR falls outside the minimum and maximum thresholds for a given time window.

The HR thresholds and time window are configured by the caregivers. We developed an algorithm that uses the thresholds to detect abnormal events, that is, to identify HR readings that are outside the configured thresholds for a longer time period than the time window configured.

Figure 2 shows the diagram for detecting abnormalities. The diagram starts with an abnormality flag in false and continuously receives HR samples. It identifies whether the incoming sample is outside the thresholds. The identification of a normal reading will reset the flag but an abnormality triggers a time stamp that is stored and accordingly sets the abnormality flag to true. Beyond this point, a normal reading resets the flag but further abnormal readings cause a comparison between the time stamp stored and the current time of the abnormal reading. If the period is longer than the time window, the algorithm alerts for an AHRE. Specifically, the steps of the process are described in the Algorithm 1.

The classification of a SL has two possible values: (a) a state without stress (SNS) or (b) a state with stress (SWS). Each one can be found in two types of activities: at rest or at a recurrent activity. The set of rules defined to establish the classification is described in Table 1.

In order to classify a condition as SNS, it is important to identify the type of PA, which in turn can be classified as recurrent or rest. A recurrent activity is given when the analysis of the HR threshold set by the physician and a tolerance percentage (15%) are above the lower limit and below the upper limit with the tolerance percentage included. A rest activity is given when the previously configured thresholds are met. Finally, for the classification of SWS, the same set of rules is used; however, if one of them is fulfilled, it is sufficient to perform this classification.

The Table 1 shows the status classifications of a wandering patient, where HR represents the heart rate, LL the lower HR limit for the patient, and UL the upper HR limit for the patient.

3.5. Android Activities of the CMA-HR-SL. Figure 3 shows a selection of screenshots of the mobile application of the CMA-HR-SL platform. The figure presents the main functionalities that can be performed with the modules implemented. A brief description of such modules is given hereafter:

(i) *Login*: Figure 3(a) shows a custom login, which is intended to authenticate user data due to the confidentiality requirements.

(ii) *User and contact registration*: this module is presented in Figures 3(b) and 3(c). It allows to register the user (i.e., patient) information (name, user, password, age, and sex) and contact information (name, kinship, e-mail, and phone number).

(iii) *Central activity*: Figure 3(d) shows a menu of submodules: (a) Start of service: it runs the service in the background of the HR monitoring, location sensing, and patient activity. (b) Pairing with the sensor: it performs the search and pairing with the external sensor. (c) Parameter setting: it allows

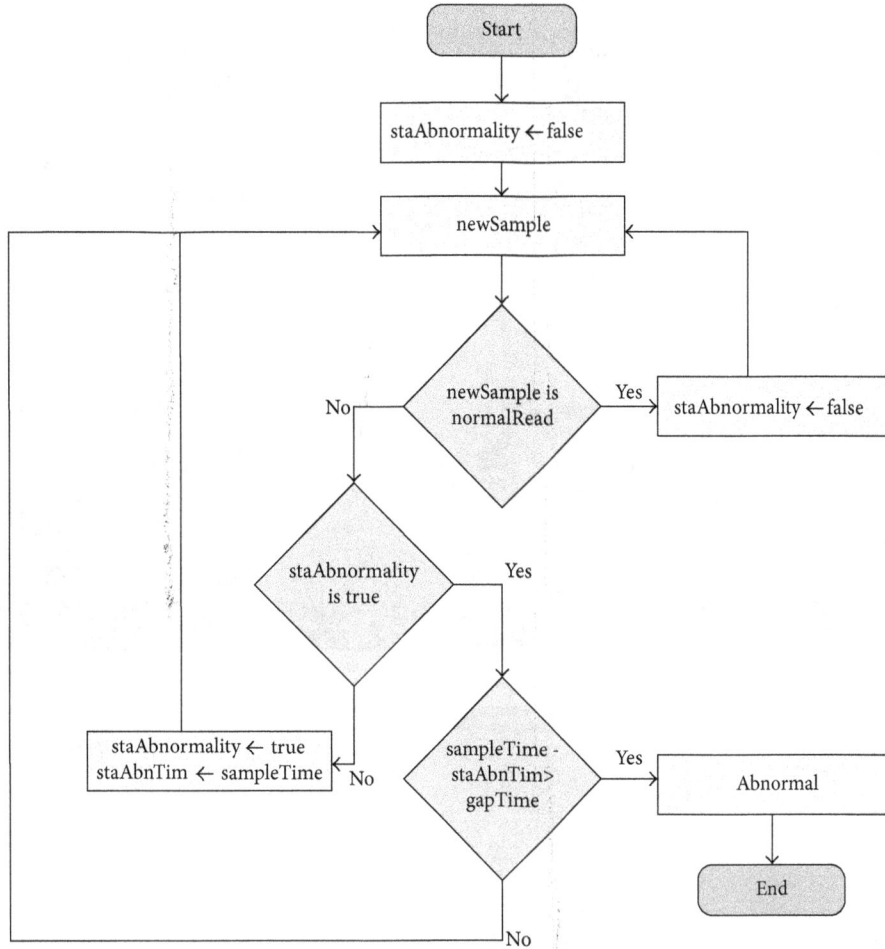

FIGURE 2: Diagram for identifying AHREs.

Require: newSample of HR
Ensure: Identification of abnormality events
(1) **if** newSample is normalRead **then**
(2)   staAbnormality is **false**
(3)   **if** staAbnormality is **true then**
(4)   **if** sampleTime − staAbnTim > gapTime **then**
(5)   Abnormal Episode
(6)   **else**
(7)   newSample
(8)   **end if**
(9)   **else**
(10)   staAbnormality es **true**
(11)   staAbnTim it is equal to sampleTime
(12)   **end if**
(13) **end if**

ALGORITHM 1: Identification of AHREs.

TABLE 1: Classification of stress levels.

State	Activity	Rules
SNS	Exercise	$(LL \leq HR) \wedge (HR \leq UL * 1.15)$
	Rest	$(LL \leq HR) \wedge (FC \leq LS)$
SWS	Exercise	$(HR < LL) \vee (HR > UL * 1.15)$
	Rest	$(HR < LL) \vee (HR > UL)$

4. Phase 2: CMA-HR-SL Functions for Storing Data in the Cloud

This section describes the CMA-HR-SL functions for storing data. It is a Cloud-based solution for storing the medical and location information of wandering patients in a single logical entity (the Cloud). The Cloud-based storage is an interface between the wandering patients and the assistance service (web service described in Section 5). The contextual framework and the most important elements of the Cloud-based framework and the assistance services are graphically shown in Figure 4. In addition, the algorithmic representation of the overall monitoring and localization process is presented in Algorithm 2.

The main functionality of the cloud storage is to secure the information of the wandering patients for further analysis. It enables storing large data sets for heart rate, stress, and

setting HR thresholds, as well as a time window value, for identifying abnormal readings as shown in Figure 3(e). (d) Configuring energy-efficient handling and transmission policies: Figure 3(f) shows the setting of the type of policy with which the data will be transmitted. (e) Exit: to exit from the application.

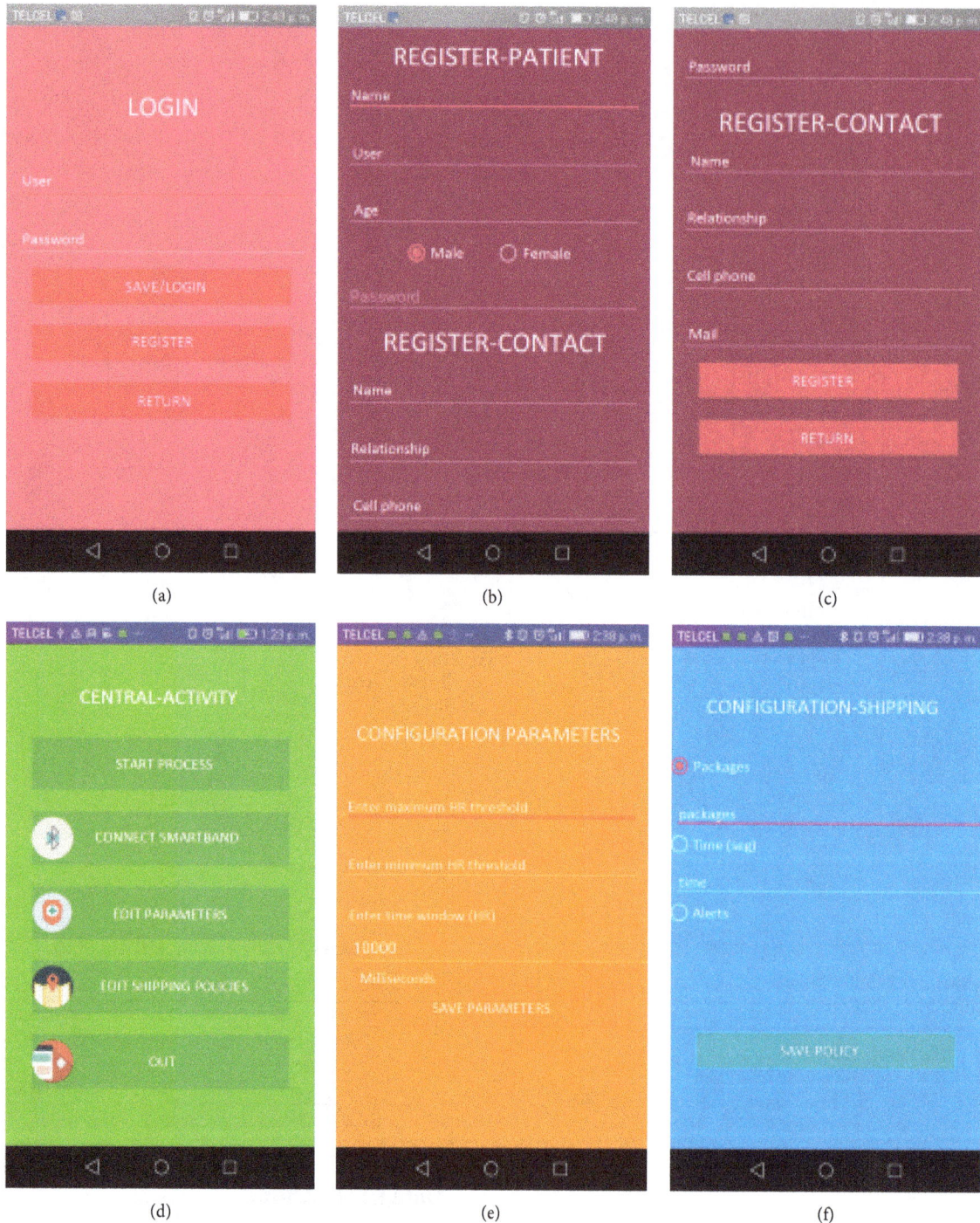

FIGURE 3: Mobile application modules. (a) Login. (b) Register I. (c) Register II. (d) Central Activity. (e) Configuration Parameters. (f) Configuration Policies.

location monitoring services, which can be accessed when required. The storage server is used for sending, handling and visualization of information, and offering availability and practicality in data presentation. Access to the online database is authenticated with both the patient's smartphone device and the web service. It is important to mention that data management security is outside the scope of the main objective of this research. The storage manager for the CMA-HR-SL storage functions is MySQL. In order to assess the required functionality, the cloud database has five tables that are described in Table 2.

The elements defined above allow storing and organizing the information provided by the monitoring carried out by the mobile application, specifically, for the purpose of processing and analyzing the vital signs and location information stored in the system database.

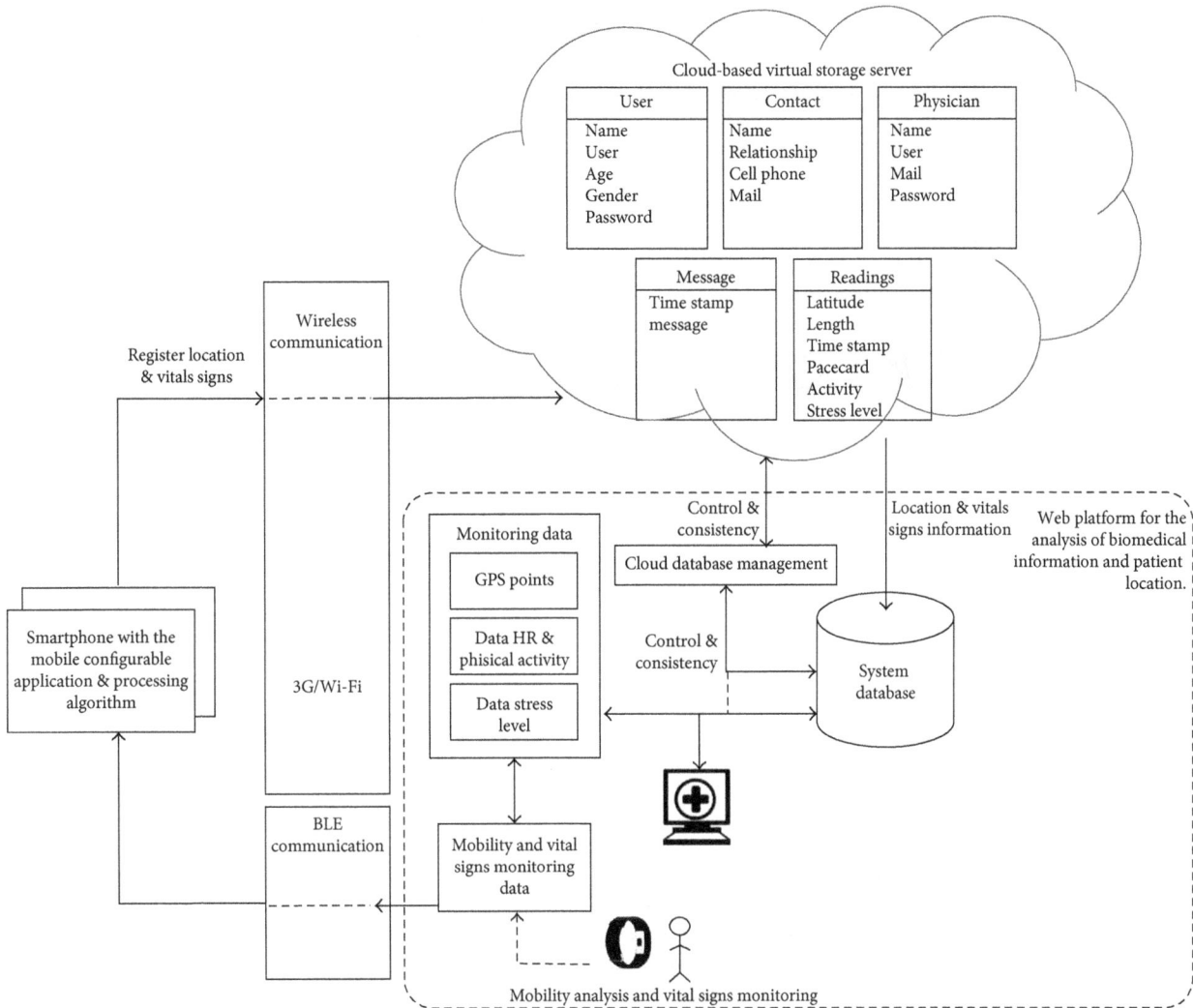

FIGURE 4: Monitoring and location service framework.

Require: Data to be monitored I = {GPS, HR, SL} not empty
Ensure: Identification of abnormality episodes
(1) **while** $I \neq$ null **do**
(2)　**if** connection BLE is **true then**
(3)　　CMA-HR-SL receives I and executes anomaly identification algorithm
(4)　**else**
(5)　　connection BLE is **false**
(6)　**end if**
(7)　Establish WiFi connection and send I to external storage service
(8)　External storage service provides I to web platform
(9)　Carer visualizes episodes of abnormality of the patient from web platform
(10) **end while**

ALGORITHM 2: Monitoring and location process.

5. Phase 3: CMA-HR-SL Support for Caregivers

The CMA-HR-SL platform allows caregivers to manage personal information from wandering patients, including reviewing and following up biomedical readings (TS, HR, GPS, SL, and PA), alert statuses, and abnormal locations so that they can assist the patients whenever the need arises. The service manages alerts about abnormal readings in order to take precaution or reactive actions as to avoid setbacks and serious alterations in patients' health.

The web platform was developed to browse through the stored information so that caregivers can have access to

TABLE 2: Tables in the system database.

Type of alert	Definition
User	Stores information related to the patient: user ID, name, user, age, gender, and password
Contact	Stores data related to the patient's direct contact: contact ID, user ID, name, relationship, cell phone, and e-mail
Physician (caregiver)	This table contains personal physician's (i.e., caregiver) information: physician ID, patient ID, name, user, e-mail, and password
Message	Table storing information of identified alert messages: Message ID, user ID, time stamp of the alert message identification, and the sent message
Readings	Stores information from identified abnormal readings: read ID, user ID, latitude, time stamp length of the reading, HR data, PA status, and SL of the patient

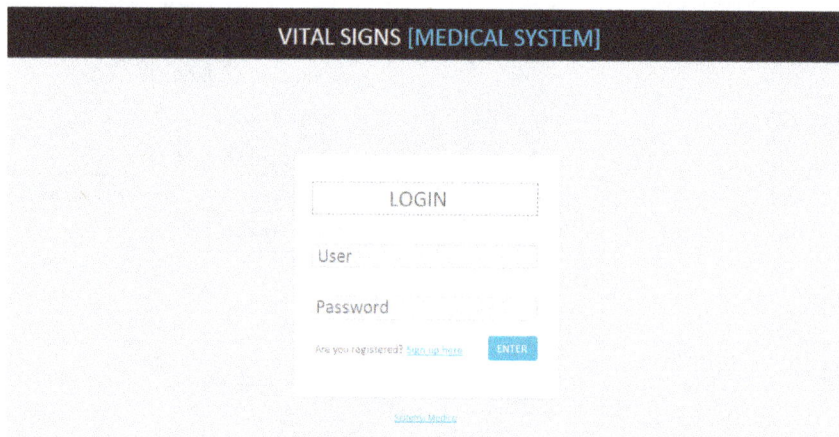

FIGURE 5: Login.

patients' information at all times. The platform was developed with PHP language, and it includes the following four configuration modules: (a) Map: it is responsible for creating the map to view the location of the patient and consult the corresponding alert notification records for a given patient or a set of patients; (b) Configuration: this module is intended to protect (through authentication) the data when establishing connections to the DB; (c) GraphFiles: this module is intended to display data from heart rate readings during a short period of time; (d) Files: this module is responsible for making queries to the DB, for registering and displaying patients' data and contacts. The web platform has two graphical interfaces: (1) Login interface and (2) Central activity interface. The Login interface is intended to register caregivers as well as to control the access to the monitored data of the patients. Figure 5 shows the Login interface of the web platform. The Central activity interface allows the caregiver to check the readings, location, and graphical representations of the heart rates of the patients. Figure 6 shows a list of abnormal readings identified for each patient, where each reading includes the heart rate, time stamp, GPS location, and stress state of the patient. Figure 7 shows a patient location map, and Figure 8 shows a graphical representation of a patient's HR data, date, and time of registration. Alerts can be retrieved or sent from the web server to the caregiver's mobile device.

6. Results and Discussion

This section evaluates the Configurable Mobile Application and System for Heart Rate and Stress Level Monitoring

(CMA-HR-SL) for wandering patients. Section 6.1 evaluates the capabilities of our CMA-HR-SL approach to detect accurate heart rates and to determine whether high HR events are due to incremental physical activity or due to arrhythmic episodes. Next, Section 6.2 evaluates the effectiveness of our CMA-HR-SL handling and transmission policies as to save energy in smartphone devices while monitoring vital signs and mobility information of the wandering patients. Following on, Section 6.3 provides an overall monitoring assessment of eleven wandering patients during their everyday activities. For this purpose, patients with chronic diseases as well as healthy patients were monitored in indoor and outdoor areas. This experimental assessment was intended to demonstrate the capabilities of our CMA-HR-SL platform to identify abnormal heart rate, stress levels, and the GPS location of the wandering patients when alert episodes occur due to such alterations. Finally, Section 6.4 presents the typical response times of our CMA-HR-SL approach when abnormal behaviors are registered in the wandering patients. Namely, this section evaluates the response latency of our CMA-HR-SL approach to typical abnormal heart rate behavior.

6.1. Evaluation of Heart Rate-Aware Capabilities. This section compares the performance of our CMA-HR-SL approach with six state-of-the-art applications with the intention to put in context the holistic solution presented in this paper with respect to current heart rate monitoring applications. Following on, we present the capabilities of the CMA-HR-SL approach to detect whether a high HR event is due to incremental physical

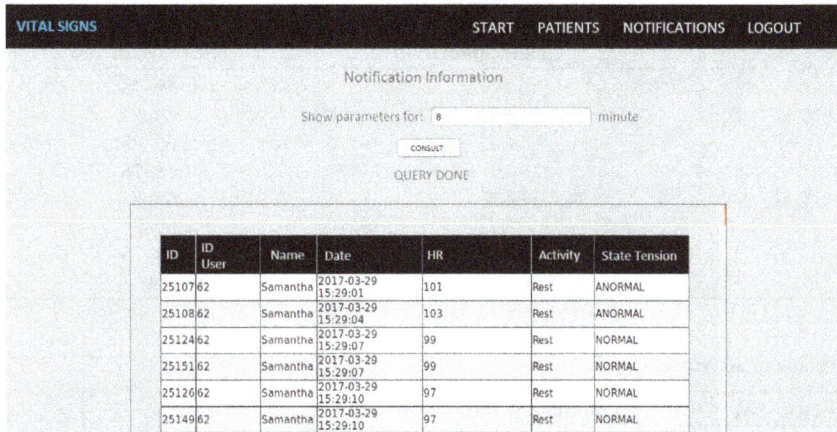

FIGURE 6: Central activity: shows list of abnormal readings identified with HR, GPS, TS, SL, and PA.

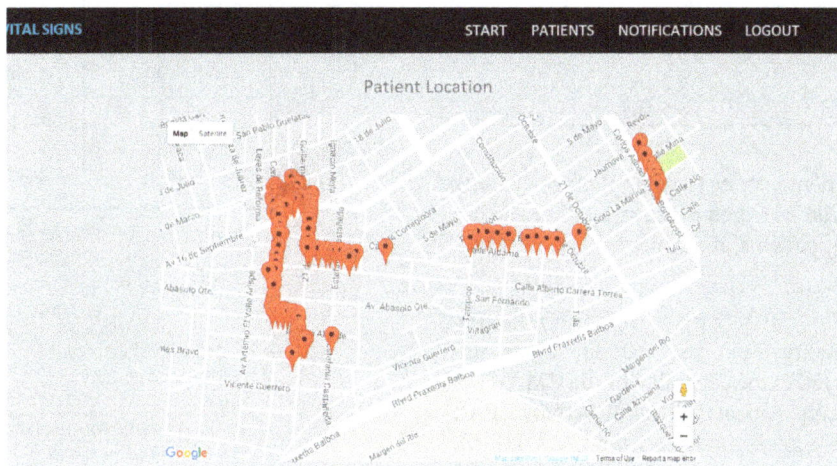

FIGURE 7: Location of the patient's GPS.

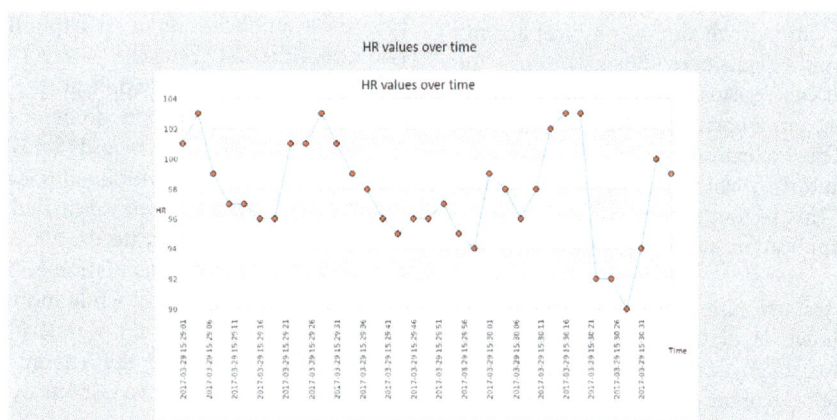

FIGURE 8: HR chart analysis.

activity or due to arrhythmic episodes, namely high HR of patients in resting state.

6.1.1. Comparing HR Monitoring with State-of-the Art Applications.
There are several mobile applications in the market; however, many of them are focused on sports, and they lack continuous and long-term analysis functionality.

This first experiment compares six of the best-rated HR monitoring applications in the market with our CMA-HR-SL mobile application. For this purpose, we measured the HR of a wandering patient in a resting state during two minutes with each application. Table 3 shows the name, qualification, HR range, and minute of monitoring of the application. The data collected by our application was taken from the web service

TABLE 3: Comparison between own app against best-rated HR applications.

Application name	Qualification	HR range at rest	Minute
Own application	—	69–75	h:01
Health-care monitor [11]	4.5	68–75	h:03
Single pulsometer [12]	4.4	66–72	h:05
Pulse heart rate monitor [13]	4.3	68–76	h:07
Runtastic heart rate: pulse [14]	4.3	68–76	h:09
Cardiograph [15]	3.8	69–72	h:11
SmartBand 2 [16]	3.4	68–75	h:13
Average HR		68–74	

TABLE 4: Average in activity response.

Activity	Expected start	Monitored start
RE	15:18:00	15:18:45
RS	15:28:00	15:28:36
RE	15:38:00	15:38:44
RS	15:48:00	15:48:17

Activity	Average in response
RE	44.5 sec
RS	26.5 sec

application. The results demonstrate that our platform collects accurate results which can be stored and analyzed for further assistance in wandering patients, if required.

6.2.2. High HR due to Physical Activity or High HR in Resting State. An HR abnormality event may be due to several factors. This section evaluates the capability of the CMA-HR-SL application to identify patients' physical activity (PA). Specifically, this section demonstrates how the CMA-HR-SL application can determine whether the HR readings outside the configured range are due to physical activity or if the HR readings outside the thresholds occur in when the patient is at rest. This is an important assessment because sometimes a patient's HR reading can be high due to physical activity (e.g., running, which could be considered normal), and, some others instead, a patient can register high HR during resting periods, the latter being an abnormality that is highly desirable to be detected. This evaluation was carried out during 50-minute activity monitoring, which was divided into rest and mobility episodes. This section determines the accuracy of the CMA-HR-SL application while detecting rest and mobility episodes.

The results of this test are summarized in Table 4, and they are graphically shown in Figure 9. RE represents an episode of high HR due to physical activity, and RS represents an episode of high HR of the patient in a resting state. Table 4 shows the time at which the RE and RS episodes start (expected start) and the time at which each episode is monitored (monitored start) by the CMA-HR-SL application. The difference between the expected and monitored is the time that the CMA-HR-SL application takes to determine the state of the patient while registering the HR. In average, the CMA-HR-SL application takes 44.5 seconds to determine that a given HR is due to physical activity and 26.5 seconds to determine that the HR is due to potential arrhythmic behavior.

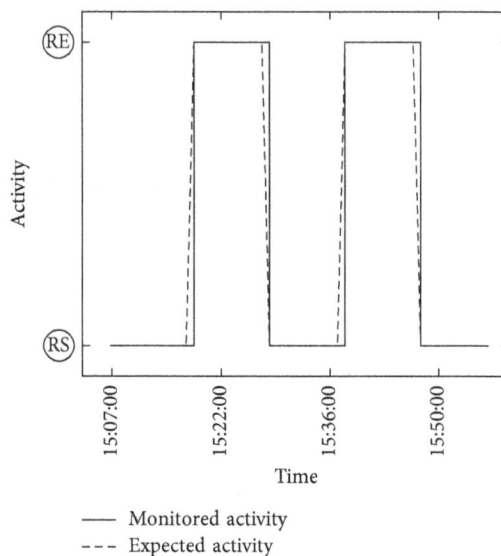

FIGURE 9: Activity monitoring over time.

6.2. Energy Consumption with Energy-Efficient Handling and Transmission Policies. In order to compare the energy load of the CMA-HR-SL application when it runs with other (common) applications in smartphone devices, we defined two experimental test sets: (Set A) tests performed with additional background applications in the smartphone, causing these to affect in some indirect way the rapid consumption of the battery and (Set B) tests performed only with the CMA-HR-SL mobile application without any additional applications. Next, we carried out experimental test and analysis to demonstrate the effectiveness of our CMA-HR-SL handling and transmission policies for saving energy in the smartphone devices while monitoring vital signs and mobility information of the wandering patients.

In order to analyze the energy consumption of our monitoring application, two test sets were designed as follows:

(i) Test Set A: tests including our CMA-HR-SL application running with social applications (Facebook, WhatsApp, Gmail, Spotify, Instagram, Twitter, etc.).

(ii) Test Set B: tests including our CMA-HR-SL application running only with essential factory applications, namely, with Android applications needed for functional Smartphone operation.

The rest of this section analyses the impact of different energy-efficient handling and transmission policies on the

TABLE 5: Package Policy over time for Test Set A.

Packages	Hour-start	Hour-end	Duration	Readings	Abnormal	Delta
50	12:05:12	21:57:33	09:52:21	11,900	1,555	10,345
70	10:36:43	21:03:16	10:26:33	12,685	1,098	11,587
90	11:29:37	21:36:40	10:07:03	13,912	1,809	12,103

TABLE 6: Time Policy over time for Test Set A.

Time	Hour-start	Hour-end	Duration	Readings	Abnormal	Delta
15 min.	11:32:22	21:37:21	10:04:59	4,592	198	4,394
30 min.	12:09:13	23:05:49	10:56:36	4,246	1,221	3,025
60 min.	11:01:39	21:26:21	10:24:42	1,328	682	646

TABLE 7: Alert Policy over time for Test Set A.

Classification	Hour-start	Hour-end	Duration	Readings	Abnormal	Delta
Healthy without phisical activity (HPnPA)	09:22:11	21:14:03	11:51:52	345	345	0
Healthy with phisical activity (HPwPA)	12:30:26	22:35:46	10:05:20	1,646	1,646	0
Chronic disease (PwCD)	10:16:32	20:40:30	10:23:58	705	705	0

performance of our CMA-HR-SL application for Test Set A and Test Set B described earlier. The policies allow the configuration of the following parameters: size of packages transmitted, time of transmission intervals, and alerts. The tests were executed starting with 100% battery, running only our mobile monitoring application on the smartphone.

We have used three energy-efficient management and transmission policies for each set of tests (A and B). For the sake of clarity, we define the three policies included in our experimental tests: Package Policy considers an amount of packages that will be collected by the application and a trigger event to send the package set to the cloud storage service, where each packet is composed of the following data: HR, GPS, TS, SL, and PA; Time Policy drives the collection and sending of packages based on a fixed time. The time parameter of this Policy is previously configured. The Time Policy drives the application to send the packages periodically to the service storage in the cloud; and lastly, Alert Policy considers the identification of only packages with abnormal biomedical readings so that they are sent to the storage service regardless of time and quantity.

Based on the information given by a caregiver where monitoring for wandering patients with chronic illnesses should be at least 12 hours, we carried out our experimental tests taking such time-frame into account. The energy consumption results (i.e., battery life time) for Test Set A is provided in Section 6.2.1. Tables 5–7 present the data collected from the experimental cases, which show the duration of the monitoring, as well as the total and abnormal readings identified for each test set corresponding to each of the shipping policies. In turn, Figures 10–12 present graphically the impact of the battery consumption for each of the energy-efficient handling and transmission policies, which are Package Policy, Time Policy, and Alert Policy, respectively. Similarly, the energy consumption results (i.e., battery life time) for Test Set B are provided in Section 6.2.2. Tables 8–10 similarly present the data collected from the experimental cases, showing the time of the biomedical monitoring, as well

as the total and abnormal readings identified, this for Package Policy, Time Policy, and Alert Policy, respectively. In turn, Figures 13–15 represent graphically the impact of the battery consumption of the mobile device for each of the proposed shipping policies, which are Package Policy, Time Policy, and Alert Policy, respectively

The results of enforcing our energy-efficient handling and transmission policies in Test Set A and Test Set B, and a discussion of these results are provided hereafter in Sections 6.2.1, 6.2.2, and 6.2.3.

6.2.1. Results of Energy-Efficient Handling and Transmission Policies for Test Set A. This section presents the results of battery energy consumption of Test Set A with the three energy-efficient handling and transmission policies: Package Policy, Time Policy, and Alert Policy. For Package Policy, the monitoring application was configured to transmit groups of 50, 70, and 90 packages. For Time Policy, the mobile application was configured to transmit information within time intervals of 15 min, 30 min, and 60 min. For Alert Policy, we defined three types of alerts: (1) healthy wandering patient without physical activity (HPnPA); (2) healthy wandering patient with physical activity (HPwPA); and (3) wandering patient with chronic disease (PwCD).

Tables 5–7 show the results of the energy consumption for the three policies described above. Each table gives the start time, end time, test duration, number of inserted readings, number of abnormal readings, and the difference between these two. Figures 10–11 present graphically the impact of the battery consumption for each of the energy-efficient handling and transmission policies, namely from 100% to 0% of battery charge. The figures also give the difference between the readings inserted and the abnormal readings identified for each type of policy.

6.2.2. Results of Energy-Efficient Handling and Transmission Policies for Test Set B. In this experiment, the tests were

(a)

(b)

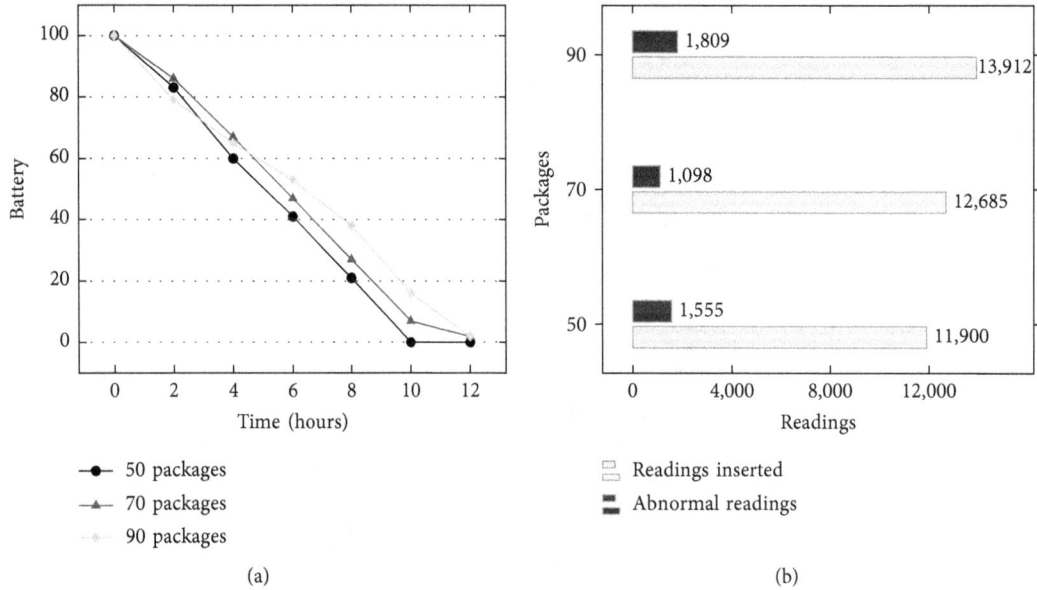

FIGURE 10: Test Set A: Package Policy. (a) Energy consumption over time. (b) Readings inserted between identified abnormal readings.

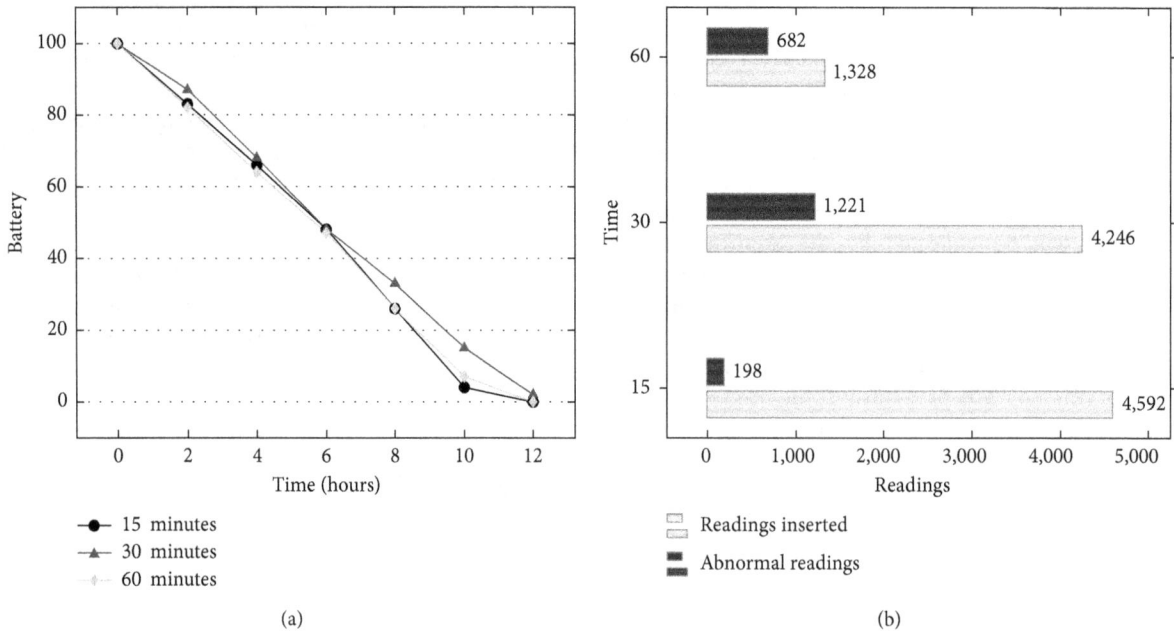

(a)

(b)

FIGURE 11: Test Set A: Time Policy. (a) Energy consumption over time. (b) Readings inserted between identified abnormal readings.

carried out in a controlled scenario, where those must-have applications were eliminated, that somehow realize updates in the system altering the behavior of the battery discharge. Likewise, the tests performed on set A and suspended mode screen were executed but now in this new controlled environment and with the same shipping policy configuration sizes. Tables 8–10 also contain seven fields specifying the send variable, start time, end time, test duration, inserted readings, abnormal readings identified, and the difference between these two, also showing graphically the discharge of the battery, as well as the difference between the readings inserted and the abnormal readings identified, all this for each type of policy.

6.2.3. *Discussion of Results for Energy-Efficient Policies.* The results presented in Section 6.2.1 and Section 6.2.2 represent an important source of information to analyze the evolution of smartphone battery consumption as function of the size of packages, time transmit readings, and type of alert classification and to analyze battery consumption when the CMA-HR-SL application is executed in a controlled or not controlled environment.

Experimental results for Package Policy: In Test Set A, when the size of the package was 70, the battery lasted 10 hours, 26 minutes, 33 seconds, and a number of 1,098 abnormal readings were identified from a total of the 11,587

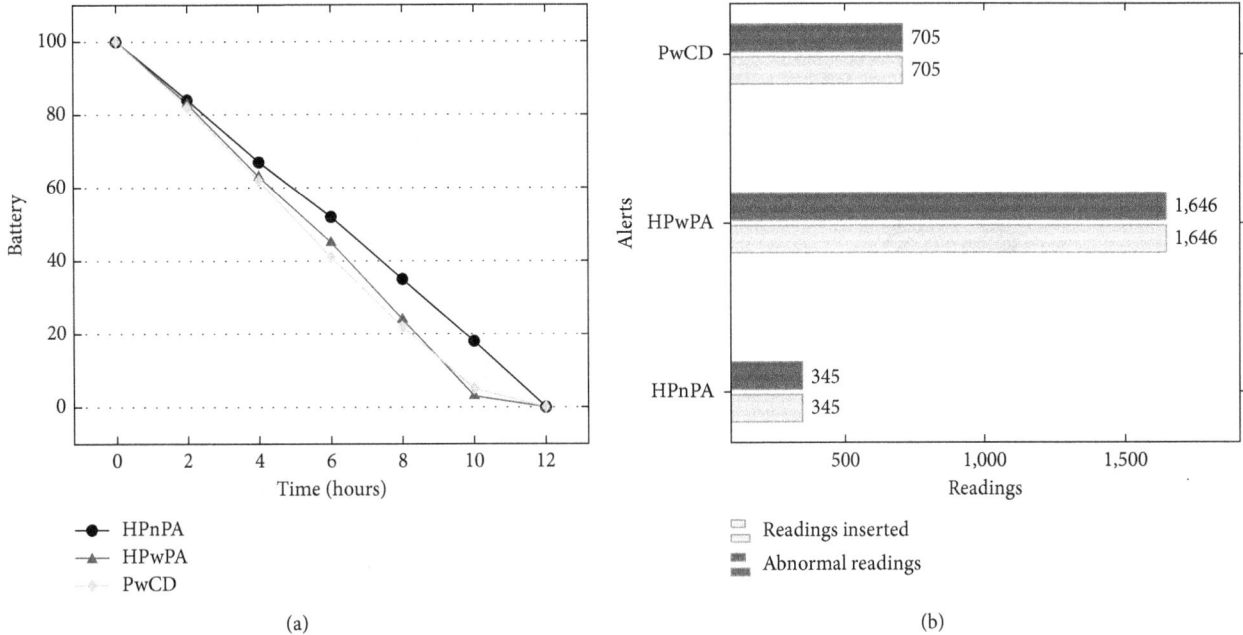

FIGURE 12: Test Set A: Alert Policy. (a) Energy consumption over time. (b) Readings inserted between identified abnormal readings.

TABLE 8: Package Policy over time (B).

Packages	Hour-start	Hour-end	Duration	Readings	Abnormal	Delta
50	11:28:43	22:04:51	10:36:08	12,365	1,010	11,355
70	09:53:53	20:31:43	10:37:50	10,058	2,467	7,591
90	09:34:44	20:48:48	11:14:04	13,485	1,187	12,298

TABLE 9: Time Policy over time (B).

Time	Hour-start	Hour-end	Duration	Readings	Abnormal	Delta
15 min.	10:46:54	20:49:53	10:02:59	4,664	1,429	3,235
30 min.	12:11:53	23:36:27	11:24:34	4,178	1,307	2,871
60 min.	10:30:30	21:28:40	10:58:10	1,445	1,212	233

TABLE 10: Alert Policy over time (B).

Classification	Hour-start	Hour-end	Duration	Readings	Abnormal	Delta
Healthy without phisical activity (HPnPA)	10:42:40	21:24:58	10:42:18	1,559	1,559	0
Healthy with phisical activity (HPwPA)	09:02:53	20:34:02	11:31:09	578	578	0
Chronic disease (PwCD)	10:10:34	20:39:15	10:28:41	1,133	1,133	0

readings inserted. When the packet size was 90, the battery lasted 10 hours, 07 minutes, 03 seconds with 1,809 abnormal readings identified from the 13,912 readings inserted. From this result, we can conclude that the more abnormal readings registered, the higher the battery consumption. This is because abnormal readings use more data transmission and, therefore, the battery life is lower, regardless of the size of the package transmission. For Test Set B, when the size of the packages was 90, the battery life was 11 hours, 14 minutes, 04 seconds with 1,187 abnormal readings identified from the 13,485 inserted. This is because apart from being in a controlled environment, the number of abnormal readings was lower than that of Test Set A; consequently, the duration of the battery was longer.

Experimental results for Time Policy: In Test Set A, when the time to transmit readings was set to 30 minutes, the battery consumption time was 10 hours, 56 minutes, 36 seconds with 1,221 identified abnormal readings from the 4,246 readings inserted. However, when the time to transmit readings was set to 60 minutes, the battery consumption time was 10 hours, 24 minutes, 42 seconds with 682 abnormal readings identified from the 1,328 readings inserted. There are two reasons for this behavior: (1) The test was executed with Test Set A, where the system environment of the mobile device is not controlled which means, it has applications that raise the battery consumption; and (2) The abnormal reading identified in the 60 minutes transmission is above the average of the total of

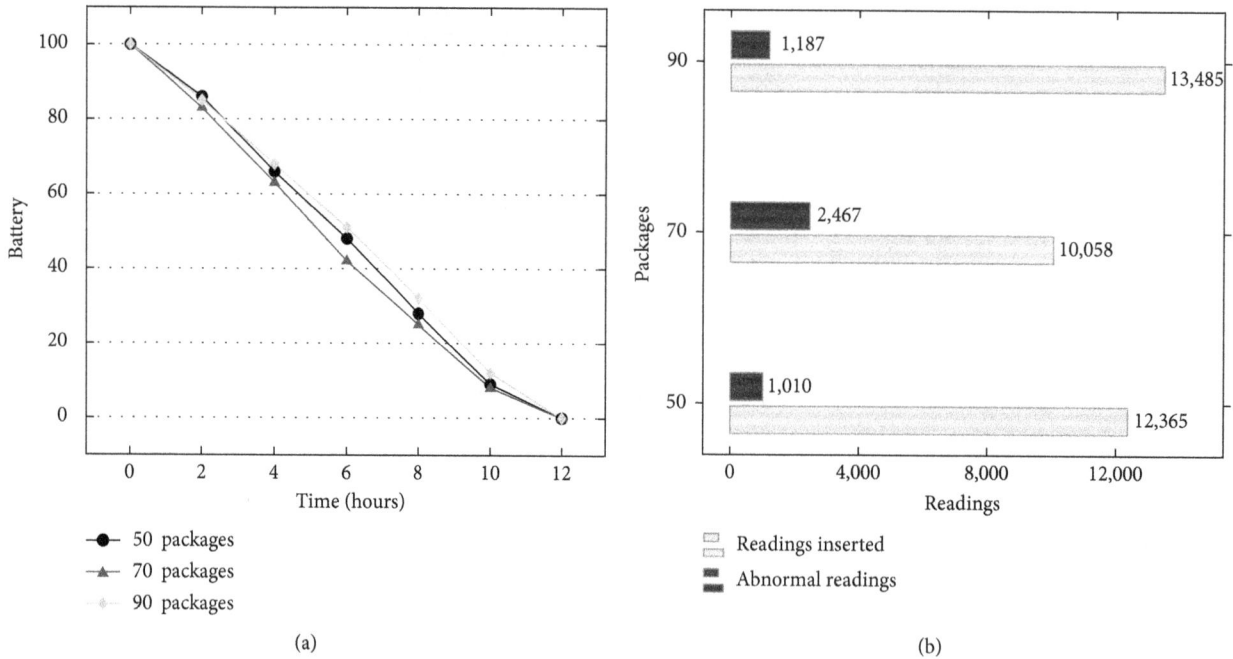

(a)

(b)

FIGURE 13: Package B: Package Policy. (a) Energy consumption over time. (b) Readings inserted between identified abnormal readings.

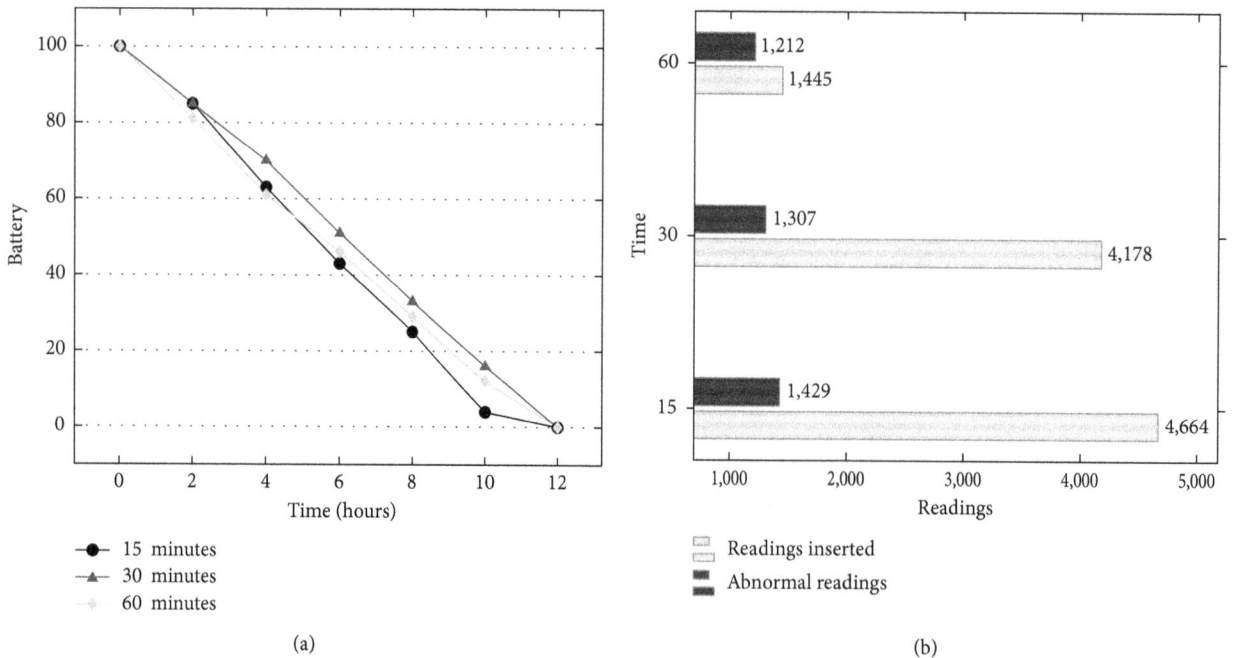

(a)

(b)

FIGURE 14: Set B: Time Policy. (a) Energy consumption over time. (b) Readings inserted between identified abnormal readings.

readings inserted, leading to higher battery consumption. On the other hand, for Test Set B, when the time to transmit readings was set to 30 minutes, the battery duration was 11 hours, 24 minutes, 34 seconds with 1,307 abnormal readings identified out of a total of 4,178 inserted readings; this was due to the higher number of abnormal readings with respect to Test Set A. In general, Test Set B produced less total readings because of the controlled environment nature of the test, that is, with less applications running on the mobile device.

Alert Policy is an important tool for finding trade-offs between intelligent monitoring, efficiency of abnormality notifications, and smartphone battery duration. In this regard, Figures 12(a) and 15(a) show results of battery performance for the patient's medical states considered. Figure 12(a) shows that lower abnormal readings produced less information for upload to the external storage service, resulting in a monitoring time of 10 continuous hours, registering 20% of battery level at the end of the test. On the other hand, the PwCD and HPwPA states registered battery

(a)

(b)

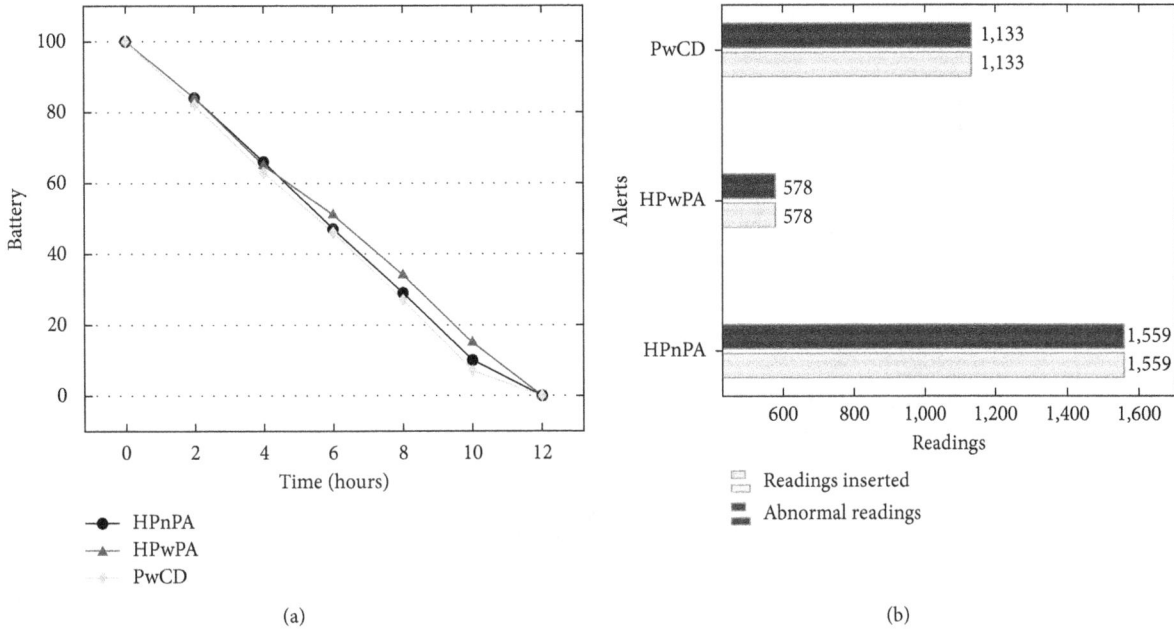

FIGURE 15: Set B: Alert Policy. (a) Energy consumption over time. (b) Readings inserted between identified abnormal readings.

utilization of 1 to 5%, preventing continuous monitoring. It is important to mention that these tests were performed with Test Set A, where the smartphone environment was not controlled. Figure 15(a) presents the results of Alert Policy with Test Set B. The figure shows that HPwPA medical state produced better performance in terms of energy utilization, registering a battery utilization level of 18% after 10 hours of continuous monitoring due to the low number of abnormal readings registered. Instead, the PwCD and HPnPA states registered battery utilization lower than 10% battery in 10 hours of continuous monitoring.

Finally, regarding the tests of Alert Policy, the battery duration is a function of the type of health classification of the patient, the patient's medical state, and number of alert readings identified. The results demonstrate the robustness of our approach to adapt and register abnormalities in the wandering patients. Nevertheless, a more elaborate evaluation of continuous monitoring with wandering patients with several medical states is presented in the following section.

6.3. Overall Monitoring Assessment for Patients with Different Medical States. This section presents monitoring results of our CMA-HR-SL approach while monitoring wandering patients with different medical and physical states. We monitored eleven wandering patients grouped as follows: (1) three healthy wandering patients without physical activity (HPnPA); (2) three healthy wandering patients with physical activity (HPwPA); and (3) five wandering patients with chronic disease (PwCD). For these three classifications, we tested the same number of Alert Policies. The details of the type of alert and the patients are provided in Table 11.

The monitoring for each wandering patient was performed in their daily activities in order to identify alert episodes of heart rate (HR) and stress levels (SL). The test was

carried out using Alert Policy, which consists of identifying abnormal SL episodes according to the SL rules presented in Table 1 and abnormal HR episodes. The HR thresholds are defined by the caregiver based on normal heart beats per minute (bpm). For our tests, the caregiver selected the threshold $60\,\mathrm{bpm} \leq \mathrm{HR} \leq 100\,\mathrm{bpm}$. In this regard, a HR outside of this threshold is considered abnormal, and the web system would issue alerts for appropriate patient's assistance.

Since alerts are issued according to the physical and medical state of each patient, the battery consumption times vary for each patient. We measured the battery consumption time until the charge of the battery was 0%. Four of the eleven tests lasted around eleven hrs and seven lasted around ten hours. Tables 12 and 13 provide the duration time for each patient while monitoring HR and SL in our overall assessment.

It is important to mention that the physiological increase in HR of all readings was due to various external events (daily activities, emotions, stress, effort, etc.) as well as due to REs and RSs activities. The lowest and highest physiological HR registered in our tests was 45 bpm and 140 bpm, respectively. Our tests confirmed that the higher the number of alerts, the lower the battery life; moreover, the CMA-HR-SL platform is able to register all of these alerts and, more importantly, is able to issue alerts to caregivers to take care of the wandering patients accordingly. Tables 14–16 show the results (abnormal readings, activity REs readings, stress level, and HR average) of our overall assessment. It is worth mentioning that for the results of the stress level, all patients registered abnormal levels; this is because during the whole test, the patients registered a number of abnormal episodes (registered as abnormal in Tables 14–16). Also, the physical activity and HR were effectively used to determine appropriate stress levels.

6.4. Response Time of the CMS-HR-SL Platform. The CMS-HR-SL platform presented in this paper has been designed to

TABLE 11: Types of alert for patient.

Type of alert	Definition
HPnPA	Healthy wandering patient without physical activity: the patients do not suffer from any chronic disease, and they do not perform physical activities. Three patients
HPwPA	Healthy wandering patient with physical activity: the patients do not suffer from any chronic disease, and they regularly perform physical activities. Three subjects
PwCD	Wandering patient with chronic disease: this classification includes wandering patients with chronic disease without physical activity (PwCDnPA) and wandering patients with chronic disease with physical activity (PwCDwPA). The chronic diseases are: diabetes, high blood pressure (HBP), Wolff Parkinson White (WPW), and nonspecific chronic ulcerative colitis (NSCUC). Five patients

TABLE 12: Functional test with approximate duration of 11 hours.

Monitoring	Duration
Patient User A	11:15:36
Patient User B	11:21:30
Patient User D	11:07:06
Patient User G	11:35:45

TABLE 13: Functional test with approximate duration of 10 hours.

Monitoring	Duration
Patient User C	10:48:41
Patient User E	10:42:56
Patient User F	10:30:54
Patient User H	10:57:16
Patient User I	10:50:26
Patient User J	10:12:22
Patient User K	10:28:41

alert caregivers when wandering patients need assistance. In this sense, timely reaction to sudden issues like abnormal heart rates, abnormal stress levels, and abnormal location of wandering patients is a pivotal requirement. This section evaluates the response time of the CMS-HR-SL platform when patients are in the need for assistance. Table 17 presents information of three patients (see patients ID 70, 54, and 60 in the leftmost column in Table 17) that presented alterations in their heart rates during our tests. The column TS-FRA-HR shows the time stamp when the first abnormal heart rate is present in the wandering patient; this information has been obtained from the mobile application. The column TS-NC shows the time stamp when the caregiver receives the alert in the web system, this information has been obtained from the web system directly. The difference between the time when the alert is produced (TS-FRA-HR) and the time when the alert is received by the caregiver (TS-NC) is presented in the Latency column in Table 17. The column ATR-HR presents the amount of alterations of heart rate in the corresponding patient received after the first abnormal reading. From the above, we determined that the order of time needed to receive an alert when an abnormality is recorded in a wandering patient is in the range of 0–3 seconds (minimum) and 10–12 seconds (maximum).

7. Concluding Remarks and Future Work

This paper has presented the design and implementation of a platform aimed at providing support in e-health control and provision of location services for wandering patients through real-time medical and mobility information analysis. The paper has focused on heart rate and stress level monitoring and analysis for wandering patients. We have demonstrated that through external, nonintrusive sensors, it is possible to perform smartphone-based continuous monitoring of vital signs and, more importantly, to determine abnormal medical states accurately by means of correlation of vital parameters in runtime and on demand.

The experimental results demonstrate that through our platform, it is possible to identify chronic episodes of a wandering patient through alerts as well as provide the location of the patient for appropriate assistance. It is also worth mentioning that through our platform, it is possible to provide assistance to caregivers when wandering patients register abnormal episodes of heart rate and stress levels for reactive or proactive health-care assistance. The platform gives support to caregivers allowing them to avoid physical and continuous evaluations of wandering patients; instead, it allows them the flexibility to browse through patients' state and analyze the evolution of vital signs remotely from a web-based service for remote clinical checkups of the patient.

We have demonstrated that it is possible to manage battery consumption by means of energy-efficient handling and transmission policies for more efficient transmission of medical information from the sensor-based smart device to the web service. The platform system gives the flexibility to adopt the more appropriate policy for taking care of the patients according to their medical follow-up requirements. In this regard, our platform conceals continuous monitoring with energy-efficient applications in favor of more granular and more focused e-health control of wandering patients.

We strongly believe that systems like our CMA-HR-SL platform are important for enhancing medical care, as they give caregivers the possibility of offering quality and prompt attention to patients with diverse needs. We performed our platform's evaluations on eleven wandering patients with different medical conditions. Our experimentation trials lasted approximately 690 hours considering the evaluation of all the monitoring and analysis parameters. We are aware that a more diverse population of wandering patients would enhance the statistical validation of our results. Moreover, the results obtained through the performed evaluation tests have allowed us to determine the effectiveness of the

TABLE 14: HPnPA classification.

User	Classification	Age	RE activity	SL	Abnormals	HR average	Duration (hh:mm:ss)
User A1	HPnPA	56	71	Abnormal	71	93	11:15:36
User B2	HPnPA	49	58	Abnormal	871	95	11:21:30
User C3	HPnPA	32	201	Abnormal	787	82	10:48:41

TABLE 15: HPwPA classification.

User	Classification	Age	RE activity	SL	Abnormal	HR average	Duration (hh:mm:ss)
User D4	HPwPA	24	175	Abnormal	729	72	11:07:06
User E5	HPwPA	27	84	Abnormal	2,125	72	10:42:56
User F6	HPwPA	24	233	Abnormal	236	84	10:30:54

TABLE 16: PwCD classification.

User	Classification	Age	HR activity	SL	Abnormal	HR average	Duration (hh:mm:ss)
User G7	PwCDwPA (HA y Pre Diabetes)	69	0	Abnormal	4	86	11:35:45
User H8	PwCDwPA (WPW)	26	29	Abnormal	1,055	95	10:57:16
User I9	PwCDwPA (HA)	50	40	Abnormal	191	82	10:50:26
User J10	PwCDwPA (Diabetes)	42	61	Abnormal	1,834	77	10:12:22
User K11	PwCDwPA (CUCI)	24	29	Abnormal	578	82	10:28:41

TABLE 17: Time in response.

ID patient	TS-FRA-HR (hh:mm:ss)	TS-NC (hh:mm:ss)	Latency	ATR-HR
70	21:03:05	21:03:05	0 seg.	23
70	21:09:14	21:09:17	3 seg.	6
54	20:11:31	20:11:43	12 seg.	5
54	20:18:58	20:19:08	10 seg.	2
60	12:15:35	12:15:40	5 seg.	2
60	13:23:03	13:23:03	0 seg.	22

approach, and they have been useful to draw the above conclusions. Therefore, future work will be devoted to analyzing the performance of our platform in more patients with other diseases, as well as extending its functionality to include body sensors for monitoring other types of biological information in accordance with patients' requirements. Another area of future work is the integration of other sensors to extend the medical care capabilities of our CMA-HR-SL platform. Finally, due to the proposed design, another fundamental aspect is the definition of self-adjustable energy-efficient handling and transmission policies, which can adjust the time and amount of package sizes according to context or type of alerts, with the intention of finding a balance between battery usage and the needs of the mobile application.

Acknowledgments

Samantha Yasivee Carrizales-Villagómez thanks the Science and Technology Council of Mexico (CONACYT) for the postgraduate studies scholarship number CVU 706194 awarded towards the realization of this research titled "A Platform for e-Health Control and Location Services for Wandering Patients." The author also appreciates the support and collaboration of the Center for Research and Advanced Studies (CINVESTAV) and the Polytechnic University of Victoria (UPV) for guiding the development of the research. This paper is partially supported also by project TEC2015-71329-C2-2-R (MINECO/FEDER).

References

[1] T. G. V. on Dementia, "Alzheimer's disease international, "types of dementia"," July 2015, http://www.alz.co.uk/.

[2] P. León-Ortiz, M. L. Ruiz-Flores, J. Ramírez-Bermúdez, and A. L. Sosa-Ortiz, "Estilo de vida en adultos mayores y su asociación con demencia," *Gaceta Médica de México*, vol. 149, no. 1, pp. 36–45, 2013.

[3] S. R. Steinhubl, E. D. Muse, and E. J. Topol, "Can mobile health technologies transform health care?," *JAMA*, vol. 310, no. 22, pp. 2395-2396, 2013.

[4] L. Xu, D. Guo, F. E. H. Tay, and S. Xing, "A wearable vital signs monitoring system for pervasive healthcare," in *Proceedings of the 2010 IEEE Conference on Sustainable Utilization and Development in Engineering and Technology (STUDENT)*, pp. 86–89, Kuala Lumpur, Malaysia, November 2010.

[5] L. Zhang, C. Yu, C. Jin et al., "A remote medical monitoring system for heart failure prognosis," *Mobile Information Systems*, vol. 2015, Article ID 406327, 12 pages, 2015.

[6] E. A. P. J. Prawiro, C.-I. Yeh, N.-K. Chou, M.-W. Lee, and Y.-H. Lin, "Integrated wearable system for monitoring heart rate and step during physical activity," *Mobile Information Systems*, vol. 2016, Article ID 6850168, 10 pages, 2016.

[7] T. Yilmaz, R. Foster, and Y. Hao, "Detecting vital signs with wearable wireless sensors," *Sensors*, vol. 10, no. 12, pp. 10837–10862, 2010.

[8] Y. Hao and R. Foster, "Wireless body sensor networks for health-monitoring applications," *Physiological Measurement*, vol. 29, no. 11, pp. R27–R56, 2008.

[9] J. Rubio-Loyola, G. Ramírez, T. Velin, J. Sánchez, and E. Burgoa, "e-health control and location services for wandering patients through cloud-based analysis," in *Proceedings of the Eighth International Conference on Advances in Human-Oriented and Personalized Mechanisms, Technologies, and Services*, pp. 12–21, Barcelona, Spain, November 2015.

[10] R. E. P. Márquez, "Sistema de monitorización de ritmo cardiaco para el soporte de aplicaciones de sensado personal con telefonos inteligentes," Ph.D. thesis, Centro de Investigación y de Estudios Avanzados del Instituto Politécnico Nacional, Mexico City, Mexico, 2016.

[11] iCare, "iCare monitor de la salud (bp), (version 3.6.0)," September 2017, https://play.google.com/store/apps/details?id=comm.cchong.BloodAssistant.

[12] Pulsometer, "Single pulsometer, (version 3.0.0)," August 2017, https://play.google.com/store/apps/details?id=com.supersimpleapps.heart_rate_monitor_newui&hl=es.

[13] H. Monitor, "Pulse heart rate monitor, (version dependent on device)," August 2017, https://play.google.com/store/apps/details?id=si.modula.android.instantheartrate&hl=es.

[14] Runtastic, "Runtastic heart rate, (version dependent on device)," September 2017, https://play.google.com/store/apps/details?id=com.runtastic.android.heartrate.lite&hl=es.

[15] Cardiograph, "Cardiograph, (version dependent on device)," July 2017, https://play.google.com/store/apps/details?id=com.macropinch.hydra.android&hl=es.

[16] Sony, "SmartBand-swr12, (version 4.4)," January 2017, https://play.google.com/store/apps/details?id=com.sonymobile.hostapp.everest&hl=es.

8

Gait Analysis using Computer Vision based on Cloud Platform and Mobile Device

Mario Nieto-Hidalgo (ID)**, Francisco Javier Ferrández-Pastor, Rafael J. Valdivieso-Sarabia** (ID)**, Jerónimo Mora-Pascual, and Juan Manuel García-Chamizo**

Department of Computing Technology, University of Alicante, Campus San Vicente del Raspeig, Alicante, Spain

Correspondence should be addressed to Mario Nieto-Hidalgo; mnieto@dtic.ua.es

Academic Editor: Pino Caballero-Gil

Frailty and senility are syndromes that affect elderly people. The ageing process involves a decay of cognitive and motor functions which often produce an impact on the quality of life of elderly people. Some studies have linked this deterioration of cognitive and motor function to gait patterns. Thus, gait analysis can be a powerful tool to assess frailty and senility syndromes. In this paper, we propose a vision-based gait analysis approach performed on a smartphone with cloud computing assistance. Gait sequences recorded by a smartphone camera are processed by the smartphone itself to obtain spatiotemporal features. These features are uploaded onto the cloud in order to analyse and compare them to a stored database to render a diagnostic. The feature extraction method presented can work with both frontal and sagittal gait sequences although the sagittal view provides a better classification since an accuracy of 95% can be obtained.

1. Introduction

This work is part of a project called Gait-A whose main objective is the early detection of frailty and senility syndromes using gait analysis. Physical activity is one of the main components involved in frailty syndrome evaluation [1, 2]. Gait is identified as a high cognitive task in which attention, planning, memory, and other cognitive processes are involved [3, 4].

Through gait analysis, quantification of measurable information of gait, and its interpretation [5], frailty and dementia syndromes can be diagnosed. This process is carried out by specialists and is based on estimations through visual inspection of gait.

In this work, we propose a computer vision approach that could aid the specialists providing them with objective measurements of gait and, thus, gain in objectivity of the gait analyses performed.

We propose the use of smartphone cameras to record the subject's gait and also provide computer vision algorithms able to analyse those sequences to extract spatiotemporal gait parameters. These parameters are then sent to the cloud to be analysed by a classifier for the purpose of determining whether abnormalities are present or not.

A lot of works dealing with gait analysis using computer vision are found in the literature. However, most of them focus on gait biometrics for human identification, and few of them address gait analysis for detection of abnormalities.

The main goal of this study is to provide a nonexpensive and easy-to-deploy solution to obtain the spatiotemporal parameters of gait, which will be fed to classification algorithms that will discriminate between normal and abnormal gait. It needs to be mentioned that the process of obtaining spatiotemporal parameters for abnormal gait compounds the task as the number of assumptions that can be made over gait patterns is drastically reduced. In such cases, neither cyclic patterns nor the totality of the gait phases can be assumed to be present. In this work, for study purposes, Parkinsonian gait, knee pain, and foot dragging among other patterns that deviate from what we consider normal gait will be taken as abnormal gait.

A set of different gait features is analysed in [6] for person identification. The process starts by extracting the silhouette with a background subtraction technique to then obtain the contour. After the contour is obtained, they extract four time-series features: width/height ratio, bounding box width, silhouette area, and center of gravity (COG). These four features follow a cyclic pattern that match the gait cycle and are used to identify a person through deterministic learning.

Xu et al. examined the suitability of the Kinect sensor to measure gait parameters while walking on a treadmill in frontal view [7]. They compared the heel strike (HS) and toe off (TO) they obtained with those obtained using a motion tracking system. HS showed less error than TO because it happens closer to the sensor.

Choudhury and Tjahjadi [8] proposed a method composed of three modules: silhouette extraction, subject classification using Procrustes shape analysis (PSA) and elliptic Fourier descriptor (EFD), and combination of both results. For silhouette extraction, they use background subtraction and morphologic operations to remove noise. PSA module analyses a group of shapes using matching of geometrical locations of a silhouette. The stride length is computed using the width of the bounding box. Finally, EFD allows to characterize the contour of the subject in key points of a gait phase.

Leu et al. proposed a method to extract skeleton joints from sagittal and frontal views [9]. The method proposed uses the horizontal and vertical projection of the silhouette pixels to obtain the neck joint. Then they apply an anatomical model to obtain hip, knees, and ankles. Yoo and Nixon [10] also extract skeleton joints using an anatomical model to segment the silhouette but they obtain the mean points of each segment and then apply linear regression to obtain a line that represents the bones. During double support gait phase, they apply motion tracking to estimate the location of the occluded points. Khan et al. [11], similarly obtain the skeleton by computing the mean points of each body segment. They obtain leg movement and posture inclination and compare it with a normal gait model to recognise Parkinsonian gait.

In addition, we find the following proposals for classifying gait patterns. In Wang [12], the method is based on optical flow that calculates a histogram of silhouette flows to which an eigenspace transformation applies. The data obtained are compared with a normal gait template to calculate deviation. In Bauckhage et al. [13], homeomorphisms apply between 2D lattices and binary shapes to obtain a vector space in which the silhouette is encoded. They performed several silhouette bounding box splittings to obtain different lattices that are then classified using support vector machine (SVM).

Apparently, most of the vision-based gait analysis proposals use sagittal view for the reason that it provides more information with which to work. However, there are obtainable benefits out of a frontal gait analysis. According to Whittle [14], more gait abnormalities can be observed from a sagittal view than from a frontal view. However, we do also undertake frontal gait analysis for the following reasons:

(i) Some abnormalities can only be observed from a frontal point of view. Whittle [14] mentions that circumduction gait, hip hiking, abnormal foot contact, and rotation among others are better observed from a frontal view.

(ii) In terms of the physical space necessary for recording, sagittal gait sequences require much more than those of frontal gait, for which only a small hall or corridor will serve.

A way to reduce the space needed for sagittal view recording is to use a treadmill, but it could alter gait patterns, especially with frail people. Another workaround is to use a motorised camera that follows the subject, but it is expensive and could complicate the background subtraction as it is moving as well. Both workarounds complicate the acquisition of gait sequences making it difficult to be processed by a smartphone.

Sagittal images show a clear view of feet displacement and enough information to locate heel and toe of each foot. In frontal view, on the other hand, it is not easy to determine where the heel and toe are located in each foot. Therefore, a different approach is required for frontal sequences.

In sagittal view, the size of the subject's silhouette is maintained along the whole of its trajectory. However, in frontal view, the size of the silhouette increases along its trajectory, so a normalization might be required.

The paper is organized as follows. Section 2 describes the sagittal and frontal methods to obtain spatiotemporal parameters of gait, their implementation in a smartphone, and the classification of normal and abnormal gait in a cloud platform. Section 3 shows the results in which the spatiotemporal gait parameters are subjected to normal and abnormal gait classification. Finally, Section 4 provides the conclusion of this work.

2. Methods

In this paper, we present a platform for gait analysis using computer vision where a smartphone records and processes a gait sequence to obtain spatiotemporal parameters to be sent to the cloud for a classification between normal and abnormal gait. The layout of the platform is shown in Figure 1. In the following subsections, each module of the platform will be described.

2.1. Sagittal Approach. The sagittal approach takes gait sequences recorded from the side as input. The method presents four phases: preprocessing, feet location, feature extraction, and skeleton extraction. Figure 2 shows the diagram of the sagittal approach. The classification phase is performed in the cloud.

2.1.1. Preprocessing. In this phase, a background subtraction is performed to obtain the silhouette of the subject using mixture of Gaussians [15] background subtraction. After that, a morphology operator is applied to remove noise. Finally, the bounding box of the remaining silhouette is extracted by computing the x, y positions using (1), and then those points are made to correspond to a rectangle $(x, y, \text{width}, \text{height})$ using (2).

$$\min x = \arg \min_{x,y} \left(\forall_{x,y} \in \text{silhouette} : x \right)$$
$$\max x = \arg \max_{x,y} \left(\forall_{x,y} \in \text{silhouette} : x \right)$$
$$\min y = \arg \min_{x,y} \left(\forall_{x,y} \in \text{silhouette} : y \right) \tag{1}$$
$$\max y = \arg \max_{x,y} \left(\forall_{x,y} \in \text{silhouette} : y \right),$$

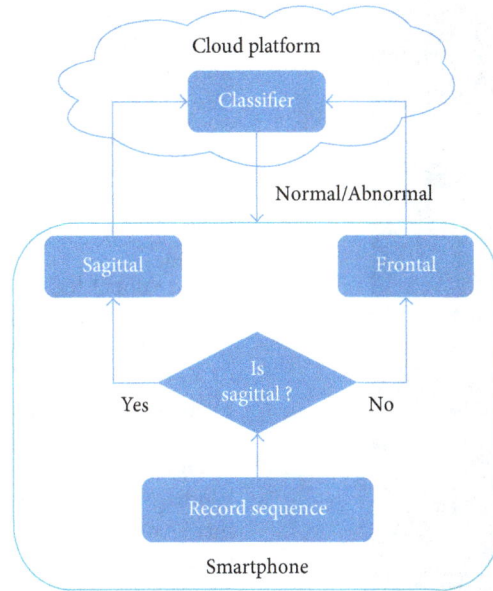

FIGURE 1: Diagram of the proposed platform. Extraction of gait features is performed on a smartphone and classification in the cloud.

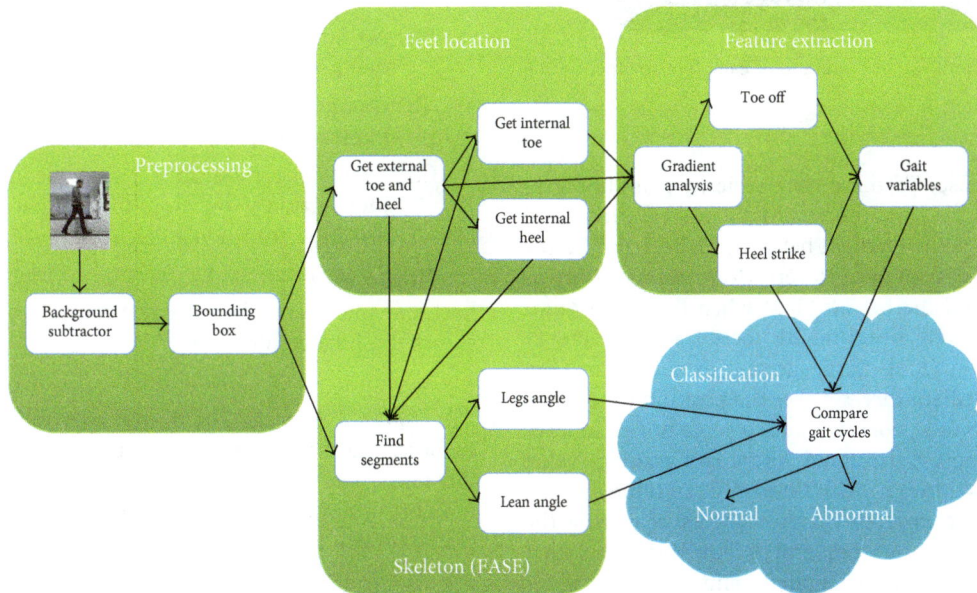

FIGURE 2: Diagram of the sagittal gait approach.

$$\text{Bounding box} = (\min x, \min y, \max x - \min x, \max y - \min y). \tag{2}$$

2.1.2. Feet Location.
The silhouette obtained by background subtraction is then enclosed in its bounding box and split into four regions, namely, *head (13% of bounding box height), torso (34%), upper legs (24%), and lower legs (29%),* according to an anthropometric model [16] as shown in Figure 3. The lower leg region is then brought to focus. We search the silhouette pixel with maximum X component to obtain the toe of the front foot (FF) using (3) and the pixel with minimum X to obtain the heel of the back foot (BF) using (4). Then, the lower leg region is split into halves vertically to separate each foot. In the BF half, we search for the lower right pixel (assuming displacement from left to right) to obtain the BF toe. In the FF half, we search for the lower left pixel to obtain the heel. The final result is shown in Figure 3.

$$\arg\max_{x,y} \left(\forall_{x,y} \in \text{silhouette} : x \right), \tag{3}$$

$$\arg\min_{x,y} \left(\forall_{x,y} \in \text{silhouette} : x \right). \tag{4}$$

2.1.3. Feature Extraction.
For each frame of the sequence, the position of the heel and toe of both feet was obtained in

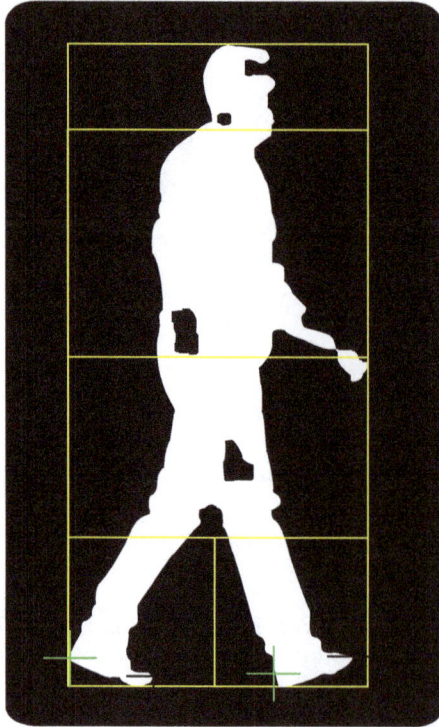

FIGURE 3: Location of the heel and toe of each foot for the sagittal approach.

the previous phase. To these time series, we applied gradient analysis of the X component to obtain heel strike (HS) when the mean point gradient between FF heel and FF toe goes from greater than zero to zero (foot stops moving as shown in (5)) and the toe off (TO) when the mean point gradient between BF heel and BF toe goes from zero to greater than zero (foot starts moving as shown in (6)). Applying the gradient directly over the position time series produces a lot of false positives due to some noise. To filter the noise, we apply a threshold where any gradient value less than that is set to zero. This threshold can remove small oscillations due to an error in the process of getting the silhouette and locating toes and heels. It follows that a Gaussian smoothing is applied, and isolated values greater than zero or equal to zero are removed using (7).

$$\text{heel\_strike}(i) \rightarrow \text{grad}(x)_i \leq 0 \wedge \text{grad}(x)_{i-1} > 0, \quad (5)$$

$$\text{toe\_off}(i) \rightarrow \text{grad}(x)_i > 0 \wedge \text{grad}(x)_{i-1} = 0, \quad (6)$$

$$v_i' = \begin{cases} \text{if } v_i = 0 \wedge v_{i-1} \neq 0 \wedge v_{i+1} \neq 0 \rightarrow \dfrac{(v_{i-1} + v_{i+1})}{2} \\[2mm] \text{if } v_i \neq 0 \wedge v_{i-1} = 0 \wedge v_{i+1} = 0 \rightarrow 0. \end{cases} \quad (7)$$

2.1.4. Skeleton Extraction. The skeleton extraction phase provides a fast way of obtaining an approximation of the locations of the head, neck, hip, knees, and feet. It uses the

four regions of the silhouette described in the feet location phase. The head and torso regions are divided in half horizontally, and the COG of each half is computed. The COG of the upper region is moved to the top, and the COG of the lower region is moved to the bottom. Then, the head lower COG and the torso upper COG are averaged to obtain a common point which is the neck. The head location corresponds to the upper COG of the head region.

The upper leg region is also split horizontally in half, and both COGs are obtained. In addition, a vertical split is also performed, and another two COGs are obtained. The upper COG is moved to top and averaged with the lower torso COG to obtain the hip location. Lower COG is discarded. Then right and left COGs are moved to bottom, those two points being the location of the knees. The knees are adjusted to simulate bending. The process to adjust the knees consists in tracing three circles: one with center at the hip and thigh length radius (which is the height of the upper leg segment) and two other circles with center at each foot and radius equal to the tibia length (which is the height of the lower leg segment). Then, an intersection between the hip circle and each of the foot circles is performed. There are three possibilities:

(i) *No intersection.* In this case, the knee point is the one given by the COG.

(ii) *One intersection.* In this case, the knee point is the intersection point.

(iii) *Two intersections.* In this case, the knee point is the intersection point more to the right (assuming gait direction from left to right).

Finally, the location of each foot is the mean point of the heel and toe obtained in the feet location phase. Figure 4 shows the final result.

2.2. Frontal Approach. The frontal approach is very similar to the sagittal one proposed in the previous subsection. It has the same phases: preprocessing, feet location, feature extraction, and skeleton detection. The diagram of the frontal gait approach is shown in Figure 5.

2.2.1. Preprocessing. This phase is exactly the same as for sagittal. The silhouette is obtained using Mixture of Gaussians as background subtraction, and then morphology operators are applied to remove noise.

2.2.2. Feet Location. In frontal view, both toes are always visible but heels are constantly occluded, so heels cannot be properly located. Therefore, we can only rely on toe information.

To obtain toes, we proceed by dividing the silhouette in four regions according to the anthropometric model established in [16]. We focus only on the lower leg segment. Then, we calculate its bounding box and split it vertically into half to separate both feet. It is important to recalculate the bounding box of this part so the vertical split separates both feet accurately; otherwise, any misalignment can cause problems. Note that the

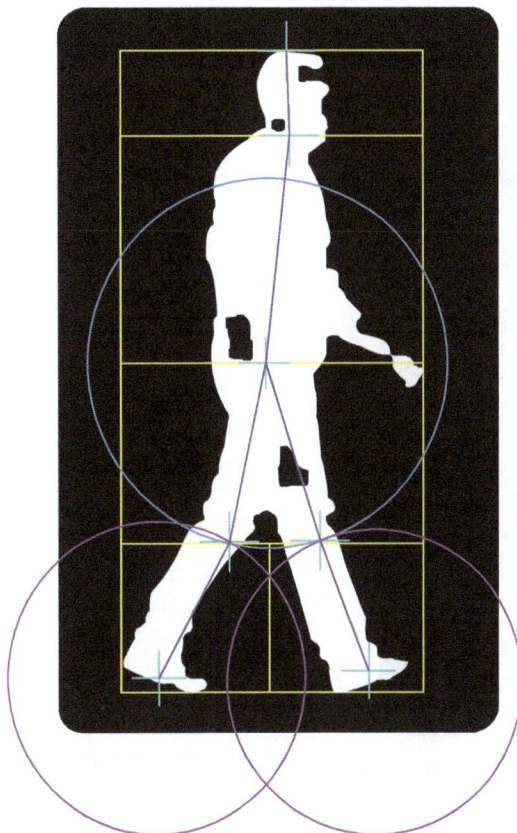

FIGURE 4: Knee adjustment for the sagittal approach.

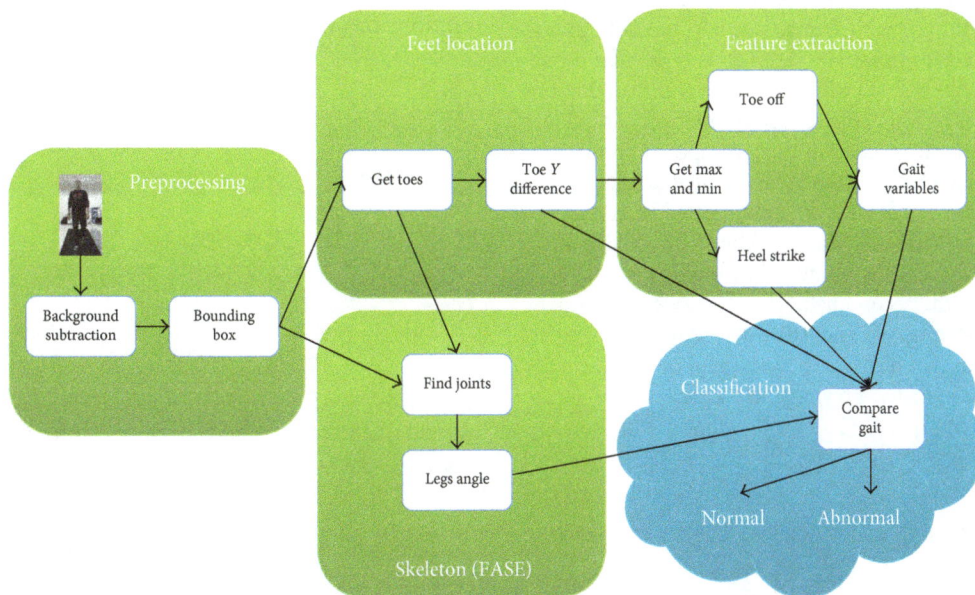

FIGURE 5: Diagram of the frontal gait approach.

process of splitting the bounding box for the purpose of separating both feet will never be accurate with gait patterns that place one foot in front of the other. We will assume that this particular gait pattern is not present in our dataset. We obtain the left and right foot toe by locating the pixel with minimum y component in the left and right half, respectively (8) (Figure 6).

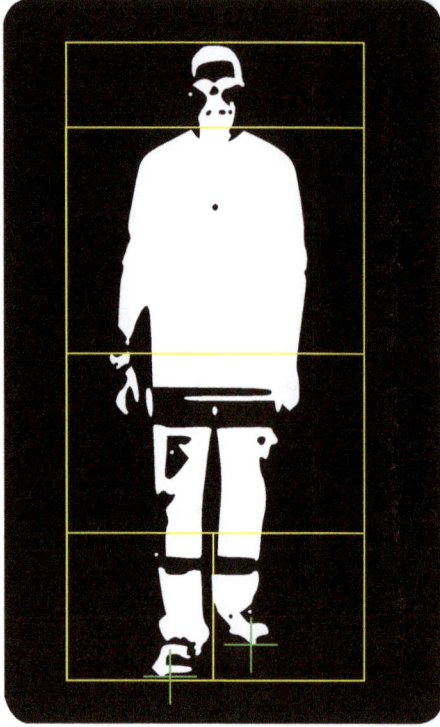

FIGURE 6: Toe location of each foot for the frontal approach.

$$\arg\min_{x,y} \left(\forall_{x,y} \in \text{silhouette} : y \right). \tag{8}$$

2.2.3. Feature Extraction. The previous phase provides the position of each toe for each frame, which is precisely the information we need to derive HS and TO. We propose an approach to obtain HS and TO with frontal gait based on the time series derived by subtracting the vertical component of both feet.

We will use the subtraction of the y component of the toes to obtain a curve in which zero crosses indicate the feet adjacent gait phase. HS and TO of each foot are located between each zero cross. We can estimate HS and TO by assuming that HS is produced before TO; HS is produced in the first half of each region and TO in the second half. Therefore, we can estimate HS and TO following (9) and (10), respectively, where zc_i relates to the frame in which a zero cross point occurs and zc_{i-1} relates to the frame of the previous zero cross point.

$$\text{HS} = zc_{i-1} + \frac{(zc_i - zc_{i-1}) \times 3}{4}, \tag{9}$$

$$\text{TO} = zc_i - \frac{zc_i - zc_{i-1}}{4}. \tag{10}$$

This approach poses some problems with some abnormal gait patterns, as shown in [17], in which some events could not be detected, for example, when a foot is always behind the other or is dragged due to some injury or pain. Figure 7 shows foot dragging where, in some cases, the curve does not cross zero during the swing phase. To solve the problem, we devise another method. Using the same curve from the previous approach (the difference of y component

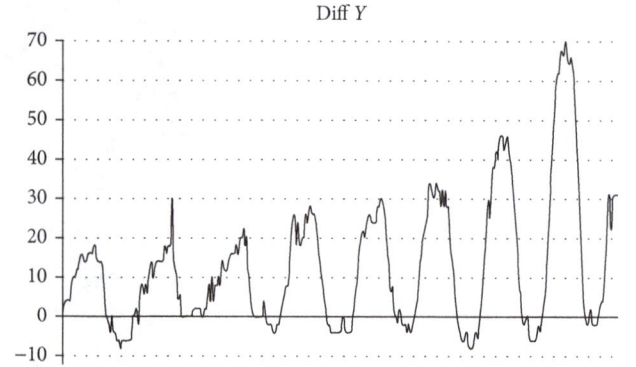

FIGURE 7: Difference of component Y of each foot with abnormal foot dragging.

of each foot), we proceed by applying Gauss filters to remove noise (Figure 8 shows the curve of Figure 7 after applying Gauss filters), then we obtain the local maxima and minima, which are located more or less at the center of each pair of zero crosses. But, in this case, the curve does not have to cross zero to produce a maximum or minimum, and the problem is solved.

HS are located before a maximum or minimum, and TO after. We know that both events are located in that region. Empirically adjusting them, we derived that the HS is located at 1/4 the distance between one maximum (or minimum) and the previous one (12), and TO is located at 1/8 the distance between one maximum (or minimum) and the next one (13).

Being M an ordered set of maxima and minima in ascending chronological order:

$$M = \{m_1, m_2, m_3 \cdots m_n\}. \tag{11}$$

HS of m_i is obtained as

$$\text{HS}_i = m_i - \frac{m_i - m_{i-1}}{4}, \tag{12}$$

and TO is obtained as

$$\text{TO}_i = m_i + \frac{m_{i+1} - m_i}{8}. \tag{13}$$

2.2.4. Skeleton Detection. The process is the same as the one described for the sagittal approach, but for frontal approach, the adjustment of knees is not necessary.

2.3. Smartphone Implementation. Sagittal and frontal approaches were implemented on Android using OpenCV native functions. We allowed two ways of processing a dataset:

(i) On a real-time video: the smartphone camera records the subject walking and processes it at the same time.

(ii) On a previously recorded video: the smartphone records the subject walking and stores it in memory, and then the stored video is processed.

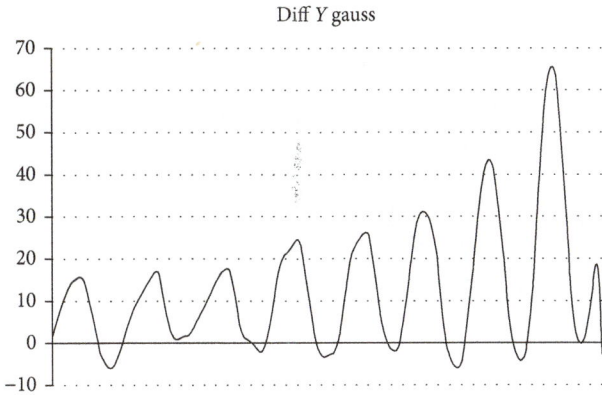

FIGURE 8: Difference of component Y of each foot after applying Gauss filters.

FIGURE 9: Dataset background showing a room with closets and windows.

To achieve real-time processing, we use the pyramidal multirresolution approach described in [18]. We achieve 10 fps using a quad core at 1.4 GHz smartphone with 1 GB memory and 25 fps using a tablet with a Tegra K1 quad core processor at 2.2 GHz and 2 GB memory. The size of the input image was reduced to 480×270 pixels. However, results shown in Section 3 are obtained using full resolution using the dataset.

2.4. Cloud Platform. To develop the cloud platform, we used the Microsoft Azure Machine Learning platform. This is a cloud platform for designing and developing predictive models. Azure provides a REST Web Service to access the Machine Learning tools.

For our purposes, we develop a K-nearest neighbour (KNN) algorithm with Dynamic Time Warping (DTW) as a distance function accessed through the REST Web Service provided by Azure. To perform a classification between normal and abnormal gait, we use the stride (bounding box width for sagittal approach, and subtraction between y component of each foot for frontal approach) and leg-angle time series (provided by the skeleton extraction algorithm computed as the angle formed by the hip and each foot).

3. Results and Discussion

We will now describe the experiments performed and the results obtained. The dataset recorded for the experiments is also described in this section.

3.1. Dataset. To test the proposed approaches, we recorded two datasets of subjects walking: one using sagittal view and the other using frontal view. Both datasets were recorded in a room with a nonhomogeneous background including windows where the light made it difficult to extract the silhouette. This was intentional because we wanted to test our approaches in real conditions, and so the silhouette is often incomplete. Figure 9 shows the room in which the recordings were performed.

To record the frontal dataset, we placed a camera at one end of an 8 m corridor and asked the subject to walk towards it.

We captured a total of 23 samples of normal gait and 20 samples of abnormal. To record the sagittal dataset, we used the same environment, but we placed a camera at a distance of 4 m from the perpendicular of the gait direction to obtain a side view. In this case, a total of 15 samples of normal gait and 15 of abnormal gait were recorded. Even if the number of recorded samples is low (43 for frontal gait and 30 for sagittal gait), there are a total of 320 HS events and 319 TO events for frontal gait and 233 HS events and 223 TO events for sagittal gait.

We asked the subjects to walk normally along the corridor and then to walk feigning some of the following abnormalities:

(i) *Knee pain*: the subject simulated pain in one of his knees.

(ii) *Foot dragging*: the subject dragged one foot.

(iii) *Parkinsonian gait*: the subject made some small steps with variable speed.

(iv) *Other*: the subject depicted random patterns.

To guarantee the privacy of the subjects, we published only the silhouettes extracted during the silhouette extraction phase. These silhouettes are stored as an ordered set of images, and a file with the elapsed milliseconds for each image is also included. For each recorded sample, we manually mark the frames in which a HS or TO event occurs to use it as a ground truth. We also include information related to pixel width to be able to calculate distances and the sample class (normal = 0 or abnormal = 1). In addition, a file with the output of the feet location and feature extraction phases is included which contains the positions of heel and toe of each foot, their gradients, and the events of HS and TO detected. These results are the output of the HS and TO detection algorithm using full resolution (1920×1080), which do not correspond to those provided by the smartphone using a quarter of that resolution.

Both datasets are accessible through the URL provided by [19].

3.2. Experiments. We performed experiments using our own datasets for sagittal and frontal gait. We used the manual marking of the HS and TO events of each gait sequences of the dataset as ground truth. The error margin of this manual marking was set to ± 1 frame because that is the minimum value. We also assumed an error of ± 1 frame in the algorithm

TABLE 1: Results of the sagittal HS and TO detection algorithm showing the amount of correct detections (less than 2 frames of difference between algorithm and manual marking), undetected cases, wrong detection (more than 2 frames of difference), and the root mean square error of both correct and wrong cases.

Approach	Correct	Undetected	Wrong	RMSE
DAI dataset normal gait heel strike				
Sagittal	90.2%	1.1%	7.6%	1.44 frames (48 ms)
Frontal	89.4%	0%	10.60%	1.88 frames (63 ms)
Toe off				
Sagittal	93.3%	2.2%	2.2%	1.08 frames (36 ms)
Frontal	89.4%	0%	10.6%	1.63 frames (54 ms)
DAI dataset abnormal gait heel strike				
Sagittal	89%	2.1%	6.9%	1.79 frames (60 ms)
Frontal	72.1%	0%	27.9%	2.42 frames (81 ms)
Toe off				
Sagittal	82.1%	3.6%	10.7%	1.59 frames (53 ms)
Frontal	75%	0%	25%	2.17 frames (72 ms)
Total heel strike				
Sagittal	89.5%	1.7%	7.2%	1.66 frames (55 ms)
Frontal	78.8%	0%	21.3%	2.23 frames (74 ms)
Toe off				
Sagittal	86.5%	3.0%	7.4%	1.41 frames (47 ms)
Frontal	80.6%	0%	19.4%	1.98 frames (66 ms)

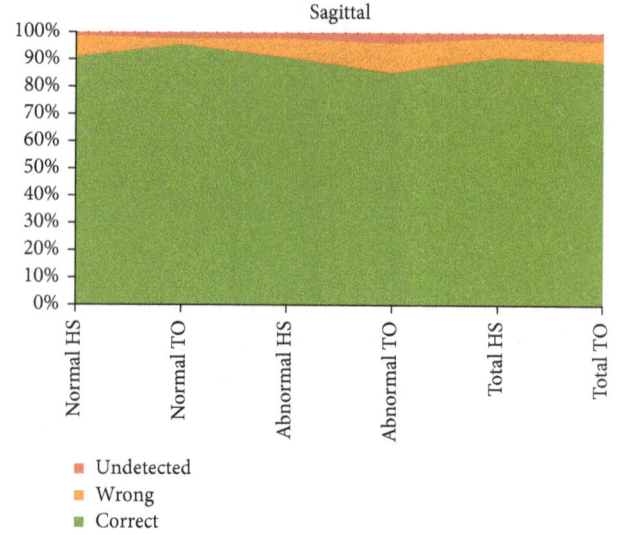

FIGURE 10: Results of the sagittal approach.

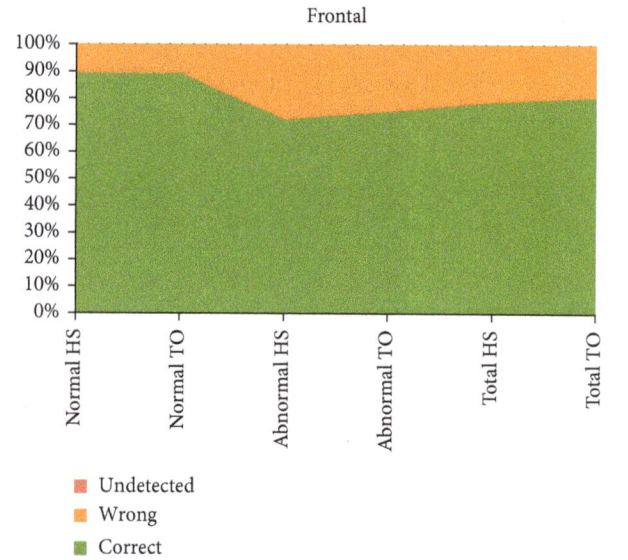

FIGURE 11: Results of the frontal approach.

output. So, the global error margin was set to ± 2 frames. Then, the difference in frames between the ground truth and the proposed algorithm was analysed. Any difference less or equal to the global error margin was considered acceptable. Then, the root mean square error (RMSE) of the differences was computed using

$$\text{RMSE} = \sqrt{\frac{1}{n} \sum_{i=0}^{n} \left(m_i - a_i\right)^2}, \qquad (14)$$

where n corresponds to the number of events (HS or TO in this case), m_i the frame of the event i in the manual marking, and a_i the frame of the event i in the algorithm output.

3.3. Sagittal Approach. In Table 1, we show the results after applying the HS and TO detection algorithm with the filtering method described in the previous section for sagittal view. The table shows the amount of correct detections (less than 2 frames of difference between algorithm and manual marking), undetected cases, wrong detection (more than 2 frames of difference), and the root mean square error of both correct and wrong cases. As observed, the RMSE of both HS and TO events is lower than the error margin of 2 frames. TO events are more accurately delimited than HS events. But, HS events show less undetected cases. Therefore, it will be HS, the event we will use to obtain the spatiotemporal parameters to perform classification. Figure 10 shows graphically the correct, wrong, and undetected cases.

3.4. Frontal Approach. Table 1 also shows the results after applying the frontal approach. As shown in there, the RMSE of both HS and TO in normal gait is smaller than the error margin of 2 frames, but it is slightly bigger for abnormal gait. Therefore, results are acceptable for both normal and abnormal. Error is mainly produced in the first steps when the silhouette is smaller (the subject is farthest from the camera). Figure 11 shows graphically the results of Table 1.

The results obtained with our sagittal view approach are similar for normal gait. We obtained 1.44 frames for HS and 1.08 for TO, which were slightly more precise than the ones we extracted from frontal approach (1.88–1.63) but close to each other. However, in the case of abnormal gait, we obtained 1.79 frames for HS and 1.59 for TO, which were more precise than those obtained with the frontal approach (2.42–2.17).

TABLE 2: 10-fold and leave-one-out cross-validation results of each classifier for the sagittal approach.

Classifier	10-Fold	Leave-one-out
Stride each cycle	77%	75%
Legs angle each cycle	92%	92%
Stride each subject	77%	77%
Legs angle each subject	100%	100%

TABLE 3: 10-fold and leave-one-out cross-validation results of each classifier for the frontal approach.

Classifier	10-Fold	Leave-one-out
Stride each cycle	75%	79%
Legs angle each cycle	71%	77%
Stride each subject	80%	84%
Legs angle each subject	88%	88%

3.5. Classification. To perform a classification between normal and abnormal gait, we use KNN to compare the stride length and leg-angle time series of the different gait cycles. To calculate the distance between two time series, we apply DTW. We perform the classification test with two different methods:

(i) Testing each gait cycle separately. The time series corresponding to each gait cycle is treated separately as if it belonged to different subjects.

(ii) Testing each gait cycle of each recording sample and outputting the mode class for each subject. In this case, a prediction for each gait cycle follows, and then another prediction is computed by outputting the mode class for the same recording sample.

To validate the proposed classification, we use 10-fold and leave-one-out cross-validations to finely measure the accuracy of each classifier.

Table 2 shows the results of the stride and leg-angle time series for the sagittal approach. We obtained an accuracy rate of 100% using leg-angle time series when outputting the mode class for each recording sample. Least accurate results, however, are the ones offered by the stride width.

The results of the classification experiments for frontal approach are shown in Table 3. As shown in there, testing each recording sample produces better results as it tends to eliminate outliers.

We have focussed on obtaining a classification between normal and abnormal gait to assess the suitability of the proposed algorithm to differentiate between the two of them. For this test, we considered knee pain and foot dragging as abnormal gait. The results obtained suggest that the classifier can differentiate between normal and abnormal gait. Therefore, future work will focus on classifying different abnormal gaits.

4. Conclusion

The main contribution of this paper is a nonexpensive and easy-to-deploy approach to obtain HS and TO and some skeleton joints using both sagittal and frontal gait sequences. Frontal view poses some problems when obtaining heels position, so we focus on toes instead. Results show acceptable precision in providing HS and TO in both the sagittal and the frontal methods. Comparing both approaches, results were similar but sagittal proved to be more accurate. The dataset recorded to test the proposed approaches is for anyone to use it [19]. To maintain the privacy of the subjects, we published only the silhouette.

We also provide a cloud platform-based web service to perform a classification between normal and abnormal gait for both sagittal and frontal views. Results show a classification rate greater than 80% in frontal view and more than 90% in sagittal view.

The ability to perform gait analysis using frontal view reduces the physical space required for the tests. In addition, this method does not rely on silhouette displacement (the sagittal approach does), so it is also suitable for treadmill gait sequences. Therefore, the space could be reduced even more in cases where the alteration of gait patterns that the treadmill could cause does not significantly matter.

Future work will focus on improving the accuracy of HS and TO for abnormal gait and classifying different abnormal gait types.

Acknowledgments

This research is part of the FRASE MINECO project (TIN2013-47152-C3-2-R) funded by the Ministry of Economy and Competitiveness of Spain.

References

[1] J. Waltson and L. Fried, "Frailty and the old man," *Medical Clinics of North America*, vol. 83, no. 5, pp. 1173–1194, 1999.

[2] L. P. Fried, C. M. Tangen, J. Walston et al., "Frailty in older adults evidence for a phenotype," *Journals of Gerontology Series A: Biological Sciences and Medical Sciences*, vol. 56, no. 3, pp. M146–M157, 2001.

[3] J. M. Hausdorff, G. Yogev, S. Springer, E. S. Simon, and N. Giladi, "Walking is more like catching than tapping: gait in the elderly as a complex cognitive task," *Experimental Brain Research*, vol. 164, no. 4, pp. 541–548, 2005.

[4] T. Mulder, W. Zijlstra, and A. Geurts, "Assessment of motor recovery and decline," *Gait & Posture*, vol. 16, no. 2, pp. 198–210, 2002.

[5] F. I. Mahoney and D. W. Barthel, "Functional evaluation: the barthel index," *Maryland State Medical Journal*, vol. 14, pp. 61–65, 1965.

[6] W. Zeng, C. Wang, and F. Yang, "Silhouette-based gait recognition via deterministic learning," *Pattern Recognition*, vol. 47, no. 11, pp. 3568–3584, 2014.

[7] X. Xu, R. W. McGorry, L. S. Chou, J. H. Lin, and C. C. Chang, "Accuracy of the microsoft kinect™ for measuring gait parameters during treadmill walking," *Gait & Posture*, vol. 42, no. 2, pp. 145–151, 2015.

[8] S. D. Choudhury and T. Tjahjadi, "Gait recognition based on shape and motion analysis of silhouette contours," *Computer Vision and Image Understanding*, vol. 117, no. 12, pp. 1770–1785, 2013.

[9] A. Leu, D. Ristić-Durrant, and A. Gräser, "A robust markerless vision-based human gait analysis system," in *Proceedings of the 6th IEEE International Symposium on Applied Computational Intelligence and Informatics (SACI) 2011*, pp. 415–420, IEEE, Timişoara, Romania, May 2011.

[10] J. H. Yoo and M. S Nixon, "Automated markerless analysis of human gait motion for recognition and classification," *ETRI Journal*, vol. 33, no. 2, pp. 259–266, 2011.

[11] T. Khan, J. Westin, and M. Dougherty, "Motion cue analysis for parkinsonian gait recognition," *The Open Biomedical Engineering Journal*, vol. 7, no. 1, pp. 1–8, 2013.

[12] L. Wang, "Abnormal walking gait analysis using silhouette-masked flow histograms," in *Proceedings of the 18th International Conference on Pattern Recognition, ICPR 2006*, vol. 3, pp. 473–476, IEEE, Cancun, Yucatán, Mexico, August 2006.

[13] C. Bauckhage, J. K. Tsotsos, and F. E. Bunn, "Automatic detection of abnormal gait," *Image and Vision Computing*, vol. 27, no. 1-2, pp. 108–115, 2009.

[14] M. W. Whittle, *Gait Analysis: An Introduction*, Butterworth-Heinemann, Oxford, UK, 2014.

[15] Z. Zivkovic and F. van der Heijden, "Efficient adaptive density estimation per image pixel for the task of background subtraction," *Pattern Recognition Letters*, vol. 27, no. 7, pp. 773–780, 2006.

[16] F. Tafazzoli and R. Safabakhsh, "Model-based human gait recognition using leg and arm movements," *Engineering Applications of Artificial Intelligence*, vol. 23, no. 8, pp. 1237–1246, 2010.

[17] M. Nieto-Hidalgo, F. J. Ferrández-Pastor, R. J. Valdivieso-Sarabia, J. Mora-Pascual, and J. M. García-Chamizo, "Vision based gait analysis for frontal view gait sequences using rgb camera," in *Proceedings of the Ubiquitous Computing and Ambient Intelligence: 10th International Conference, UCAmI 2016, Part I*, vol. 10, pp. 26–37, Springer, San Bartolomé de Tirajana, Gran Canaria, Spain, November–December 2016.

[18] M. Nieto-Hidalgo and J. M. García-Chamizo, "Real time gait analysis using rgb camera," in *Proceedings of the Ubiquitous Computing and Ambient Intelligence: 10th International Conference, UCAmI 2016, Part I*, vol. 10, pp. 111–120, Springer, San Bartolomé de Tirajana, Gran Canaria, Spain, November–December 2016.

[19] M. Nieto-Hidalgo, R. J. Valdivieso-Sarabia, J. Mora-Pascual, J. M. García-Chamizo, and F. J. Ferrández-Pastor, "Gait-A Database: dataset of subjects walking," 2017, http://hdl.handle.net/10045/70567.

9

Malaria Vulnerability Map Mobile System Development using GIS-Based Decision-Making Technique

Jung-Yoon Kim [ID],[1] Sung-Jong Eun,[2] and Dong Kyun Park [ID][3]

[1]*Graduate School of Game, Gachon University, 1342 Seongnam Daero, Sujeong-Gu, Seongnam-Si, Gyeonggi-Do 461-701, Republic of Korea*
[2]*Health IT Research Center, Gil Medical Center, Gachon University College of Medicine, Incheon, Republic of Korea*
[3]*Department of Gastrointestinal Medicine, Gil Medical Center, Gachon University College of Medicine, Incheon, Republic of Korea*

Correspondence should be addressed to Dong Kyun Park; pdk66@gilhospital.com

Academic Editor: Jaegeol Yim

This paper aimed at improving the lack of GIS information use and compatibility of multiplatform which represented limits that existing malaria risk analysis tools had. For this, the author developed mobile web-based malaria vulnerability map system using GIS information. This system consists of system database construction, malaria risk calculation function, visual expression function, and website and mobile application. This system was developed based on Incheon region only. Database includes information on air temperature and amount of precipitation as well. With regard to malaria risk calculation, guideline provided by Korea Centers for Disease Control and Prevention was followed first and then decision-making technique was used. Calculating criteria value for risk index made it possible to estimate more precise risk. With regard to visual expression function, database constructed earlier and risk information were linked to print out graphic map and graphs so that more intuitive and visible expression can be provided based on animation technique. This system allows a user to check information in real time and can be used anywhere anyplace. Mobile push function is to enhance user's convenience. Such web map is useful to general users as well as experts.

1. Introduction

Malaria is infectious disease which is mediated and disseminated by *Anopheles sinensis*. Malaria is caused by *Plasmodium* (*Plasmodium* genus) which penetrates in red blood cell. Malaria shows symptom of fever, chill, hepatomegaly, splenomegaly, and so on. Malaria is reported to have occurred in one hundred four countries in the world. It is said that over one billion people around the world are exposed to a danger of malaria and over three hundred million people are infected with malaria and over one million people die of malaria on a yearly basis [1]. *Plasmodium* is *Plasmodium falciparum*, *P. vivax*, *P. malariae*, *P. ovale*, and *P. knowlesi*. It is said that malaria occurs with mosquito as mediator. It is reported that in South Korea, mainly *vivax* malaria caused by *P. vivax* which is mediated

by *Anopheles* genus mosquito has occurred. As from 1979, South Korea was declared as malaria-free area. However, in 1993, *vivax* malaria occurred in soldiers on the front line who served in the army near North Korea. It is reported that from 1993 to 2013, 500–4,000 malaria cases occurred centering on DMZ (demilitarized zone) annually. *Vivax* malaria which reoccurred in 1993 has appeared for about twenty years in northern part of Gyeonggi Province and varied in frequency of occurrence. Until now, *vivax* malaria occurred with three times of peaks (1998~2000, 2007, and 2010, respectively). It is reported that malaria cases have continued to occur until 2017 [1–3]. Figure 1 shows occurrence of malaria in South Korea by years.

Malaria is infectious disease that needs continuous control. Scholars have conducted studies on the cause of malaria and incidence rate by correlation considering meteorological elements and community characters [4–6]. In addition, they

[Ref. Korea Centers for Disease Control and Prevention]

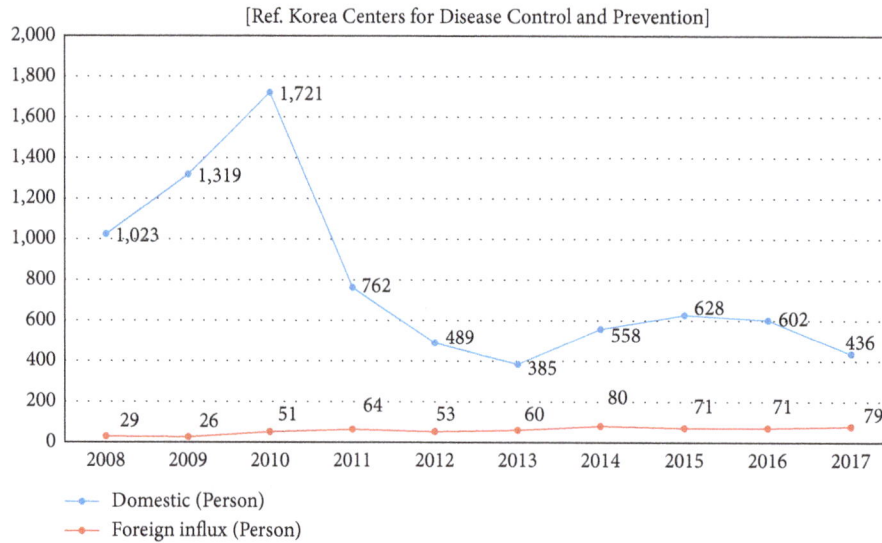

FIGURE 1: Annual malaria infection status.

analyzed route of infection in terms of epidemiology or environmental factors to prevent occurrence of malaria.

Software tool [7] that supports analysis of malaria-related information is being developed. As quantity of data becomes increased, applying technique of processing big data to correlation analysis is being raising with care. Like this, in health science, studies have been conducted in a specialized way making good use of IT-related technologies. Database in the field of health science includes a variety of data most of which is obvious in terms of space or refers to space. Therefore, database in the field of health science has the advantage of using data in GIS. Diversity, size, and complexity of health science data need more application of GIS [8]. Field of controlling infectious disease such as malaria needs GIS-based tools for risk analysis. This is to make good use of GIS-based analysis that can analyze information in an intuitive and visible way. Such tools have been developed, but they are not widely used due to problem of compatibility of independent platform, problem of exclusive institutional policy, and limited information.

This paper built malaria vulnerability map system using GIS information-based decision-making technique in order to limit problems of abovementioned tools and compatibility of multiplatform. Malaria vulnerability map system drew well-founded resulting values by making good use of various data on air temperature, humidity, amount of precipitation, lakes, and population with GIS information. Malaria vulnerability map system was built based on mobile web to enhance users' convenience. In terms of UX/UI, animation function helps enhance users' immersion and accessibility by providing intuitive information. This study consists of present condition of relevant tool, content of development system, conclusion, and investigation.

2. Related Works

Tools that analyze risk of malaria have been developed at home and abroad. In South Korea, abovementioned tools are used in some institutions for internal use only. In foreign countries, MDAST (Malaria Decision Analysis Support Tool) project is being carried out. MDAST project is developing a tool that analyzes a cause of occurrence of malaria in Kenya, Tanzania, and Uganda where malaria occurs frequently and predicts incidence rate [9].

MDAST aims to provide a platform to analyze influences in terms of health, society, and environment for institution that establishes and evaluates malaria-related policy by using various scientific methods of analyzing information in malaria. This service downloads a tool. Sets it up on local, enters relevant data, sets several factors, applies desired analysis methods, and obtains a result. Analysis methods support various statistical analysis methods. Above tools are widely used but require professional knowledge in using such as factor setting and analysis method selection and impossible to provide online-based real-time support. The major drawback of above tools is that it is difficult for general user to access.

In South Korea, above tools are used through intranet or local installation by health-care-related institutions. They are used by some institutions for their internal use. Since the early 2000, some institutions have developed analysis tools. Environmental factors such as weather and geography and weights are selected on a case-by-case basis to handle system. GIS-based analysis system predicts a risk of malaria and prints out information on a map. For above tools, setting factors of several environmental elements plays an important role. Factors are handled through user definition.

GIS-based analysis system is to cope with unexpected local occurrence of malaria and prevent malaria from spreading by detecting an area where malaria is likely to occur as soon as possible. GIS-based analysis system which is used in South Korea varies in tools among institution, which makes performance deviation for functions supported great. It is difficult to define tools which are used because they are used for internal use.

GIS-based analysis system which is used in South Korea is used by some institutions for their internal use and is based on local. There are few systems which show information on environmental elements relating to malaria on the website. This study developed map system which delivers GIS-based information on risk of malaria on mobile website in consideration of the foregoing. System was built in Incheon region. DB was built by receiving data from institutions such as Korea Meteorological Administration and K-Weather. This study built web-based malaria vulnerability map system so that general users can be provided with information on malaria risk by region in real time.

3. Developing Malaria Vulnerability Map Mobile Using GIS-Based Decision-Making Technique

The purpose of this paper was to develop a map that can analyze factors which cause malaria by using mobile web-based GIS (Geographic Information System). Decision tree analysis method was applied to model areas where malaria is more likely to occur. GIS and decision-making system to develop Korean-type malaria vulnerability map were finally built.

A map of this system was made based on overlay operation method that can provide new information which is not identified on respective map by combining several maps. Population, humidity, temperature, amount of precipitation, and lakes were considered for decision-making technique. Weight was calculated by using AHP decision-making method. Final malaria vulnerability map was made by using PROMETHEE technique.

Map system was implemented based on mobile web for compatibility with mobile platform. For web-based visual output, frame animation-based reading of areas where malaria appeared, reading of summarized information on areas where malaria appeared, reading of detailed information on areas where malaria appeared, and display of risk according to areas where malaria appeared were expressed visually.

3.1. Details of System Development. System functions consist of system database, malaria risk calculation, visual expression, and website and mobile application. Detail core functions are decision-making technique-based modelling part for malaria vulnerability calculation and mobile web map construction part for visual expression and program extension. The system configuration is shown in Figure 2.

3.1.1. Building Up System Database. Web-based system and MySQL with good compatibility and proven stability were used as database system to organize database.

Database information specific to Incheon region consists of data from 2011 to 2017. Area code was extracted based on longitude and latitude to provide information specific to Incheon so that it can provide detailed information according to region.

Information consisting of DB includes population, humidity, air temperature, amount of precipitation, and lake in Incheon. Information on lakes was used in determining vulnerability because whether there is mosquito can be

a factor on which high weight for malaria risk can be placed. With regard to information on air temperature, relative humidity, and amount of precipitation, data provided by Korea Meteorological Administration and K-Weather were used. Other information was provided by Incheon Institute of Health Environment. For relevant DB, schema was defined in one table without dividing it into several tables according to characteristics of data to cut down on expense caused by reference between tables and extract database information rapidly. Table 1 shows the structure of the system DB, which shows the structure of information on *Anopheles sinensis*, region code, sex, temperature, relative humidity, and precipitation.

For system database organized, unlike existing analysis tools where information on air temperature or amount of precipitation was not provided, various environmental factors were considered.

3.1.2. Developing a Model for Calculating Malaria Risk

(1) Calculating optimum weight using AHP (analytic hierarchy process) technique. Malaria risk is calculated based on various standards. For risk modelling work, various weights are considered. In this paper, population, humidity, air temperature, amount of precipitation, and lake were considered. Weight for each criterion was reflected based on experts' opinion as importance of standards differs in creating malarial risk. For this, AHP (analytic hierarchy process) was used. AHP [10–12] is a calculation model which was devised based on the fact that brain uses stepwise or hierarchical analysis process. AHP reaches a final decision making by dividing whole process of decision making into multiphases and analyzing it on step-by-step basis. The author of this paper carried out AHP based on flowchart as shown in Figure 3.

Order of standards which were considered based on Figure 3 was developed as shown in Figure 4. In phase 2 of order of rank, comparison of standards was made based on judgment of experts in malaria [13]. Table 2 shows comparison of pairs. Lastly, weight for standards was calculated by using eigenvector and eigenvalue in each comparison matrix.

Final weight criteria estimated through AHP decision-making technique were calculated based on temperature of 0.51, humidity of 0.34, population of 0.1, and lake of 0.15.

(2) Calculating decision-making matrix using PROMETHEE technique. Methods that select the best alternative by objectively measuring preference for alternatives considered under each criterion have been studied a lot. Typical techniques are ELECTRE and PROMETHEE. ELECTRE I, II, III, and IV technique proposed by B. Roy can solve difficult problems well and is the most reliable method, but the calculation process is very complex and requires decision maker to decide lots of parameters. On the other hand, PROMETHEE (Preference Ranking Organization Method Enrichment Evaluations) is very simple and easy for decision maker to understand and

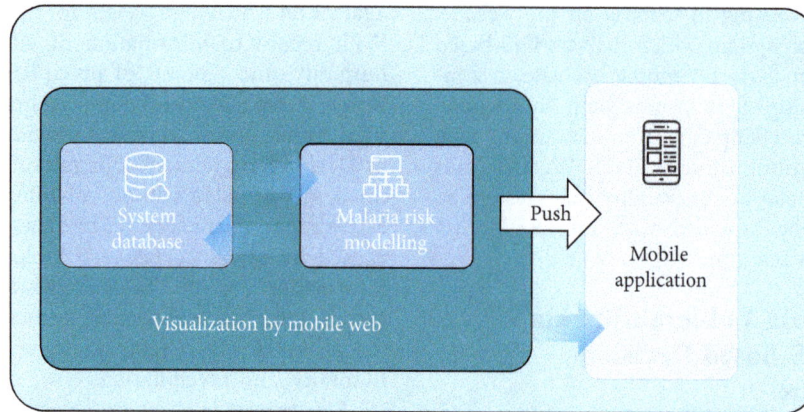

FIGURE 2: Conceptual diagram of malaria vulnerability map mobile system.

TABLE 1: External schema.

Data	Unit	Type
Anopheles sinensis	Count	Integer
Region code	—	Integer
Male	1,000 persons	Integer
Female	1,000 persons	Integer
Temperature	°C	Float
Relative humidity	%	Float
Amount of precipitation	mm	Float

preference intensity for alternatives is expressed with easy concept and requires decision maker to decide up to two parameters, which makes it preferred (Brans, 1985).

PROMETHEE proposed by Brans and Vincke [14] defines preference functions according to each evaluation criterion as Type I (Usual Criterion), Type II (Quasi-criterion), Type III (Criterion with Linear Preference), Type IV (Level Criterion), Type V (Criterion with Linear Preference and Indifference Area), and Type VI (Gaussian Criterion) [15–19]. PROMETHEE evaluates preference relationships among alternatives by using preference index $\pi(a, b)$.

$$\pi(a, b) = \frac{1}{k} \sum_{h=1}^{k} p_h(a, b), \qquad (1)$$

All evaluation criteria must be defined as one of the abovementioned six preference functions. $p_h(a, b)$ in Equation (1) represents preference function value for evaluation criteria h. Decision maker should decide a shape of preference function according to evaluation criteria and assign threshold according to preference functions.

The author of this paper implemented PROMETHEE for each level five times. In each repetition, *nxk* decision-making matrix was created. *n* is the number of points (alternatives) and *k* is the number of criteria. Decision-making matrixnumber represents criteria for alternatives (regions). Table 3 represents decision-making matrix drawn when first time was repeated.

Decision matrix elements to be used in decision-making technique are decided based on the foregoing. Applicable components are decision-making matrix. Incheon region was divided into 28 areas. Matrix values to

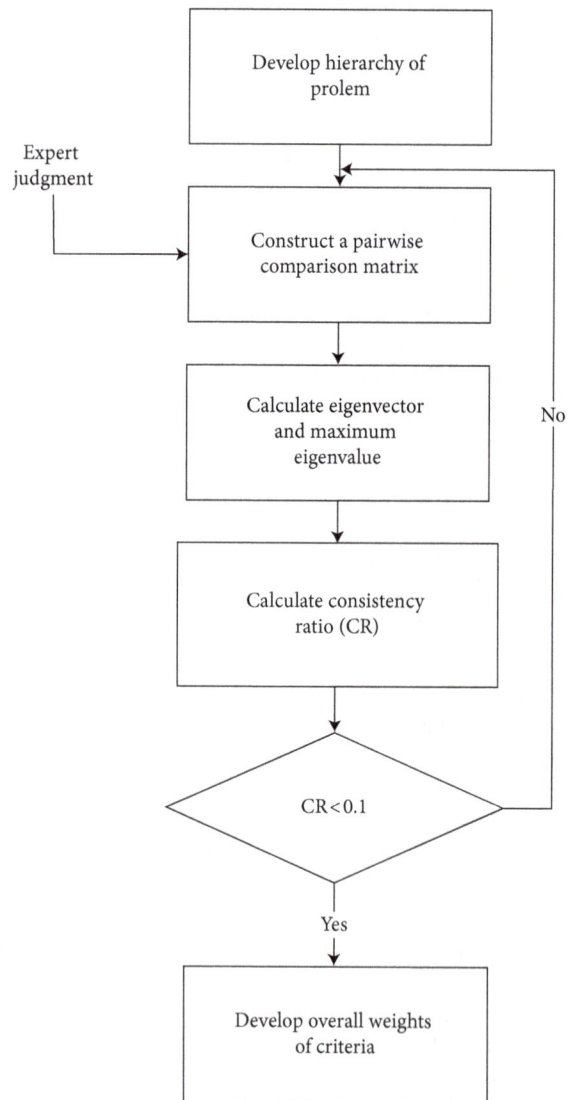

FIGURE 3: AHP for calculating criteria weight.

decide population, humidity, temperature, and lake were as shown above. Table 4 represents final result obtained from PROMETHEE. Five different rankings for 28 areas were obtained by implementing PROMETHEE five times.

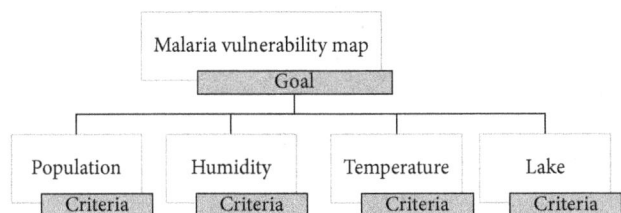

FIGURE 4: Structure of order of rank in criteria.

TABLE 2: External schema.

Criteria	Air temperature	Humidity	Lake	Population
Air temperature	1	3	3	4
Humidity	1/3	1	2	2
Lake	1/3	1/2	1	2
Population	1/4	1/2	1/2	1

TABLE 3: Decision-making matrix.

Zones	Population	Humidity	Temperature	Lake
1	4.085648	5.867935	1	4.064596
2	6.652511	6	1.685264	8.64964
3	6.11661	7.43814	8.964341	5
4	4.140974	7.444163	7.649631	7.165594
5	4	2.487076	1	0
6	3	5	1	0
7	3	5	1	0
8	3.452714	3.34022	1	0
9	5.655222	6	40951.15	6.865466
10	8.097396	6	97781.63	4.169856
11	5.438997	6.478203	1.986046	4.359985
12	7.964342	6.408581	7.865646	3.598465
13	7.019162	7.085386	6.131337	4.006419
14	5.354643	8.078567	8.643161	7.645985
15	8.989335	6	3.641969	4.036452
16	1.969482	9	4.945094	7.264995
17	4.945331	7.766397	5.619694	9.165985
18	7.842272	6.142719	6.940964	4.358643
19	3.928667	2.519822	1	0
20	4.123696	4.496581	1.690846	2.365646
21	4	3.717338	1	0
22	4	2.113713	1	0
23	3	5.402396	1	0
24	3	5	1	0
25	4	3.196849	1	0
26	3.219948	3.921217	1	0
27	3	4.640017	1	0
28	3.361399	3.569278	1	0

TABLE 4: PROMETHEE index.

Zones	Population
1	14
2	10
3	2
4	8
5	21
6	0
7	0
8	25
9	11
10	6
11	12
12	5
13	4
14	1
15	9
16	13
17	3
18	7
19	23
20	15
21	17
22	26
23	16
24	18
25	19
26	22
27	20
28	28

3.1.3. Creating Malaria Vulnerability Map

(1) Raster overlay operation. Overlay operation is applied to most GIS programs. Overlay operation provides new information by combining several maps. In overlay operation, new special elements are formed based on several maps. Overlay operation is handled based on raster and vector maps. Raster data structure is suitable for such operation because all maps used for analysis have the same georeference. They have the same number of pixels and consist of row and column and have the same pixel size and coordinate. Accordingly, when several maps are combined, a program examines each pixel and the same figures can be checked from different maps. In raster overlay, cell numbers are combined in a specific way and figures obtained are assigned to corresponding cells in output layer. Raster overlay is applied to explicit or ordinal number data. Number or characters are stored in each raster cell. Figures in each cell correspond to items of raster variables. In Figure 5, Layer A represents solid data and Layer B shows an example of raster overlay that records land use. Number of potential output items is number of combination that input items can have $(2 * 3 = 6)$.

Figure 6 shows number of population, humidity, temperature, lake, and vulnerability map result calculated by raster overlay. Population shape file based on areas of distribution of population has been converted to raster file. As all criteria maps must have common scale, figures of population raster file have been converted to figures of 0 to 10 by using reclassification analysis. Only reclassification analysis was implemented based on humidity data. Figure was obtained by reclassifying humidity raster file. Like humidity map, air temperature map was created by using reclassification analysis. This was handled by reclassifying figures of air temperature raster file. As distance from lake is greater, malaria risk decreased. A number of ring buffer analyzes were implemented to determine area around lake from diverse distances. By using this analysis, various buffers were created from distances of 100, 200, 300, 400, 600, 800, 1000,

Layer A

Geographic data		
A	B	B
A	B	B
A	A	B

Attribute data	
Type	Soil_name
A	Evard loam
B	Cecil clay

Overlay →

Layer B

Geographic data		
2	3	3
2	3	1
2	1	1

Attribute data	
ID	Land use
1	Forest
2	Urban
3	Farm

→

Output layer

Geographic data		
A2	B3	B3
A2	B3	B1
A2	A1	B1

Attribute data		
ID	Land use	Soil_name
A1	Forest	Evard loam
A2	Urban	Evard loam
B1	Forest	Cecil clay
B3	Farm	Cecil clay

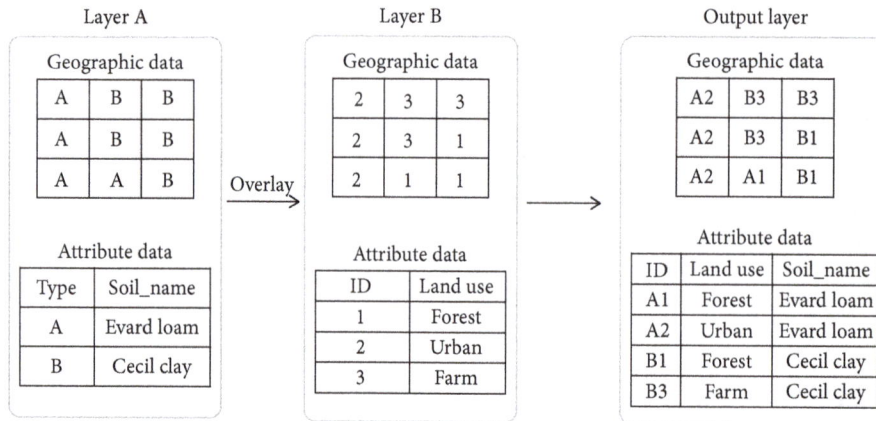

FIGURE 5: Analysis of raster overlay.

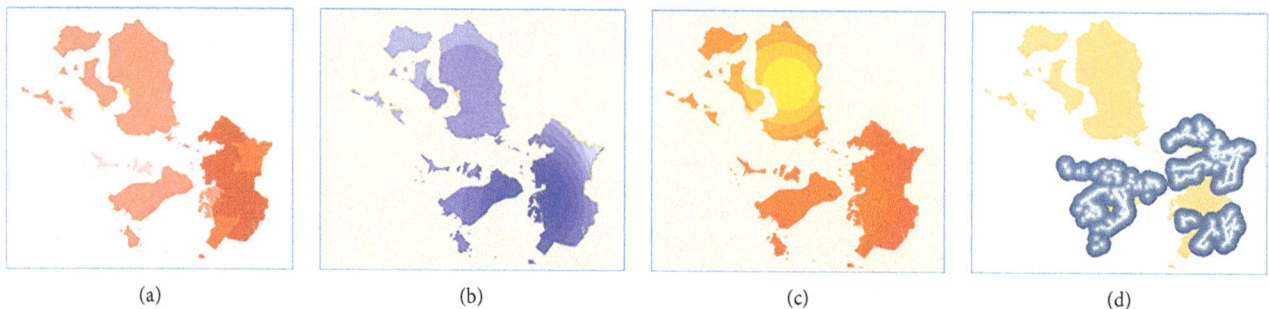

| (a) | (b) | (c) | (d) |

FIGURE 6: A result of raster overlay-based conversion and mapping. (a) Population map. (b) Humidity map. (c) Temperature map. (d) Lake map.

1500, and 2000. Buffer layers were converted to raster file based on distance area. The result that overlaps the map based on the converted raster file is shown in Figure 7.

(2) Creating malaria vulnerability map through Thiessen polygon. Incheon region was classified into 28 zones. Procedures for creating polygon were repeated five times, and 28 zones were set in Incheon Metropolitan City at random. Thiessen polygon [20] analysis was conducted by using 28 zones. Figure 7 shows Thiessen polygon created by repetition of five times.

Analysis was conducted by using mean statistics obtained by assigning mean figures in each zone to output cells. This procedure was implemented for Thiessen polygon and shows criteria map according to zones for first repetition. Accordingly, PROMETHEE was applied to criteria figures corresponding to 28 zones. Malaria risk map was created by combining risk maps obtained every time repetition was made. Repetition of applicable procedures helps in creation of more accurate malaria risk map (map closer to reality).

3.2. Developing Mobile Web-Based Malaria Vulnerability Map Application

3.2.1. Developing Mobile Web-Based Application. PHP is used as server development language for construction of

website. Client side language was implemented as HTML, CSS3, and JavaScript (Bootstrap, jQuery). Google Maps JavaScript API v3 was used as a map. Marker was put on control area with JavaScript. When marker is selected, information contained in database is presented in graph.

Mobile application was developed as android-based hybrid web app. Push function and GPS sensor were used so that location can be tracked on background. Information on malaria rick according to periods can be provided to a user based on it. Restful API was developed through PHP so that when longitude and latitude are handed over on background, a user can check whether it entered applicable area. For push implementation, GCM (Google Cloud Message) [21, 22] service was used so that push service can be used in a stable manner irrespective of terminal's connection to network.

3.2.2. Developing Animation-Based Visualization Technology. Visual expression of information was implemented for intuitive and high visibility based on mobile web map. Frame animation [23], one of the animation techniques, was used for immersion. Wind direction, wind speed, spot atmospheric pressure, and sea level pressure in addition to foregoing database information were expressed in graph. Implementation functions consists of reading of areas where malaria appear, reading of summarized information on areas where malaria appear, reading of detailed

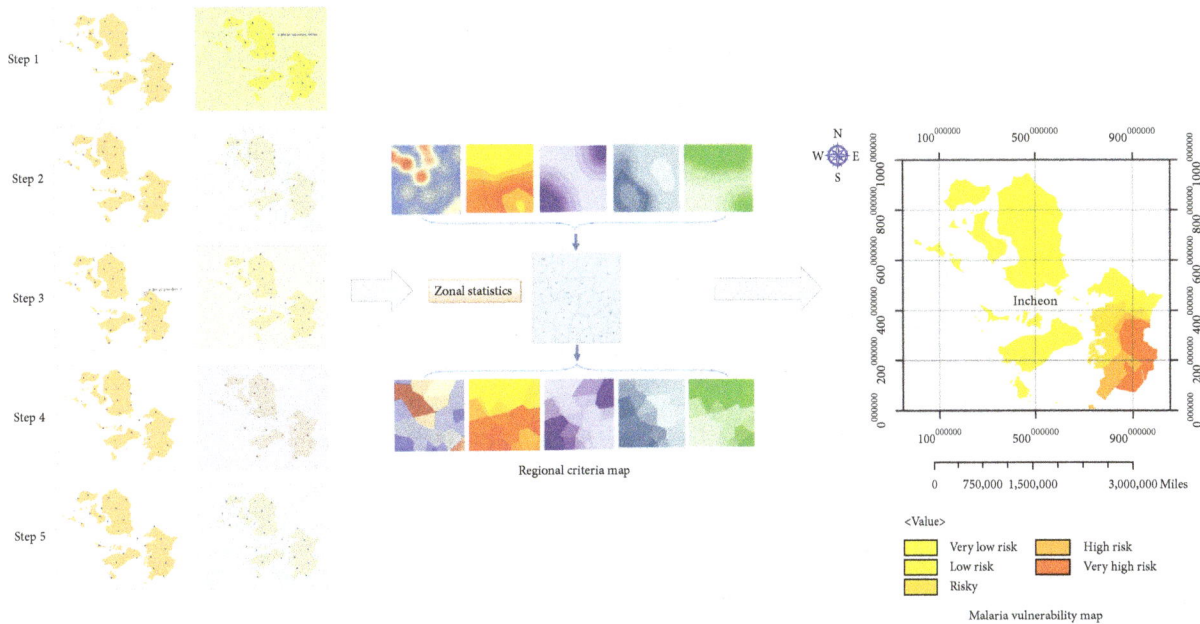

FIGURE 7: Raster overlay analysis.

information on areas where malaria appear, and risk animation function for areas where malaria appear. Visual expression function was implemented by using Google Maps API and JavaScript (Bootstrap, jQuery). Frame animation technique was used for effective view of risk information. Frame animation technique can highlight specific content and provide information to a user by changing images by time and static image to dynamic image stimulating human senses which are sensitive to dynamic motions. The author of this paper attempted to convey risk and environmental information by using the abovementioned functions and make a user have easier access by supporting intuitive UI for higher level of immersion. Applicable implementations are presented in Figure 8. The DB defines the schema as shown in Table 1.

4. Evaluation and Conclusion

4.1. System Model Evaluation. In order to evaluate the system performance, we measured accuracy of the malaria risk model. Comparisons of accuracy criteria were compared with those of the experts' risk criteria. So we use the confusion matrix and calculate the accuracy by true positive (TP), false positive (FP), false negative (FN), and true negative (TN). Table 5 represents confusion matrix result for the evaluation of the system model.

A total of 30 test cases were used for the evaluation, which resulted in a good performance of 91.7% accuracy. Equation (2) for calculating the accuracy is processed through the following confusion matrix:

$$\text{Accuracy} = \frac{(TN + TP)}{(TN + TP + FN + FP)}. \tag{2}$$

For some inaccurate results, there is a problem in weight calculation of the AHP model to create the decision matrix.

Therefore, the issue of weight is in the order of rank in criteria. This section is classified as a special case, and more accurate weighting methods are needed. As a workaround, we expect to give more conditions to the quarter.

4.2. Conclusion. PHP is used as server development language for construction of website. Client side language was implemented as HTML, CSS3, and JavaScript (Bootstrap, jQuery). Google Maps JavaScript API v3 was used as a map.

Scholars have conducted research to develop GIS-based tools that can analyze malaria risk as continuous control of occurrence of malaria is needed. However, analysis tools at home and abroad are not widely used due to the lack of GIS information use, independent platform environment, and exclusive institutional policies.

This paper aimed at improving the lack of GIS information use and compatibility of multiplatform which represented limits that existing malaria risk analysis tools had. For this, the author developed mobile web-based malaria vulnerability map system using GIS information.

This system consists of system database construction, malaria risk calculation function, visual expression function, and website and mobile application. This system was developed based on Incheon region only. Database includes information on air temperature and amount of precipitation as well. With regard to malaria risk calculation, guideline provided by Korea Centers for Disease Control and Prevention was followed first and then decision-making technique was used. Calculating criteria value for risk index made it possible to estimate more precise risk. With regard to visual expression function, database constructed earlier and risk information were linked to print out graphic map and graphs so that more intuitive and visible expression can

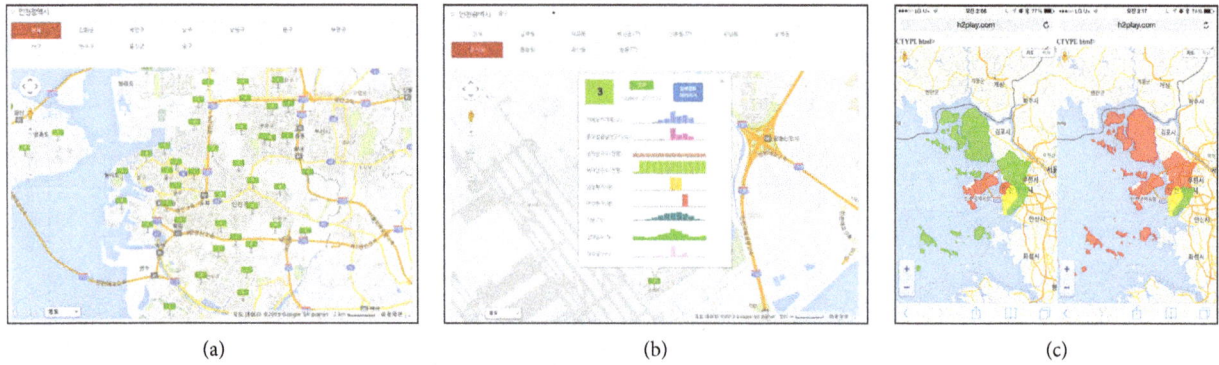

(a)   (b)   (c)

FIGURE 8: Final result of system visualization. (a) Marking of malaria generation region. (b) Summation information (tooltip) of malaria generation region. (c) Risk time-series animation of malaria generation region.

TABLE 5: Confusion matrix result for evaluation.

Confusion matrix	Malaria risk model
True positive	27
False positive	3
True negative	28
False negative	2

be provided based on animation technique. This system allows a user to have easier access via website and mobile application which print out above information. This system allows a user to check information in real time and can be used anywhere anyplace. Mobile push function is to enhance user's convenience. Such web map is useful to general users as well as experts.

As an evaluation of system model, we can check the good performance of 91.7% accuracy. However, for some inaccurate results, there is a problem in weight calculation of the AHP model to create the decision matrix. This section is classified as a special case, and more accurate weighting methods are needed. To solve the issue, we expect to give more conditions to calculate the weight value.

This system is expected to be used for institutions that operate under exclusive policy and service that prints out information based on web map. This system is expected to be used as basic technology that provides visible and intuitive information in the field of infographics which offer location-based information, GIS, industries as well as health care. The author of this paper will consider environmental elements which were not used in analyzing this system and seek additional mobile functions and a way to enhance user's accessibility. And we will also consider network security technologies for securing user data [24–27].

Acknowledgments

This research was supported by the Bio & Medical Technology Development Program of the National Research Foundation (NRF) funded by the Ministry of Science, ICT and Future Planning (2017M3A9E2072689).

References

[1] J. Y. Kim, "Assessments on trend of malaria prevalence in Republic of Korea," *Public Health Weekly Report*, vol. 7, no. 12, pp. 237–242, 2014.

[2] J.-W. Park, "Status of *vivax* malaria in the Republic of Korea," *Journal of the Korean Medical Association*, vol. 47, no. 6, pp. 521–526, 2004.

[3] Korea Center For Disease Control and Prevention, *Infectious Diseases Surveillance Yearbook, Ministry of Health and Welfare*, Korea Center For Disease Control and Prevention, Cheongju-si, South Korea, 2014.

[4] H. S. Shin, "Malaria prevalence rate and weather factors in Korea," *Health and Social Welfare Review*, vol. 31, no. 1, pp. 217–237, 2011.

[5] S. M. Chae, D. J. Kim, S. J. Yoon, and H. S. Shin, "The impact of temperature rise and regional factors on malaria risk," *Health and Social Welfare Review*, vol. 34, no. 1, pp. 436–455, 2014.

[6] J. Kwak, J. Lee, H. Han, and H. Kim, "A case study: malaria modeling based on climate variables in Korea," *Health and Social Welfare Review*, vol. 33, no. 4, pp. 547–569, 2013.

[7] Y. W. Gong, "Design and implementation of outbreak estimation system for infectious disease using GIS (for malaria)," Engineering Master thesis, Kyungwon University, Gyounggi-do, Republic of Korea, 2001.

[8] R. Laurini, *Information Systems for Urban Planning: A Hypermedia Cooperative Approach*, CRC Press, Boca Raton, FL, USA, 2014.

[9] http://sites.duke.edu/mdast/.

[10] T. L. Saaty, "Analytic hierarchy process," in *Encyclopedia of Operations Research and Management Science*, pp. 52–64, Springer, Boston, MA, USA, 2013.

[11] O. S. Vaidya and S. Kumar, "Analytic hierarchy process: an overview of applications," *European Journal of operational research*, vol. 169, no. 1, pp. 1–29, 2006.

[12] S. H. Hashemi, A. Karimi, and M. Tavana, "An integrated green supplier selection approach with analytic network process and improved Grey relational analysis," *International Journal of Production Economics*, vol. 159, pp. 178–191, 2015.

[13] Korea Center for Disease Control and Prevention, *Malaria Administrative Guideline 2013*, Korea Center for Disease Control and Prevention, Cheongju-si, South Korea, 2013.

[14] J. P. Brans and and Ph. Vincke, "Note—A Preference Ranking Organisation Method: (The PROMETHEE Method for Multiple Criteria Decision-Making)," *Management science*, vol. 31, no. 6, pp. 647–656, 1985.

[15] J. Rezaei, "Best-worst multi-criteria decision-making method," *Omega*, vol. 53, pp. 49–57, 2015.

[16] J.-J. Wang and D.-L. Yang, "Using a hybrid multi-criteria decision aid method for information systems outsourcing," *Computers & Operations Research*, vol. 34, no. 12, pp. 3691–3700, 2007.

[17] C. Kahraman, S. C. Onar, and B. Oztaysi, "Fuzzy multicriteria decision-making: a literature review," *International Journal of Computational Intelligence Systems*, vol. 8, no. 4, pp. 637–666, 2015.

[18] M. Dağdeviren, "Decision making in equipment selection: an integrated approach with AHP and PROMETHEE," *Journal of Intelligent Manufacturing*, vol. 19, no. 4, pp. 397–406, 2008.

[19] A. Mardani, A. Jusoh, K. M. Nor, Z. Khalifah, N. Zakwan, and A. Valipour, "Multiple criteria decision-making techniques and their applications: a review of the literature from 2000 to 2014," *Economic Research-Ekonomska Istraživanja*, vol. 28, no. 1, pp. 516–571, 2015.

[20] P. L. Ibisch, M. T. Hoffmann, S. Kreft et al., "A global map of roadless areas and their conservation status," *Science*, vol. 354, no. 6318, pp. 1423–1427, 2016.

[21] Y. S. Yilmaz, B. I. Aydin, and M. Demirbas, "Google cloud messaging (GCM): an evaluation," in *Proceedings of Global Communications Conference (GLOBECOM)*, pp. 2807–2812, Austin, TX, USA, December 2014.

[22] N. Saravanan, A. Mahendiran, N. V. Subramanian, and N. Sairam, "An implementation of RSA algorithm in Google cloud using cloud SQL," *Research Journal of Applied Sciences, Engineering and Technology*, vol. 4, no. 19, pp. 3574–3579, 2012.

[23] M. Achibet, G. Casiez, A. Lécuyer, and M. Marchal, "THING: introducing a tablet-based interaction technique for controlling 3d hand models," in *Proceedings of 33rd Annual ACM Conference on Human Factors in Computing Systems*, New York, NY, USA, May 2015.

[24] J. Cui, Y. Zhang, Z. Cai, A. Liu, and Y. Li, "Securing display path for security-sensitive applications on mobile devices," *Computers Materials and Continua*, vol. 55, no. 1, pp. 17–35, 2018.

[25] C. Yin, J. Xi, R. Sun, and J. Wang, "Location privacy protection based on differential privacy strategy for big data in industrial internet-of-things," *IEEE Transactions on Industrial Informatics*, vol. 14, no. 8, pp. 3628–3636, 2017.

[26] J. Wang, C. Ju, H.-J. Kim, R. Simon Sherratt, and S. Lee, "A mobile assist coverage hole patching scheme based on particle swarm optimization for WSNs," *Cluster Computing*, vol. 1, pp. 1–9, 2017.

[27] Y. Tu, Y. Lin, J. Wang et al., "Semi-supervised learning with generative adversarial networks on digital signal modulation classification," *Computers Materials & Continua*, vol. 55, no. 2, pp. 243–254, 2018.

10

User Evaluation of the Smartphone Screen Reader VoiceOver with Visually Disabled Participants

Berglind F. Smaradottir (ID),[1] **Jarle A. Håland** (ID),[2] **and Santiago G. Martinez** (ID)[3]

[1]*Department of Information and Communication Technology, University of Agder, Grimstad N-4879, Norway*
[2]*Kongsgård School Centre, Kristiansand N-4631, Norway*
[3]*Department of Health and Nursing Science, University of Agder, Grimstad N-4879, Norway*

Correspondence should be addressed to Berglind F. Smaradottir; berglind.smaradottir@uia.no

Guest Editor: Giuseppe De Pietro

Touchscreen assistive technology is designed to support speech interaction between visually disabled people and mobile devices, allowing hand gestures to interact with a touch user interface. In a global perspective, the World Health Organization estimates that around 285 million people are visually disabled with 2/3 of them over 50 years old. This paper presents the user evaluation of VoiceOver, a built-in screen reader in Apple Inc. products, with a detailed analysis of the gesture interaction, familiarity and training by visually disabled users, and the system response. Six participants with prescribed visual disability took part in the tests in a usability laboratory under controlled conditions. Data were collected and analysed using a mixed methods approach, with quantitative and qualitative measures. The results showed that the participants found most of the hand gestures easy to perform, although they reported inconsistent responses and lack of information associated with several functionalities. User training on each gesture was reported as key to allow the participants to perform certain difficult or unknown gestures. This paper also reports on how to perform mobile device user evaluations in a laboratory environment and provides recommendations on technical and physical infrastructure.

1. Introduction

Since the last decade, touchscreen technology has been increasingly used not only across multiple types of devices, such as smartphones and tablets [1–3], but also in photocopying machines, automated teller machines (ATMs), and ticket machines in bus, railway stations, and airports. Reviews from the perspective of human factors and ergonomics and studies of people with developmental disabilities pointed out the relevance of the specific context of system interaction in order to maximize safety, performance, and user satisfaction [4] and the need for more research [5]. Touchscreens require the use of fingers and a choreography of gestures for interaction between the user and the device's user interface (UI) [6, 7]. However, this type of screen interaction can represent a challenge for visually disabled users where the screens are designed for a visual feedback while using the system [8].

The World Health Organization (WHO) estimates that the number of people with visual disability is around 285 million globally and that about 2/3 of them are older than 50 years [9, 10]. Traditionally, visually disabled people have used different assistive technology devices, such as an external keyboard, a braille terminal, or a screen reader that provides speech feedback related to the visual elements on the screen. Mobile phones with physical buttons are still functional for many visually disabled people because of the surface and the rugosity of the buttons that provide palpable guidance when using the device. However, this type of communication device has become less popular in favour of smartphones with touchscreens that currently dominate the market. Smartphones with touchscreen interaction do mainly incorporate visual and sound feedback for communication with the user. This type of communication represents a challenge for the UI navigation to visually disabled people who do not see the screen with sufficient details and buttons without tactile feedback [11]. Several solutions are available in the market to improve the accessibility of smartphone technology for visually disabled people [12–14]. Some of these solutions are standalone products, and others are used in conjunction with other technology. One of the products available is VoiceOver [12], the integrated screen reader in Apple Inc. products. VoiceOver allows users to interact with the UI through gestures and with speech feedback to guide the navigation. The screen reader has been included in Apple Inc. products since April 2005 in Mac OS X 10.4, since June 2009 in iPhone 3GS OS 3.0, and in iPad OS 3.2 since its introduction in April 2010. VoiceOver has to be activated in the device's settings, and when activated, the device provides a speech feedback when a user interacts using hand gestures on the touchscreen. There are different gestures that can be performed on the UI, and they provide immediate feedback interpreted by the screen reader. For instance, tap with one finger and drag will read the item in the cursor (selected), and four-finger tap near the top of the screen will read the first item at the top. The gestures must be made with the fingers, and the screen reader does not respond to voice commands or sense motion.

In this context, the research project "Visually impaired users touching the screen—A user evaluation of assistive technology" aimed at evaluating the accessibility and usability of a screen reader for touchscreens in smartphones [15]. This paper presents the results from the evaluation of the usability and the accessibility of the screen reader VoiceOver (iOS 7.1.2), which is an integrated functionality in iPhone mobile devices. In addition, the paper provides recommendations on technical and physical infrastructure to perform an evaluation of mobile devices in a laboratory environment.

The three research questions (RQs) targeted by this study were as follows:

> RQ1: What is the user experience of visually disabled users when interacting with the VoiceOver?
>
> RQ2: How is the VoiceOver screen reader response to a set of 16 performed hand gestures during a user evaluation?
>
> RQ3: What technical infrastructure can be suitable for an evaluation of mobile assistive technology with visually disabled users?

Following this introduction, the research methodology and the technical test infrastructure are described. The results are presented based on the user evaluation outcomes and experience related to the test infrastructure. Furthermore, a discussion of the main results is provided followed by a summary of the research contributions and conclusions.

2. Materials and Methods

A mixed methods research approach was employed in the evaluation of the screen reader [16–18], with quantitative and qualitative measures. The evaluation was conducted in three phases: (1) individual user training at the participant's home and introduction to the gestures a few days before the test, supplied with a written instruction sent by e-mail;

(2) a usability test in a controlled laboratory environment including a pretest interview for collecting participant background information; and (3) a posttest interview for qualitative analysis of the test output. The research team had three members whose background was health technology, educational training with assistive technology, and clinical practice. All research team members had professional experience in working with people with visual disabilities.

In the initial preparation of the study, phone interviews were made with three key informants with expertise in visual disabilities, who worked at the Norwegian State Agency for Special Needs Education Service (StatPed) [19]. The goal of the interviews with the key informants was to gather insights on assistive technology for visually disabled people. Based on the interviews, a pilot test of the evaluation was prepared with a comparison of Android and Apple tablet devices. Two voluntary members from the Norwegian Association of the Blind and Partially Sighted [20] participated in the pilot test, running several tasks. Afterwards, a focus group interview was conducted in order to better understand the interactions and any of the problems that the users found. In the phone interviews and also in the pilot test, the informants explained that their experience was that the smartphone iPhone was the most commonly used and preferred device among their peers, also visually disabled people. Based on that information, an iPhone 4 (iOS 7.1.2) device was chosen for the study (the device can be seen in Figure 1) because it was widely available and had the VoiceOver screen reader integrated. The tasks were inspired by the standard gestures' descriptions in the VoiceOver guide manual [21].

2.1. Recruitment of Participants. The recruitment of participants was made in collaboration with the Norwegian Association of the Blind and Partially Sighted [20]. In addition, the professional network of one of the researchers with expertise in teaching and user training of assistive technology was used to support the recruitment process. The first contact made with the participants was a phone conversation to inform them about the study. The second contact was an e-mail with information about the study and a consent form to be signed by each participant. Six visually disabled people were recruited to participate in the user evaluation, see Table 1 for distribution of participants. They had a mean age of 42.8 years and an average of 1.9 years of user experience with VoiceOver. All the participants had previous experience with using a screen reader for desktop and/or laptop computers.

2.2. Test Procedure. In the first phase of the evaluation, each participant had individual user training at home (Figure 2) on 16 specific hand gestures for screen interaction. The individual user training lasted 15–30 minutes (with an average of 21.7 minutes), led by a member of the research team. The gestures that a user knew in advance and which ones were learned during the training were registered during the training session.

The second phase was executed in a usability laboratory. One of the researchers acted as the moderator and sat down

FIGURE 1: The smartphone used in the test.

TABLE 1: The background of the test participants.

Participants $n = 6$	Age	Gender	Device familiarity	Years of VoiceOver use	Self-graded skill
1	60	Female	iPhone	1.5	Medium
2	31	Male	iPhone	1.5	Advanced
3	27	Female	iPhone	1	Medium
4	30	Male	iPod	3	Advanced
5	48	Female	iPhone	2	Medium
6	61	Male	iPhone	2.5	Advanced

FIGURE 2: User training of VoiceOver gestures at a participant's home.

FIGURE 3: The moderator (left) guiding a participant (right) through the task solving in the test room.

beside the test participant. The participants were informed about the subsequent test and signed a consent form before the test began. Demographic information and user experience with specific technical devices were also collected. Each user evaluation followed the same test plan, with a set of 16 tasks related to the use of gestures for touchscreen interaction. The moderator guided through the tasks and asked the participants to speak out loudly during the task solving (Figure 3) following a think aloud protocol [22–24].

The task solving was followed by a posttest individual interview (third phase). The participants were asked to score the gesture performance and task solving, choosing among three categories: "easy," "medium," or "difficult." In addition, problems or obstacles observed or reported were discussed. The interviews also covered the general user experience with the smartphone and the first-time use of the VoiceOver.

Each test session (second and third phases) lasted between 90 and 120 minutes, and a total of six test sessions were run across three separate days.

2.3. Technical and Physical Test Infrastructure. The evaluation was executed in the usability laboratory at the Centre for eHealth of the University of Agder, Norway [25]. The usability laboratory consisted of two rooms; one test room and one control room, connected through a one-way mirror with visualisation towards the test room. In the test room, the moderator was placed together with a test participant, and in the control room, two observers followed the test from monitors and directly through the one-way mirror. The technical and physical infrastructure is described in Figure 4.

For replicability and information purposes, the technical material and equipment used during the study are presented below grouped by rooms.

Test room:

(i) Apple Inc. iPhone 4 MD128B/A iOS 7.1.2 with VoiceOver activated

(ii) Fixed camera: Sony BRCZ330 HD 1/3 1CMOS P/T/ Z 18x optical zoom (72x with digital zoom) colour video camera

(iii) Portable camera: Sony HXR-NX30 series

(iv) Apple Inc. iPad MD543KN/A iOS 8.1 for additional sound recording

(v) Sennheiser e912 condenser boundary microphone

(vi) Landline phone communication

Control room:

(i) Stationary PC: HP Z220 CMT workstation, Intel Core i7-3770. CPU @ 3.4 GHz, 24 GB RAM, Windows 7 Professional SP1 64 bit

(ii) Monitor: 3x HP Compaq LA2405x

(iii) Remote controller: Sony IP Remote Controller RM-IP10

(iv) Streaming: 2x Teradek RX Cube-455 TCP/IP 1080p H.264

(v) Software Wirecast 4.3.1

(vi) Landline phone communication

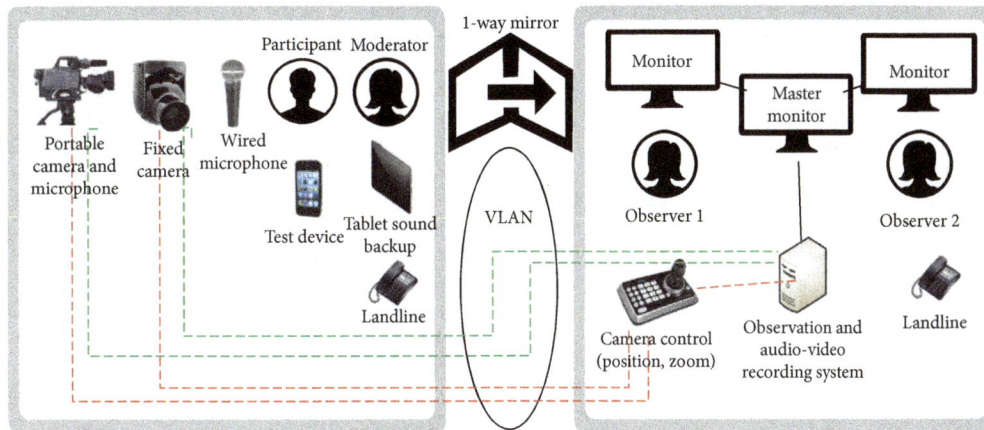

FIGURE 4: The technical and physical test infrastructure.

2.4. Data Collection. The test sessions were audio-visually recorded in a F4V video file format. The recordings from two audio-visual sources were merged into one video file using the software Wirecast v.4.3.1 [26], with multiple video perspectives and one single audio channel. The files were exported to the Windows Media Video (WMV) format and then imported to the qualitative software tool QSR NVivo 10 [27]. The recordings were transcribed verbatim and categorized for a qualitative content analysis [28]. Quantitative measurements of the time and number of attempts in the task solving were made as a part of the analysis of the recordings. In addition, the research team made annotations during the test sessions that were included in the data collection (Figure 5).

2.5. Ethical Approval. The Norwegian Centre for Research Data [29] approved this study with the project number 40636. All participants received verbal and written information about the project and confidential treatment of their collected data. They were informed that their participation was voluntary, and each participant signed a consent form. The participants were aware that they could withdraw at any time without reason. In that case, their data would be consequently withdrawn and deleted. For health and safety reasons, each test participant was thoroughly informed about the physical environment before entering the test room and the participants were never left alone in the laboratory facilities.

3. Results

All six participants went through the laboratory test. The test results are presented divided into three categories: user training, quantitative metrics from the user tests, and qualitative outcome of the posttest interviews.

3.1. Pretest User Training. The familiarity with the Voice-Over gestures registered in the user training is presented in Table 2. The registration showed that all participants knew the double tap gesture (number 4) and three-finger flick to the left or right (number 10). 5 were 6 were familiar with the

FIGURE 5: The control room showing the visual access to the test room through the one-way mirror.

one-finger tap gestures (numbers 1–3). For gesture numbers 6 and 7, the four-finger tap at the top or the bottom of the screen, 5 out of 6 participants did not know them in advance.

3.2. User Evaluations. The quantitative measurements from the user evaluations are presented in Table 3, separated in six columns. The first column describes the 16 VoiceOver standard gestures that were used to solve the associated task. The tasks are described in the second column. The third column displays the average number of attempts needed for the task solving. The fourth column shows the task solving average time that was used, measured in seconds. The fifth column presents the system response to the gesture interaction differentiated in the categories "consequent" and "inconsequent" speech feedback. Consequent speech feedback refers to sufficient and adequate information in the system response and inconsequent feedback to insufficiency or lack of information in the system response. In usability studies, the task accuracy is often categorized into completed or not completed task [23, 30]. In this particular test, there was an additional variable related to the task performance, which was the feedback that the system provided when a participant performed a specific action. The categories chosen were therefore "consequent feedback" or "inconsequent feedback" to the specific hand gesture performed. The "consequent feedback" referred to the system appropriately

TABLE 2: Familiarity per participant with the VoiceOver gestures in the pretest user training.

Gesture	P#1	P#2	P#3	P#4	P#5	P#6
(1) Tap with one finger, lift and tap again	Yes	Yes	Yes	Yes	Yes	Yes
(2) Tap with one finger and drag	Yes	Yes	Yes	Yes	Yes	Yes
(3) Tap with one finger and swipe to right or left	Yes	Yes	No	Yes	Yes	Yes
(4) One-finger double tap	Yes	Yes	Yes	Yes	Yes	Yes
(5) Split-tap: touch one finger and then tap with a second finger	No	Yes	No	Yes	No	No
(6) Four-finger tap at the top of the screen	No	No	No	No	No	Yes
(7) Four-finger tap at the bottom of the screen	No	No	No	No	No	Yes
(8) Two-finger flick up	Yes	Yes	No	Yes	No	No
(9) Two-finger flick down	Yes	Yes	No	Yes	No	No
(10) Three-finger flick to the left or right	Yes	Yes	Yes	Yes	Yes	Yes
(11) Three-finger tap	No	Yes	No	Yes	No	No
(12) Two-finger rotate	Yes	Yes	No	Yes	Yes	Yes
(13) Flick up and down with one finger	Yes	Yes	No	Yes	No	Yes
(14) Three-finger double tap	Yes	No	Yes	Yes	Yes	Yes
(15) Three-finger triple tap	Yes	Yes	No	Yes	Yes	Yes
(16) Two-finger double tap	Yes	Yes	No	Yes	Yes	Yes

P = participant.

TABLE 3: Quantitative metrics of the user evaluations.

Gesture	Task	Average no. of attempts	Average time for task solving in seconds	System speech feedback	Specification
(1) Tap with one finger, lift and tap again	Speak the item in the cursor: find the app Map	24.2	28	Consequent	
(2) Tap with one finger and drag	Speak the item in the cursor: find the app Clock	14.3	13.8	Consequent	
(3) Tap with one finger and swipe to right or left	Speak the item in the cursor: find the app Calendar	22	16.5	Consequent	
(4) One-finger double tap	Open the app Calendar	1	1	Consequent	
(5) Split-tap: touch one finger and then tap with a second finger	Open the app Weather	1	1	Consequent	
(6) Four-finger tap at the top of the screen	Read the item at the top	3.7	16.8	Consequent	
(7) Four-finger tap at the bottom of the screen	Read the item at the bottom	9.2	27	Consequent	
(8) Two-finger flick up	Read the current page starting at the top	1.2	1.2	Consequent, except once	The screen reader did not read
(9) Two-finger flick down	Read from the cursor to the end of the current page	1	1	Consequent	
(10) Three-finger flick to the left or right	Change to the next page in the start screen and back	2.8	6.7	Inconsequent	The screen reader did not consequently read the next page
(11) Three-finger tap	Read where the cursor is	1.8	3.5	Inconsequent	The screen reader did not read the cursor (application's name)
(12) Two-finger rotate	Rotor: find the setting for the speed of the speech feedback	10.3	13.3	Consequent	
(13) Flick up and down with one finger	Rotor: adjust the speed of the speech feedback	1.5	2.8	Consequent	
(14) Three-finger double tap	Mute VoiceOver	1.2	1.7	Consequent	
(15) Three-finger triple tap	Turn the screen curtain on	1.3	2.8	Consequent	
(16) Two-finger double tap	Terminate a phone call	3.6	1.8	Inconsequent	The system did not terminate the phone call

providing feedback that corresponded to the hand gesture performed by a participant. The "inconsequent feedback" referred to a system feedback that did not correspond to the hand gesture performed by a participant of absence of any feedback. The sixth column specifies the type of inconsequent response occurred.

TABLE 4: The grading of the task solving made by the participants in the posttest interview (n = 6).

Gesture	Associated task	Easy	Medium	Difficult
(1) Tap with one finger, lift and tap again	Speak the item in the cursor: find the app Map	4	1	1
(2) Tap with one finger and drag	Speak the item in the cursor: find the app Clock	5	1	
(3) Tap with one finger and swipe to right or left	Speak the item in the cursor: find the app Calendar	6		
(4) One-finger double tap	Open the app Calendar	6		
(5) Split-tap: touch one finger and then tap with a second finger	Open the app Weather	4	2	
(6) Four-finger tap at the top of the screen	Read the item at the top	3	1	2
(7) Four-finger tap at the bottom of the screen	Read the item at the bottom	4	1	1
(8) Two-finger flick up	Read the current page starting at the top	3	2	1
(9) Two-finger flick down	Read from the cursor to the end of the current page	6		
(10) Three-finger flick to the left or right	Change to the next page in the start screen and back	5	1	
(11) Three-finger tap	Read where the cursor is	5	1	
(12) Two-finger rotate	Rotor: find the setting for the speed of the speech feedback	5	1	
(13) Flick up and down with one finger	Rotor: adjust the speed of the speech feedback	5	1	
(14) Three-finger double tap	Mute VoiceOver	6		
(15) Three-finger triple tap	Turn the screen curtain on	6		
(16) Two-finger double tap	Terminate a phone call	3	1	2

The performance of three different one-finger tap gestures (tasks 1–3) for speaking the item in the cursor required many attempts to succeed. The system response was consequent. The double tap and slit-tap gestures (tasks 4-5) were easy and fast to perform for the participants. The gesture four-finger tap at the top and bottom of the screen (tasks 6-7) were reported as technically difficult to perform by the participants, which was also indicated by the time for the task solving. The gestures two-finger flick up and down to read the page from top or bottom (tasks 8-9), were easy to perform and showed consequent speech feedback. The three-finger flick and tap gestures (tasks 10-11) were reported as easy to perform, but there was inconsequent system response related to insufficiency in the speech feedback when trying to inform about the current page. For the rotor-related tasks, 12 and 13, two of the participants needed several attempts (7 and 41) for finding the rotor settings, but adjusting the speed of the speech feedback was easier. The gestures three-finger double and triple tap (tasks 14-15) were easy to perform and with a quick task solving. The two-finger double tap in task 16, to terminate a phone call, was easy to perform but there was inconsequent feedback from the system and the phone call was not terminated in three out of six tests.

3.3. Posttest Interviews. The participants graded the performance of gestures and task solving (Table 4) during the individual posttest interview.

Five of the gestures in the task solving were categorized "easy" to perform, such as the one-finger double tap and the three-finger double and triple taps. Six gestures were categorized as "easy" or "medium," such as the one-finger flick up and down and three-finger tap. There were gestures that were categorized as "difficult" by two participants, such as the four-finger tap at the bottom and the top of the screen and the two-finger double tap. The task for the two-finger double tap was termination of a phone call, and in the interviews, the participants confirmed that during the test

but also in general, the gesture was associated with inconsistency from the system. For the rotor-related gestures, one participant emphasised the importance of user training to succeed with the specific use of the rotor function.

Regarding the first-time user experience, all participants needed user training to be able to start using the smartphone and for activation of the screen reader VoiceOver. Three had family or friends that helped them with the first-time use: one went to a course organized by the Norwegian Association of the Blind and Partially Sighted and two found it out by themselves explaining that VoiceOver as such provides user training and guidance by informing about which gesture to perform for an action. Four participants stated: *It was a bit complicated with first-time set up of the new phone with apple-id and activation of VoiceOver, besides that it is easy to use. [. . .] After user training, when I understood how the system worked, I found it easy to use. [. . .] The functions make sense, and there is a logical structure. [. . .] It was terrible in the beginning, because I knew none of the gestures and I wanted to throw the phone away, but the price stopped me from doing it . . . now I find it fantastic!*

Two participants highlighted the benefits of the smartphone: *I like that I can buy it myself in the store, I did not need to apply for and receive assistive technology from the municipal services. [. . .] This is the first device I use with built-in accessibility, as the screen reader is included.*

Two participants described how the use of the screen reader had increased their self-management: *I feel more included in the society, now I can use the Internet and check the same apps as other people do, such as Facebook, weather forecast and reading news. [. . .] It is a feeling of freedom when the phone can read messages for you when you are outdoors, before I had to ask people I did not know about reading from the screen if I received a message, I can now manage it myself and that is a new world for me.* In addition, one participant expressed: *VoiceOver has made my life much easier and I have become much more independent. Everyone with a visual impairment should use a phone with it.*

However, user text input with the VoiceOver keyboard was reported as complicated by four participants, and, for this reason, those participants preferred to use an external keyboard. Another participant stated: *It was hard in the beginning with the virtual keyboard, but with some training I overcame the difficulties.* Five participants told that they preferred to use at home a desktop or laptop computer with reading list because the text input was quicker than in the smartphone and relying on the latter when they were out of home. Two participants expressed that it was easier for them to navigate on a small screen when compared to a larger tablet screen.

4. Discussion

This paper has presented a user evaluation of the Apple screen reader VoiceOver (iOS 7.1.2) with six visually disabled participants. The aim was to identify challenges related to the performance of the standard VoiceOver gestures and evaluate the associated system response. Considering the sensory limitation of the target user group, the screen reader was expected to be intuitive with an optimal presentation of the functionality and distribution of the UI. The study showed that most of the gestures were easy to perform for the participants; however, some gestures were unfamiliar to the participants, especially those connected to the rotor function. The possibility of receiving individual user training before the evaluation was an advantage to succeed with the practical use of those gestures. The system appropriately responded to the users' hand gestures, but inconsistent responses and lack of information were reported in the two-finger flick up, three-finger flick to the left or right, three-finger and double-finger taps. The three research questions (RQs) formulated at the beginning of this paper are answered below based on the results from the study.

RQ1 asked about the user experience when interacting with the VoiceOver. The user experience with VoiceOver in general was positive, as the function was described to increase the self-management and support independence. Most of the gestures were both reported and observed as easy to perform, with some exceptions. The two most difficult ones reported by the participants were the four-finger tap and the two-finger double tap gestures. The gesture made using four-finger tap on the bottom or on the top of the screen to, respectively, read the content of the UI from either side was explicitly reported as difficult to perform.

RQ2 asked about the system response to the 16 hand gestures made on the touchscreen mobile device. The speech feedback appropriately responded during the test with useful information for participants to navigate through the UI, but a few inconsistent responses on correctly performed gestures were registered such as with the two-finger double tap to terminate a phone call. The phone call was terminated correctly only in 3 out of 6 tests and can be considered as a weakness in the system with a negative consequence for the users since speaking on the phone is one of the most frequently used functions. Other user problems identified were related to the gesture made by three-finger flick to the left or right for swiping between screens where the speech feedback was inconsistent and lacked information.

RQ3 asked about recommended technical infrastructure in evaluations of mobile assistive technology with visually disabled users. A suitable infrastructure would be the one that optimizes the data collection and allows an effective retrospective analysis under more demanding conditions than other user evaluations. In addition, the comfort, safety, and trust of the visually disabled test participants are crucial to avoid interference and distortion with the test results. The described technical and physical infrastructure in Figure 4 serves as an example of a controlled scenario for an evaluation with the same type of technology and participants. The video recordings require a sufficient quality allowing us to zoom in the user interface and the finger interactions in details. A professional software video program is needed to substantially reduce the speed for optimal viewing and retrospective analysis. In addition, the data should be collected with synchronized audio and video signals because streaming over a network usually incorporates latency. The synchronization is of high importance for the retrospective analysis, as the gestures and finger interactions with a mobile device's screen are often made at high speed. Another issue experienced and specific for tests with visually disabled participants was that that the sound from the VoiceOver interfered and overlapped with the sound from the test participant and the moderator in the recordings from the table microphone unit. This might complicate the retrospective analysis, and based on that experience, we recommend using several microphones to record the sound sources separately.

This study of the screen reader VoiceOver had some limitations such as the number of test participants ($n = 6$) and tests were conducted only in a usability laboratory setting. However, the number of the participants with a distribution in their ages and smartphone skills meaningfully represented the user group of visually disabled users of smartphones. Other studies have shown that a small number of participants in usability studies can be sufficient for having valid results [31–33]. The laboratory setting allowed the collection of detailed research data under controlled conditions. The collected data material was thoroughly analysed in detail to study the interaction between the visually disabled user and the UI touchscreen. Furthermore, the application of mixed method research, combining laboratory tests with detailed interviews, provided insights into the user experiences, as well as benefits and barriers of using the VoiceOver function.

5. Conclusions

This study was made as a part of the project "Visually impaired users touching the screen—A user evaluation of assistive technology" that aimed at evaluating the usability and accessibility of the screen reader VoiceOver. The main contribution of this study lies in the detailed analysis of the interaction with gestures between the visually disabled participants and the screen reader, preceding the responses from the system. In general, most of the hand gestures were easy to perform for the participants, although user training played a key role for the understanding and successful performance of specifically complex gestures. Without training, participants could not have been able to perform such gestures. The system response and speech feedback were in most cases correct, but some functionalities of the system might be improved. The results presented are in line with other studies on assistive technologies and visually disabled users [34–36]. The methodological procedures with the use of mixed methods, combining quantitative laboratory test with qualitative interviews and observations, can be recommended to other studies of similar characteristics. The test procedure with user training on the specific hand gestures in advance reduced the memory load in the laboratory test situation, as all the participants were familiar with the gestures and could focus on performing the tasks. The application of a think aloud protocol in the usability laboratory together with posttest interviews is strongly recommended for other studies related to touchscreen assistive technology because they may provide a more comprehensive result.

In terms of future work, it is proposed to validate the laboratory results in the field and address research with a larger sample size focusing on text input and navigation using VoiceOver on a smartphone or tablet device. A comparison between the screen readers VoiceOver from Apple Inc. and TalkBack, which is mainly developed for Android devices, could illustrate differences across different platforms. The integration of VoiceOver in the Apple Watch provides new opportunities of studying user-friendliness and accessibility for visually disabled users. A comparison of the use of VoiceOver on a desktop or laptop computer which are generally more command based could be easily made in a similar usability laboratory. Finally, newer models of iPhone to date, such as 8 and Xs, provide more tactile feedback through vibration during interactions than previous versions and the impact of those functions for visually disabled users would be interesting to evaluate.

Acknowledgments

The authors would like to thank all the participants in the study and the informants prior to the study for their disinterested contribution. Thanks are due to Hildegunn Mellesmo Aslaksen for professional photographing and also to the Centre for eHealth at the University of Agder, Norway, for close collaboration and adaptation of the test infrastructure facilities.

References

[1] R. Ling, *The Mobile Connection: The Cell Phone's Impact on Society*, Morgan Kaufmann, Burlington, MA, USA, 2004.

[2] S. L. Jarvenpaa and K. R. Lang, "Managing the paradoxes of mobile technology," *Information Systems Management*, vol. 22, no. 4, pp. 7–23, 2005.

[3] G. Goggin, *Cell Phone Culture: Mobile Technology in Everyday Life*, Routledge, Abingdon, UK, 2012.

[4] A. K. Orphanides and C. S. Nam, "Touchscreen interfaces in context: a systematic review of research into touchscreens across settings, populations, and implementations," *Applied Ergonomics*, vol. 61, pp. 116–143, 2017.

[5] J. Stephenson and L. Limbrick, "A review of the use of touchscreen mobile devices by people with developmental disabilities," *Journal of Autism and Developmental Disorders*, vol. 45, no. 12, pp. 3777–3791, 2015.

[6] P. A. Albinsson and S. Zhai, "High precision touch screen interaction," in *Proceedings of the ACM SIGCHI Conference on Human Factors in Computing Systems*, pp. 105–112, Fort Lauderdale, FL, USA, April 2003.

[7] A. Butler, S. Izadi, and S. Hodges, "SideSight: multi-touch interaction around small devices," in *Proceedings of the 21st Annual ACM Symposium on User Interface Software and Technology*, pp. 201–204, Monterey, CA, USA, October 2008.

[8] L. Hakobyan, J. Lumsden, D. O'Sullivan, and H. Bartlett, "Mobile assistive technologies for the visually impaired," *Survey of Ophthalmology*, vol. 58, no. 6, pp. 513–528, 2013.

[9] World Health Organization (WHO), October 2018, http://www.who.int/mediacentre/factsheets/fs282/en/.

[10] D. Pascolini and S. P. Mariotti, "Global estimates of visual impairment: 2010," *British Journal of Ophthalmology*, vol. 96, no. 5, pp. 614–618, 2012.

[11] S. K. Kane, J. P. Bigham, and J. Wobbrock, *Fully Accessible Touch Screens for the Blind and Visually Impaired*, University of Washington, Tacoma, WA, USA, 2011.

[12] Voiceover, October, 2018, https://www.apple.com/accessibility/iphone/vision/.

[13] Window-Eyes, October, 2018, http://www.windoweyesforoffice.com.

[14] TalkBack, October, 2018, https://support.google.com/accessibility/android/answer/6283677?hl=en&ref_topic=3529932.

[15] B. F. Smaradottir, S. G. Martinez, and J. A. Håland, "Evaluation of touchscreen assistive technology for visually disabled users," in *Proceedings of IEEE Symposium on Computers and Communications (ISCC)*, pp. 248–253, Heraklion, Greece, July 2017.

[16] J. W. Creswell and V. L. P. Clark, *Designing and Conducting Mixed Methods Research*, SAGE Publications Inc., Thousand Oaks, CA, USA, 2007.

[17] R. B. Johnson, A. J. Onwuegbuzie, and L. A. Turner, "Toward a definition of mixed methods research," *Journal of Mixed Methods Research*, vol. 1, no. 2, pp. 112–133, 2007.

[18] C. Teddlie and A. Tashakkori, "Mixed methods research," in *The SAGE Handbook of Qualitative Research*, N. K. Denzin and Y. S. Lincoln, Eds., vol. 4, pp. 285–300, SAGE Publications Inc., Thousand Oaks, CA, USA, 2011.

[19] StatPed, October, 2018, http://www.statped.no/Spraksider/In-English/temp/Professional-services-and-areas-of-expertise/Expertise-within--the-specialisation-felds-for-special-needs-education/#6.6.

[20] The Norwegian Association of the Blind and Partially Sighted, October, 2018, https://www.blindeforbundet.no/om-blindeforbundet/brosjyrer/muligheter-til-et-aktivt-liv-engelsk.

[21] Learn VoiceOver gestures on iPhone, October, 2018, https://help.apple.com/iphone/12/?lang=en#/iph3e2e2281.

[22] M. W. M. Jaspers, "A comparison of usability methods for testing interactive health technologies: methodological aspects and empirical evidence," *International Journal of Medical Informatics*, vol. 78, no. 5, pp. 340–353, 2009.

[23] A. W. Kushniruk and V. L. Patel, "Cognitive and usability engineering methods for the evaluation of clinical information systems," *Journal of Biomedical Informatics*, vol. 37, no. 1, pp. 56–76, 2004.

[24] K. A. Ericsson and H. A. Simon, "Verbal reports as data," *Psychological Review*, vol. 87, no. 3, pp. 215–251, 1980.

[25] Centre for eHealth at University of Agder, October 2018, https://www.uia.no/en/centres-and-networks/centre-for-ehealth.

[26] Wirecast, October 2018, http://www.telestream.net/wirecast/overview.htm.

[27] QSR NVIVO 10, October 2018, http://www.qsrinternational.com/products_nvivo.aspx.

[28] J. Lazar, J. H. Feng, and H. Hochheiser, *Research Methods in Human-Computer Interaction*, John Wiley & Sons, Hoboken, NJ, USA, 2010.

[29] Norwegian Social Science Data Services, October 2018, http://www.nsd.uib.no/personvernombud/en/about_us.html.

[30] J. M. C. Bastien, "Usability testing: a review of some methodological and technical aspects of the method," *International Journal of Medical Informatics*, vol. 79, no. 4, pp. e18–e23, 2010.

[31] J. Nielsen and T. K. Landauer, "A mathematical model of the finding of usability problems," in *Proceedings of ACM Conference on Human Factors in Computing Systems*, pp. 206–213, Amsterdam, Netherlands, April 1993.

[32] C. W. Turner, J. R. Lewis, and J. Nielsen, "Determining usability test sample size," *International Encyclopedia of Ergonomics and Human Factors*, vol. 3, no. 2, pp. 3084–3088, 2006.

[33] J. Nielsen, "Why you only need to test with 5 users," Alertbox, October 2018, https://www.nngroup.com/articles/why-you-only-need-to-test-with-5-users/.

[34] K. Park, T. Goh, and H. J. So, "Toward accessible mobile application design: developing mobile application accessibility guidelines for people with visual impairment," in *Proceedings of HCI Korea*, pp. 31–38, Hanbit Media, Inc., Seoul, Republic of Korea, 2014.

[35] B. Leporini and M. C. Buzzi, "Interacting with mobile devices via VoiceOver: usability and accessibility issues," in *Proceedings of the 24th Australian Computer-Human Interaction Conference*, pp. 339–348, ACM, Melbourne, VIC, Australia, November 2012.

[36] D. McGookin, S. Brewster, and W. Jiang, "Investigating touchscreen accessibility for people with visual impairments," in *Proceedings of the 5th Nordic Conference on Human-Computer Interaction: Building Bridges*, pp. 298–307, ACM, Lund, Sweden, October 2008.

11

An Education Application for Teaching Robot Arm Manipulator Concepts Using Augmented Reality

Martín Hernández-Ordoñez ⓘ,[1,2] Marco A. Nuño-Maganda ⓘ,[3]
Carlos A. Calles-Arriaga ⓘ,[3] Omar Montaño-Rivas ⓘ,[4] and Karla E. Bautista Hernández ⓘ[3]

[1]Instituto Tecnológico Superior de Alvarado, Alvarado, VER, Mexico
[2]Instituto Tecnológico de Veracruz, Veracruz, VER, Mexico
[3]Universidad Politécnica de Victoria, 87138 Cd. Victoria, TAMPS, Mexico
[4]Universidad Politécnica de San Luis Potosí, 78363 San Luis Potosí, SLP, Mexico

Correspondence should be addressed to Carlos A. Calles-Arriaga; ccallesa@upv.edu.mx

Academic Editor: Byeong-Seok Shin

Teaching robotics is a challenge in many universities due to the mathematics concepts used in this area. In recent years, augmented reality has improved learning in several engineering areas. In this paper, a platform for teaching robotic arm manipulation concepts is presented. The system includes a homemade robotic arm, a control system, and the RAR@pp. The RAR@pp is focused on learning robotic arm manipulation algorithms by the detection of markers in the robotic arm and displaying in real time the values based on the data obtained by the control system. Details on the design of the platform are presented, and the related results are discussed. Experimental data about the usability of the application are also shown.

1. Introduction

In the last decade, Information Technologies and Communication (ITC) have spread to all areas of the society. An example of this is the area of learning. The constant changes in the traditional methodology of teaching seek more productive methods, with the goal to improve the learning experience and increase the intellectual level of the students. Currently, the use of mobile devices has continuously been increased. The growth in the demand for cell phones and tablets is due to the evolution of technologies and the implementation of many functionalities. One of the new technologies in development is augmented reality (AR). This technology allows the co-existence of real and virtual objects in the same environment.

Among the different qualities of reality, technology increased its adaptability to a high number of scenarios. Augmented reality, due to its portability and usability, may be implemented in various equipment such as personal computers, mobile devices, and smartphones, among others.

Besides, reality technology can be combined with other techniques to enrich the applications, emerging thus a tool to improve the teaching-learning process inside the classroom and laboratories [1]. An area with the potential to implement technology augmented reality is that of robotics since there are issues that require a 3D perception [2]. Nevertheless, traditional teaching methods make use of material didactic in 2D, which makes it difficult to understand these themes. For this reason, the present work shows the implementation of a robotic platform equipped with augmented reality technology. The purpose is to achieve in an educational setting an understanding of robotics topics which are more interactive.

In this work, a platform for teaching robotic arm manipulation concepts is presented. The system includes a homemade robotic arm, a control system, and the RAR@pp (Robotics through Augmented Reality Application). The RAR@pp is focused on learning robotic arm manipulation algorithms by the detection of markers in the robotic arm and the real-time visualization of the angles of each articulation of the homemade robotic arm using augmented reality. These angles are obtained from the robot based on the data obtained by the encoders of each motor and transmitted to the RAR@pp using Bluetooth communication. These applications allow students to send value to the robotic arm and visualize in real time the response of the arm to the com mands using the RAR@ pp. This application is specifically designed for mechatronics, information technology, and manufacturing undergraduate programs. The incorporation of augmented reality technology into the mobile application allows real-time viewing of joint angles through virtual objects corresponding to the real angles in the physical platform. The robotic platform is programmed with a reference profile using a desktop application. And by means of a Proportional Integral Derivative controller applied to each joint, it is possible that the robot follows the provided trajectory, allowing the joints to reach the required angles. This paper is structured as follows. In Section 2, a concise review of the state of the art is given. In Section 3, the design and implementation of the proposed system is described. In Section 4, the experimental setup and results are described. Finally, in Section 5, final recalls are established, and future work is outlined.

2. State of the Art

Nowadays, augmented reality represents a potential solution to problems in several areas such as robotics [3], teaching [4], mobile apps [5], or medicine [6]. For instance, Clemente et al. [7] proposed an improvement in sensing feedback of a robotic hand by visual elements. Although other options are typically used as sensors in prostheses such as vibro- or electrotactile stimulations, one advantage of AR implementation is the increase in sensibility resolution and easiness to adapt in the real world. Medical procedures could also be benefited of AR and robotics combination. In [8], a robotic system was proposed for treatment of tumors with the ablation technique. AR implementation helped to increase precision and consistency, which could be reflected in better medical outcomes. Augmented reality can also be used for safety purposes. Quercioli [6] developed a new approach to visualize lasers without any risk to eye damage. The mechanism consists of a modified smartphone camera without the infrared filter and a Google cardboard-type viewer. The system was successfully implemented to obtain real-time images of a Nd:YAG laser and a Ti:sapphire oscillator, both at near-infrared emission which is widely used in medical procedures.

In [2], a desktop application named *Build-A-Robot* was developed to help students to understand the forward kinematics of serial robots arms. The app improves the visualization and configuration of a 3D robotic arm in a controlled (virtual) environment. Jara et al. [9] developed an e-learning system based on a robotic platform and a graphical user interface with an integrated AR module. This platform can be used to control a robot arm through the Internet. The system is capable of planning a path remotely utilizing augmented reality.

In a work oriented to improve students learning, Ibañez et al. [10] developed an electromagnetism experiment using augmented reality (AR). Questions associated with fundamental topics, for example, Coulomb's law, electric field, and Ohm's law, were prepared. A test was carried out in high school students comparing a web-based tool and the AR app in an experimental/control group. Although results showed that the students' outcomes were statistically similar in both cases, the motivation to study was better for the AR app. This factor is also mentioned, along with creativity by Wei at al [4]. This is a very interesting finding since the students' attitude is fundamental during the learning process. Higher education also has a big potential to implement augmented reality as a learning tool. Martín-Gutiérrez et al. [11] developed an app for training students in real electrical labs using AR. One advantage of this work is that it was designed to promote independent as well as collaborative work in laboratories, which could be helpful mainly in engineering environments.

AR has significant growth potential in advanced manufacturing developments; for example, Ni et al. [12] created the haptic robot for programming welding paths. The main advantages of this prototype were a user-friendly interface and the possibility to improve this work through seam tracking sensors. In another work related to programming, Collett and MacDonald [13], design a system to test AR debugging.

Speed of processing is a main concern in mobile applications. Ruan and Jeong [14] proposed and implemented the substitution of traditional markers that elaborated with ARToolkit [15] for the Quick Response (QR) code in an AR system. A related work of applying QR to AR apps was developed by Kan et al. [16]. Barcode has also been used in combination with AR for commercial purposes [17]. AR had also been used for broadcasting enhancement as mentioned by Yan and Hu [18] which in general depends on AR display, AR tracking, and robotic AR broadcast.

3. Proposed System

In this section, the proposed system is described. The system includes three main components: a homemade physical robotic platform, a 2 DoF robotic arm, and the required hardware for moving the robotic arm, including a control module which receives signals from desktop application and sends back the angles obtained by the encoders of the robotic arm. The second component is the desktop application, focused on allowing students to prototype different control algorithms, and the last one is the RAR@pp, for the visualization of the angles in real time.

The dataflow of the proposed system is shown in Figure 1. The details of each component are described below:

(1) The desktop application sends commands to the robotic arm control using a USB serial protocol.

(2) The robotic arm control sends back the angles of each articulation, and these angles are used for plotting a comparison graph between desired and real path. The robotic control generates the

FIGURE 1: Main components and dataflow of the proposed system.

movement commands for each articulation, based on the information obtained from the desktop application.

(3) The mobile application sends a connection request to the robotic arm using a Bluetooth channel. The robotic arm receives this request and grants the connection. Once the connection is established, the mobile application identifies the markers located in the robot arm and sends a request to the robotic arm controller.

(4) Each robot articulation has a marker. In this specific case, due to the design of the robotic arm, only two markers were used, but this design can be extended to robotic arms with more than two articulations.

(5) The mobile application receives the degree of each articulation and displays them using a virtual degree protractor.

3.1. Main Blocks of the Desktop App.

The desktop application is designed for sending commands to the control system of the robotic arm using a serial USB connection and gets feedback about the angle of each articulation. This application was developed using Matlab/Simulink®, and the position of each articulation was independently programmed using a proportional-integral-derivative (PID) control. The desktop application also allows the user to perform simulation and generates comparison plots between real and desired trajectory, taking as input the times and angles for each articulation.

3.2. Main Blocks of the Control Module.

In Figure 2, a flow diagram including the software routines of the control module is shown. These software modules are hosted on an Arduino board, with an HC-05 Bluetooth shield. Each one of these modules is described below:

(i) *Configuration routine.* This module establishes the configuration parameters of both serial connections (Bluetooth and USB), the transfer speeds, and the pin modes required for external communication.

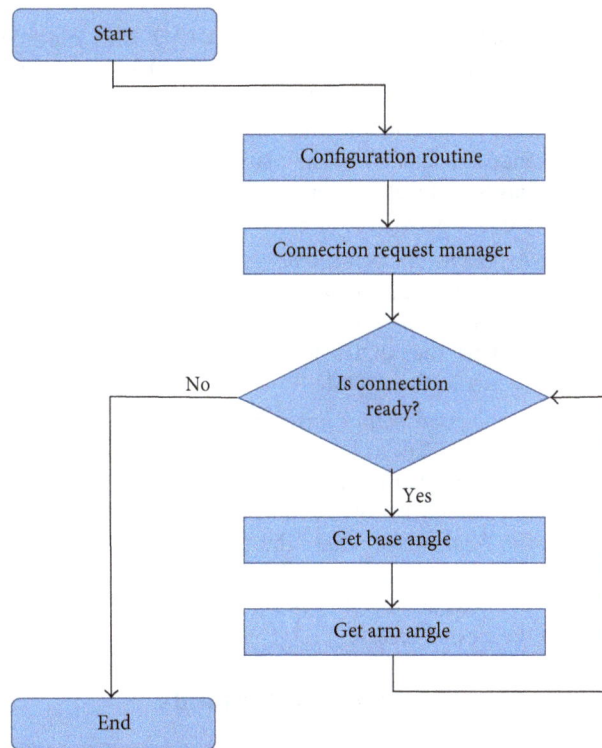

FIGURE 2: Main blocks of the robotic arm control module.

(ii) *Connection request manager.* This module detects if there is at least one connection request by the mobile application or by the desktop application.

(iii) *Get base angle routine.* This module translates the lecture obtained by the encoder located in the base articulation to digital values in order to generate the angle of the base. This angle is sent to the mobile application using the previous Bluetooth connection.

(iv) *Get arm angle routine.* This module translates the lecture obtained by the encoder located in the arm articulation to digital values in order to generate the angle of the arm. This angle is sent to the mobile application using the previous Bluetooth connection.

3.3. Main Blocks of the Mobile App. The main blocks of the mobile application are shown in Figure 3. The details of each block are described below:

(i) *Wireless connection module (WCM).* This module must allow users to detect if any compatible robot arm Bluetooth connection is available. If it is the first time connection, this module asks the user for linking the device with the platform. Once the platform has been linked, the module is ready to get data from the platform to the application, activating the rest of the application functionalities.

(ii) *Image acquisition module (IAM).* This module serves as a bridge between the physical sensor and the application. The IAM accesses the camera buffer for accessing the image obtained by the camera sensor.

(iii) *Marker localization module (MLM).* This module obtains the input video from the IAM and performs the marker localization using the previously obtained data in the training phase for the specific markers located in the platform. Once the marker has been located, then the obtained coordinates are stored and sent to the ARIM.

(iv) *Articulation degree module (ADM).* This module performs the articulation degree requests depending on the located marker by the MLM. The WCM returns the angles for each located marker and passes to the ARIM.

(v) *Augmented reality integrator module (ARIM).* This module takes as input the dimensional localization of the marker obtained by the MLM and the degree of each marker obtained by the ADM and generates the final image where the protractor with the obtained angles is overlaid on the image obtained by the IAM. Finally, this image is shown to the user on the device screen.

4. Experimental Setups and Results

4.1. Hardware Tools

(i) Two 12 V DC motors with optical encoders.

(ii) A L298N dual H-bridge motor driver.

(iii) A 12 V 10 A power supply.

(iv) A HC-05 Bluetooth module for communicating Arduino via Bluetooth communication with the AR application.

(v) An Arduino ONE device for hosting the communication modules.

(vi) Homemade robotic arm designed using Solid-Works® and built using material form local providers. The SolidWorks design is shown in Figure 4(a), and the final design is shown in Figure 4(b).

(vii) A mobile device with back camera and Bluetooth adapter. In this device, the developed application is deployed, and several tests were performed in order to validate the localization of the markers and

image visualization. The proposed application has been validated on devices with several Android versions. In Table 1, a list of devices used and its specifications used for testing the proposed application are shown. The application worked without problems in the listed devices.

In Figure 5, connections required among the Arduino board, the driver, and the power source are shown. The Arduino board hosts the control module and sends commands to the motor driver, and each motor is connected to the dual motor driver. The Arduino board receives commands for the robot arm from the desktop application through USB connection, and the angles of each arm articulation are sent to the mobile application using Bluetooth USB connection.

4.2. Software Tools. The desktop application was developed using Matlab and Simulink, on a desktop computer with Windows OS.

The mobile application was developed using Android Studio IDE with Android Studio SDK and Java SE Development Kit. The modules of the application were written in Java. In addition, ARToolkit was used for the marker recognition and integration of virtual and real objects in live capture obtained by the camera sensor of the device [15].

4.3. Designed Application. In Figure 6, the base (Figure 6(a)) and arm (Figure 6(b)) markers are shown. These markers were used for training the ARK toolkit classification model for its recognition.

In Figure 7, the desktop application and the mobile application are shown.

In Figure 8, two frames of the obtained angles of each articulation and its visualization in the APP are shown. From the obtained images, it is possible to conclude that the recognition of each marker is performed successfully, and the display of the protractor with the obtained angles is visible in each of the tested angles.

4.4. Time and Angle Results. In order to verify the alignment precision of the two axes (base and shoulder), measurements were carried out using Matlab. Figure 9(a) shows results from base angle trajectories, where the position error ranges from 0.75° to 3.8°. In the case of shoulder axis (Figure 9(b)), variations are from 0.08° to 7.28°. These variations could be reduced by using some control strategies. With respect to time lag between sending the instructions and the display of the response, a delay from 1 to 2 seconds was observed in the experiments.

5. Final Recalls and Future Work

In this paper, an approach for teaching robotic arm manipulator concepts was proposed. The system includes a homemade robotic arm, a control system, and the RAR@pp. The RAR@pp is focused on learning robotic arm manipulation algorithms by the detection of markers in the

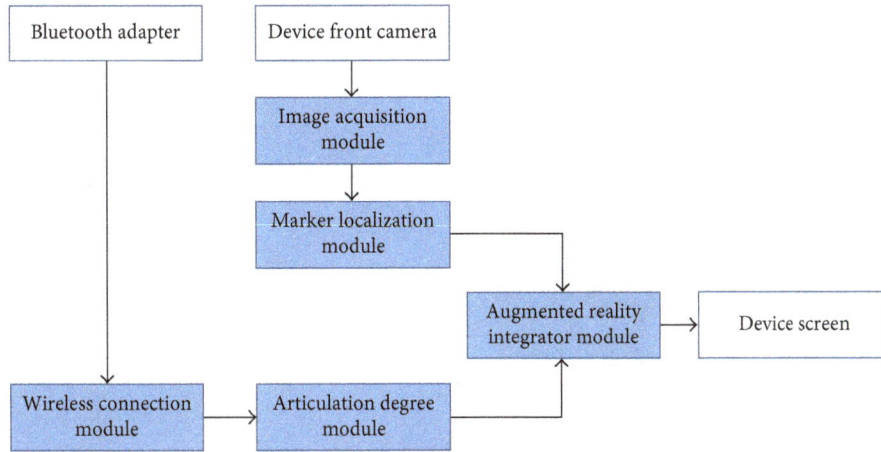

FIGURE 3: Main blocks of the mobile application.

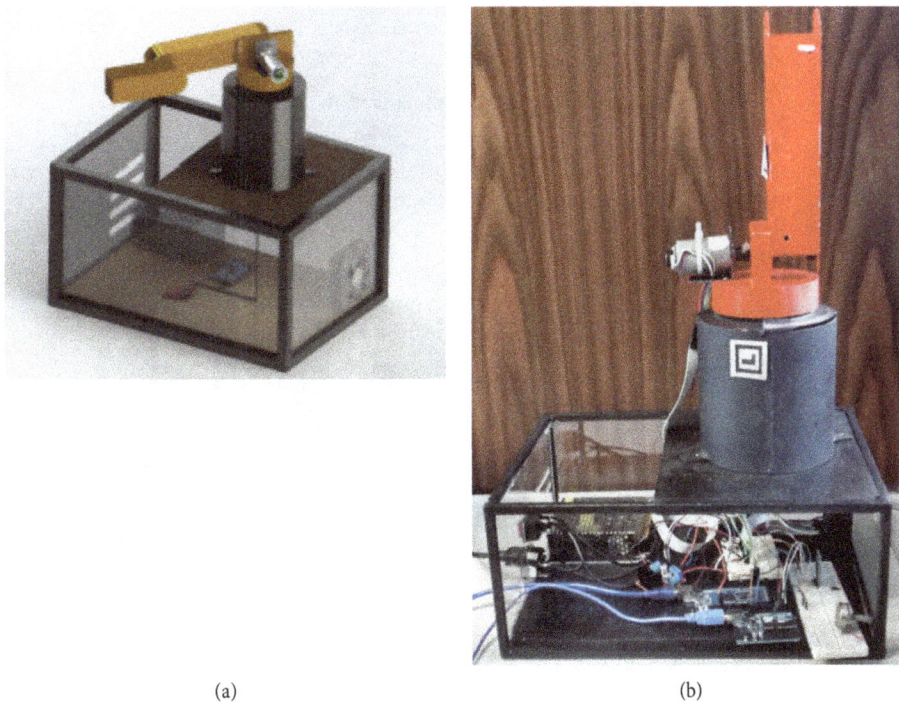

(a)

(b)

FIGURE 4: Design and realization of the robotic arm: (a) robotic arm design with SolidWorks; (b) robotic arm prototype.

TABLE 1: Devices utilized for testing the proposed application.

Device	Processor	RAM	Android version
Galaxy S2	Dual-core 1.2 GHz Cortex-A9	1 GB	4.1
Galaxy S4	Dual-core 1.7 GHz Krait 300	1.5 GB	5.0.1
Polaroid Tab	Dual-core 1.0 GHz Broadcom 21663	1 GB	4.2.2
Galaxy Tab 4	Quad-core 1.2 GHz Marvell PXA1088	3 GB	5.0.2
Galaxy Tab 10.1	Quad-core 2.3 GHz Krait 400	3 GB	5.1.1
LG G3 Stylus	Quad-core 1.3 GHz Cortex-A7	1 GB	5.0.2
Motorola Moto G	Quad-core 1.4 GHz Cortex-A53	1 GB	5.1.1

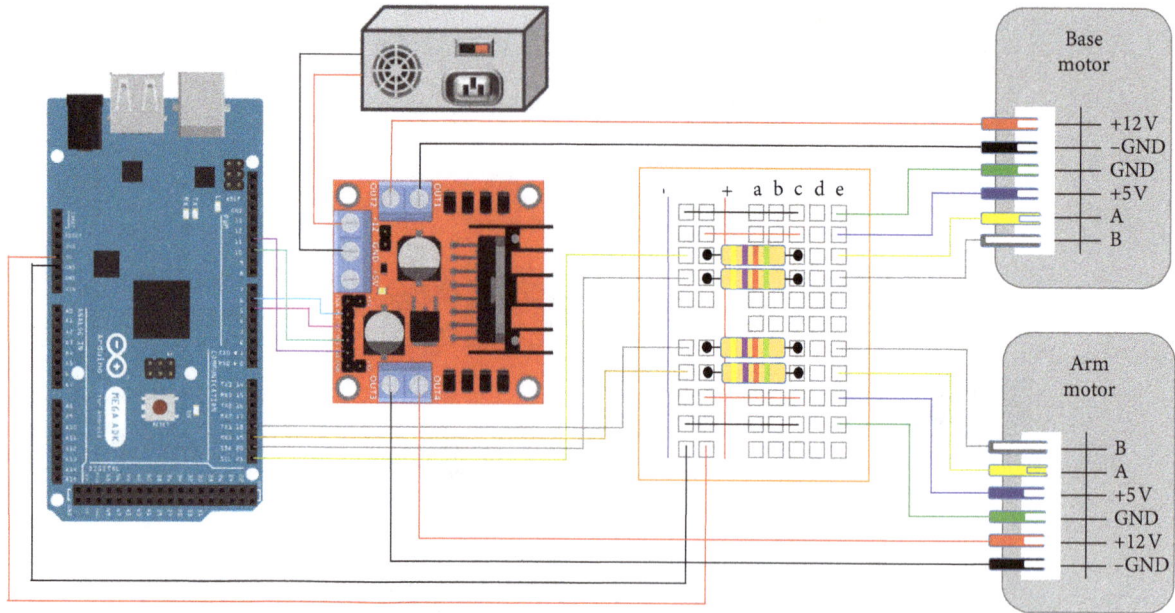

FIGURE 5: Integration of the components of the control module.

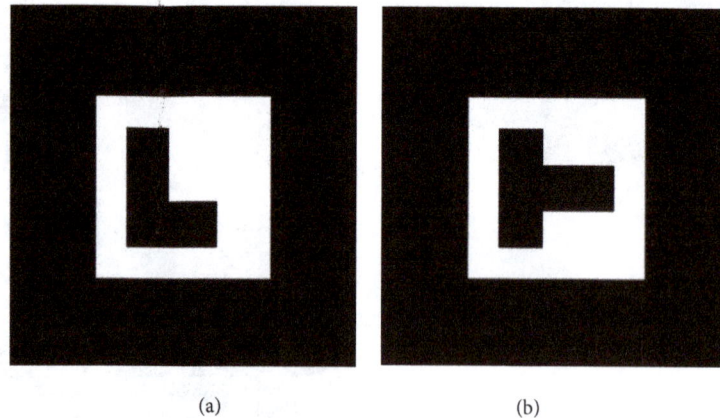

(a)   (b)

FIGURE 6: Marker for the robot arm articulations and localization of each market by the AR demo: (a) base articulation marker; (b) arm articulation marker.

robotic arm and the real-time visualization of the angles of each articulation of the homemade robotic arm using augmented reality. These angles are obtained from the robot based on the data obtained by the encoders of each motor and transmitted to the RAR@pp using Bluetooth communication. This application allows students to send configuration parameters to the robotic arm and visualize in real time the response of the arm to the commands using the RAR@pp. The application was tested in a large number of mobile devices, including smartphones and tablet devices.

The proposed platform allows capturing the student's attention, facilitating the understanding of complex robotics and kinematics concepts. The implementation of a smartphone-based application will allow a large number of students to access to this type of educational resources,

improving their performance and their understanding of key concepts of robotics and kinematics.

In this work, basic techniques combining augmented reality have been applied for the visualization of the articulation arm angles. Future work could be oriented to replace the use of markers for the direct recognition of the shape of the object for AR applications. The direct object recognition capabilities could be used for working with different types of arms and for larger degrees of freedom. Moreover, several sensors could be added to the robotic platform in order to expand its capabilities. For example, current sensors could be used to visualize the effect of torque in the axis motors. Implementation of pressure sensors in an end effector or inertial sensor to study the device behavior could also been implemented. It also can be added to the RAR@pp the

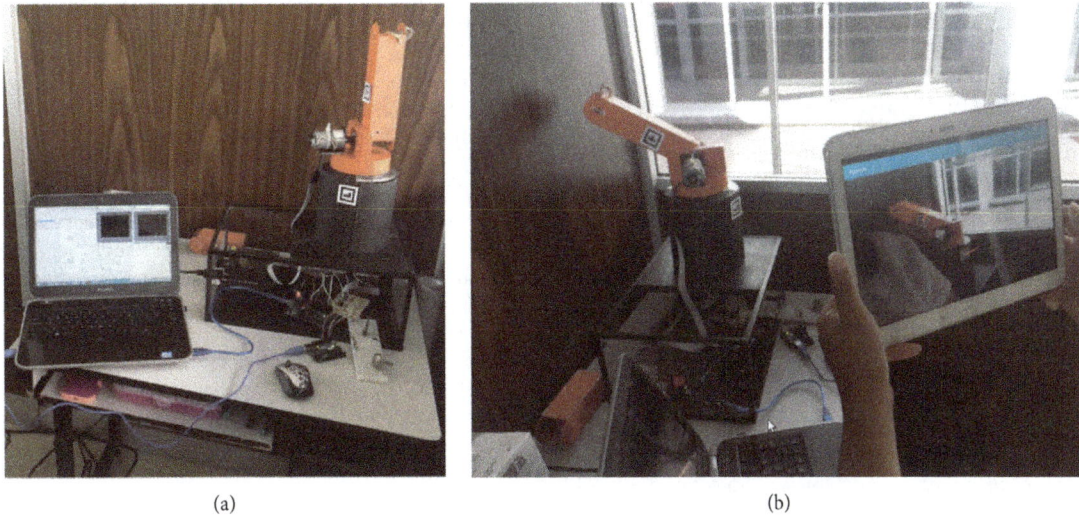

(a) (b)

FIGURE 7: Integration of the components of the proposed system: (a) robotic arm and desktop application for monitoring; (b) mobile application and its interaction with the robotic arm.

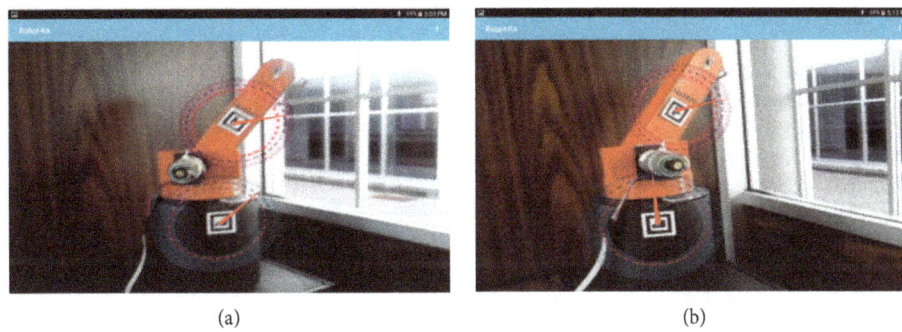

(a) (b)

FIGURE 8: Robotic arm AR interface: (a) initial position; (b) real-time angle measurements.

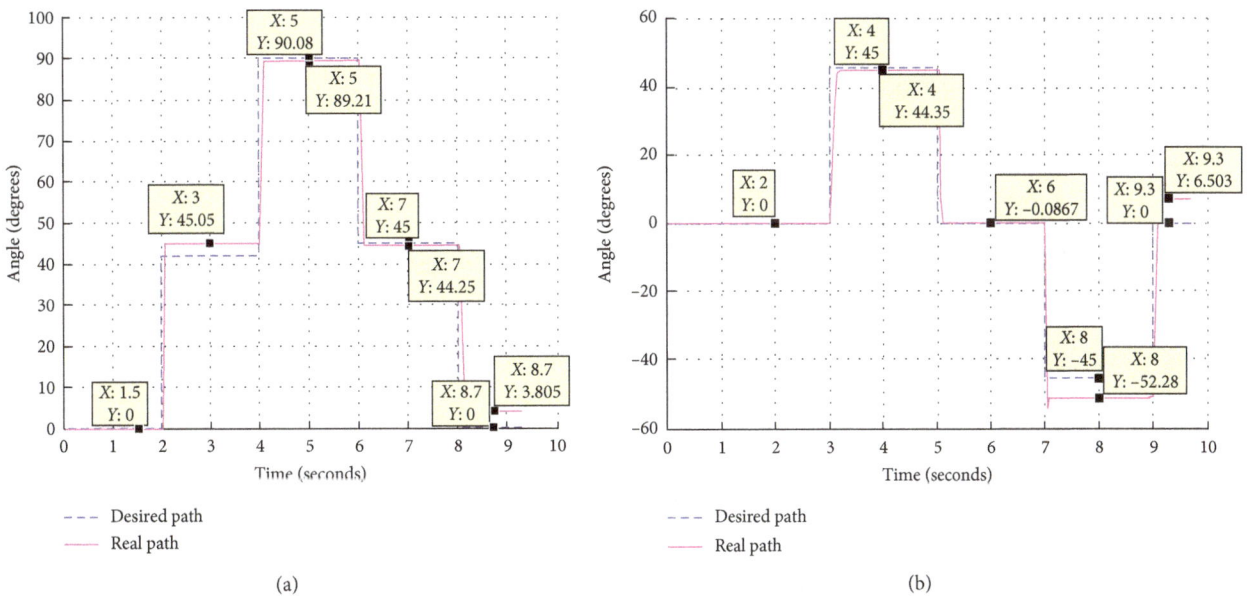

(a) (b)

FIGURE 9: Real and desired paths for robotic arm axes: (a) base axis; (b) shoulder axis.

capability to display information about motor's temperature and the measurement of execution time for a specific task.

Acknowledgments

The present project was partially funded by the National Council of Science and Technology of Mexico through a scholarship granted to Karla E. Bautista Hernández.

References

[1] G. Akçayır, H. M. Pektaş, and M. A. Ocak, "Augmented reality in science laboratories," *Computers in Human Behavior*, vol. 57, pp. 334–342, 2016.

[2] M. Flanders and R. C. Kavanagh, "Build-a-robot: using virtual reality to visualize the Denavit-Hartenberg parameters," *Computer Applications in Engineering Education*, vol. 23, no. 6, pp. 846–853, 2015.

[3] M. Stilman, P. Michel, J. Chestnutt, K. Nishiwaki, S. Kagami, and J. Kuffner, "Augmented reality for robot development and experimentation," Technical Report, Robotics Institute, Carnegie Mellon University, Pittsburgh, PA, USA, 2005.

[4] X. Wei, D. Weng, Y. Liu, and Y. Wang, "Teaching based on augmented reality for a technical creative design course," *Computers and Education*, vol. 81, pp. 221–234, 2015.

[5] W. Tarng, Y.-S. Lin, C.-P. Lin, and K.-L. Ou, "Development of a lunar-phase observation system based on augmented reality and mobile learning technologies," *Mobile Information Systems*, vol. 2016, Article ID 8352791, 12 pages, 2016.

[6] F. Quercioli, "Augmented reality in laser laboratories," *Optics & Laser Technology*, vol. 101, pp. 25–29, 2018.

[7] F. Clemente, S. Dosen, L. Lonini, M. Markovic, D. Farina, and C. Cipriani, "Humans can integrate augmented reality feedback in their sensorimotor control of a robotic hand," *IEEE Transactions on Human-Machine Systems*, vol. 47, no. 4, pp. 583–589, 2015.

[8] L. Yang, C.-K. Chui, and S. Chang, "Design and development of an augmented reality robotic system for large tumor ablation," *International Journal of Virtual Reality*, vol. 8, no. 1, pp. 27–35, 2015.

[9] C. A. Jara, F. A. Candelas-Herías, M. Fernández, and F. Torres, "An Augmented Reality Interface for Training Robotics through the Web," in *Proceedings of the 40th International Symposium on Robotics*, Barcelona, Spain, 2009.

[10] M. B. Ibáñez, Á. D. Serio, D. Villarán, and C. D. Kloos, "Experimenting with electromagnetism using augmented reality: impact on flow student experience and educational effectiveness," *Computers & Education*, vol. 71, pp. 1–13, 2014.

[11] J. Martín-Gutiérrez, P. Fabiani, W. Benesova, M. D. Meneses, and C. E. Mora, "Augmented reality to promote collaborative and autonomous learning in higher education," *Computers in Human Behavior*, vol. 51, pp. 752–761, 2015.

[12] D. Ni, A. W. W. Yew, S. K. Ong, and A. Y. C. Nee, "Haptic and visual augmented reality interface for programming welding robots," *Advances in Manufacturing*, vol. 5, no. 3, pp. 191–198, 2017.

[13] T. H. J. Collett and B. A. MacDonald, "An augmented reality debugging system for mobile robot software engineers," *Journal of Software Engineering for Robotics*, vol. 1, no. 1, pp. 18–32, 2009.

[14] K. Ruan and H. Jeong, *An Augmented Reality System Using QR Code as Marker in Android Smartphone*, IEEE, Piscataway, NJ, USA, 2012.

[15] H. Kato, "ARtoolkit," 2018, http://www.hitl.washington.edu/artoolkit/.

[16] T.-W. Kan, C.-H. Teng, and W.-S. Chou, "Applying QR code in augmented reality applications," in *Proceedings of the 8th International Conference on Virtual Reality Continuum and Its Applications in Industry (VRCAI'09)*, pp. 253–257, Yokohama, Japan, 2009.

[17] J.-C. Chien, H.-Y. Lu, Y.-S. Wu, and L.-C. Liu, "ARtoolkit-based augmented reality system with integrated 1-d barcode: combining colorful markers with remote servers of 3D data for product promotion purposes," in *Proceedings of the Second International Conference on Computational Collective Intelligence: Technologies and Application (ICCCI'10)*, vol. 3, pp. 200–209, Kaohsiung, Taiwan, 2010.

[18] D. Yan and H. Hu, "Application of augmented reality and robotic technology in broadcasting: a survey," *Robotics*, vol. 6, no. 3, p. 18, 2017.

12

VR-CPES: A Novel Cyber-Physical Education Systems for Interactive VR Services based on a Mobile Platform

Hanjin Kim ⓘD, **Heonyeop Shin** ⓘD, **Hyeong-su Kim** ⓘD, **and Won-Tae Kim** ⓘD

Smart CPS Lab, The Department of Computer Science & Engineering, KOREATECH University, Cheonan, Republic of Korea

Correspondence should be addressed to Won-Tae Kim; wtkim@koreatech.ac.kr

Academic Editor: Byeong-Seok Shin

The evolution of virtual reality technology allows users to immerse themselves into virtual environments, providing a new experience that is impossible in the real world. The appearance of cyber-physical systems and the Internet of things makes humans to understand and control the real world in detail. The integration of virtual reality into cyber-physical systems and the Internet of things may induce innovative education services in the near future. In this paper, we propose a novel, a virtual reality-based cyber-physical education system for efficient education in a virtual reality on a mobile platform, called VR-CPES. VR-CPES can integrate the real world into virtual reality using cyber-physical systems technology, especially using digital twin. We extract essential service requirements of VR-CPES in terms of delay time in the virtual reality service layer. In order to satisfy the requirements of the network layer, we design a new, real-time network technology interworking software, defined as network and time-sensitive network. A gateway function for the interworking is developed to make protocol level transparency. In addition, a path selection algorithm is proposed to make flexible flow between physical things and cyber things. Finally, a simulation study will be conducted to validate the functionalities and performance in terms of packet loss and delay as defined in the requirements.

1. Introduction

Internet of things (IoT) is a new paradigm that plugs things into the Internet in order to provide various intelligent services to the users. Most of IoT services, such as air conditioning and lighting management, detect the current states of things and environments through sensors and make reasonable decisions based on the data for the specific situations. As the number and functions of the things increase, the relationship between the things has become more complicated. This complexity makes it difficult to monitor and manage the data of things precisely, which leads to difficulties in proper control of things.

Cyber-physical systems (CPS), or *digital twin*, may be a solution for the problem [1]. The digital twin is considered an element of CPS because it means the mirror image of a physical thing. In this paper, we use the term *cyber thing* as the digital twin. CPS has some aspects of IoT in terms of using data of physical things collected through the Internet. The control devices, mechanical body, and corresponding cyber things are connected to fully operate the corresponding physical systems [2]. By virtue of the tight interworking between cyber things and physical things, CPS can be used for many critical applications, including smart city, smart factory, and smart grid, which require accurate analysis and control functionalities based on physical systems and the environments [3–5]. Since cyber things inherit the exact features and states of the physical things, they can be used for accurate estimation, prediction, and control of the physical things by means of dynamics simulations and the other computational means.

The emergence of *virtual reality* (VR) technologies makes it possible to provide services by immersion into computer-generating virtual environments [6]. Currently, most of the VR services are provided in the virtual environment that has no direct relations with the physical environment. Although *augmented reality* (AR) has more information of the real world than VR, it just overlays cyber things or cyber information on the physical things or the physical space. If the cyber thing or space is actually connected to the physical thing or space, users in the cyber space can vividly experience more immersive VR/AR services. Therefore, the integration of CPS and VR/AR empowers the legacy VR/AR services to dramatically enhance the reality and the interactivity with real world [7]. In this paper, we propose a new type of education service called virtual reality-based cyber-physical education system (VR-CPES) to support the VR/AR education services based on CPS.

1.1. Scenario. We introduce a new driver-training service scenario in order to explain, in detail, what VR-CPES can provide for efficient training in wireless/mobile environments. The VR-CPES services are executed on a mobile device, which is a sort of the VR device. A comparison between the advantages of the services implemented in VR-CPES and the existing services execution environments is shown in Table 1.

Figure 1 shows the scenario and architecture of VR-CPES driver-training service. VR-CPES driver training is very safe from risks, such as vehicle accidents. In addition, the constraints of the training caused by the lack of equipment and space are eliminated by virtue of the mobile platform. Subsequently, users can practice in a variety of driving environments, such as where they wanted to practice and where the steering-wheel position changes according to the country. On the physical side, using actual data allows user to practice in a realistic environment because conditions of the vehicle are immediately reflected by temperature, weather, and time changes in the virtual environment. The services listed above require not only generation of the virtual environment, but also interworking of cyber-physical environments. Therefore, the VR-CPES mobile platform needs to satisfy the following requirements.

1.2. Requirements. In this section, we suggest the following requirements to solve the problems shown in the scenario. In VR/AR education services, physical things must be represented in cyber space as cyber things, and the two assets must be interworking. As mentioned above, the cyber things are called the digital twin. A digital twin is a digital replica of a physical thing that conducts all of its functionalities [8]. For example, vehicles, traffic signs, and even people in physical space can be reflected by the digital twin in cyber space. In the general IoT service, the physical things usually gather environment data and use them in mobile devices. However, the physical things in VR-CPES use digital twin models so that monitoring status and management are improved compared with physical things in general IoT [9, 10].

Because VR-CPES uses a digital twin in cyber space, it needs to generate and manage various digital twin models in cyber space, according to various physical spaces. An IoT platform considering a digital twin needs to manage resources of things based on time because physical things should be reflected by the digital twin in real time. A general IoT platform does not sufficiently support such time-based resource management. Moreover, the IoT platform for VR-CPES must consider cloud and edge computing, which is

responsible for generating, simulating, and analyzing the digital twin because a single mobile device cannot manage the uncountable resources of the physical thing and digital twin, respectively. In other words, the *quality of experience* (QoE) of a user in VR-CPES is totally related to the interworking of the physical thing and digital twin.

Reliability of VR-CPES increases as the virtual environment and digital twin become more reliable. The reliability of the virtual environment and digital twin is closely related to the seamless interworking of cyber-physical environments. It also depends on a dependable transmission of physical space and physical-thing data. Table 2 shows the network *quality of service* (QoS) requirements for data which is transmitted in VR-CPES. The VR-CPES needs to satisfy QoS requirements for all applications in the cyber-physical environments. Since the main things in our scenario are vehicles and VR devices, we investigated the traffic class based on this [11–13].

In this paper, we propose a novel VR-CPES platform to provide virtual reality contents for efficient education in the mobile environment, as mentioned in the above scenario, and a network framework that can satisfy the requirements for all contents. The remainder of the paper is organized as follows: in Section 2, we summarize IoT platform, SDN, and TSN and describe insufficiency of each technology for the VR-CPES service; in Section 3, we propose the VR-CPES platform by describing each component of the platform and the real-time network framework for seamless VR-CPES service; Section 4 verifies the performance of the proposed real-time network framework; finally, Section 5 concludes this paper.

2. Related Works

2.1. IoT Platform. In order to apply IoT technologies to various domains, such as smart city and health care, IoT platforms are being developed that can adaptively operate in environments where many things need to be monitored and managed [14–16]. In standard organizations, such as oneM2M and OCF design, IoT has standards for the standardized operation of platforms, and these standard-based platforms facilitate the connection of different IoT devices and resource management [17, 18]. Various commercial platforms, such as GE's Predix platform and IBM's Watson IoT platform, are developed as platforms considering the digital twin [19, 20]. They support IoT services for the digital twin using Predix Machine, Asset, and Machine Data Analytics. In addition, multiple technologies have incorporated an IoT platform to replicate physical assets of a digital twin and utilize them [2]. Due to the characteristic of the digital twin that is a digital replica of physical assets, the IoT platforms for the digital twin must consider communication network for seamless data transmission of cyber and physical things.

2.2. Software-Defined Networking. As mentioned above, reliability of VR-CPES is based on dependable data transmission, according to the dynamically changing interconnection of cyber-physical things. *Software-defined*

TABLE 1: Comparison of driver-training services.

Conditions	Real training	VR simulator	VR-CPES driver training
Safety under training	Low	High	High
Volume of training space	Large	Small	Small
Diversity of training scenario	Low	Middle	High
Reality of training	High	Low	High

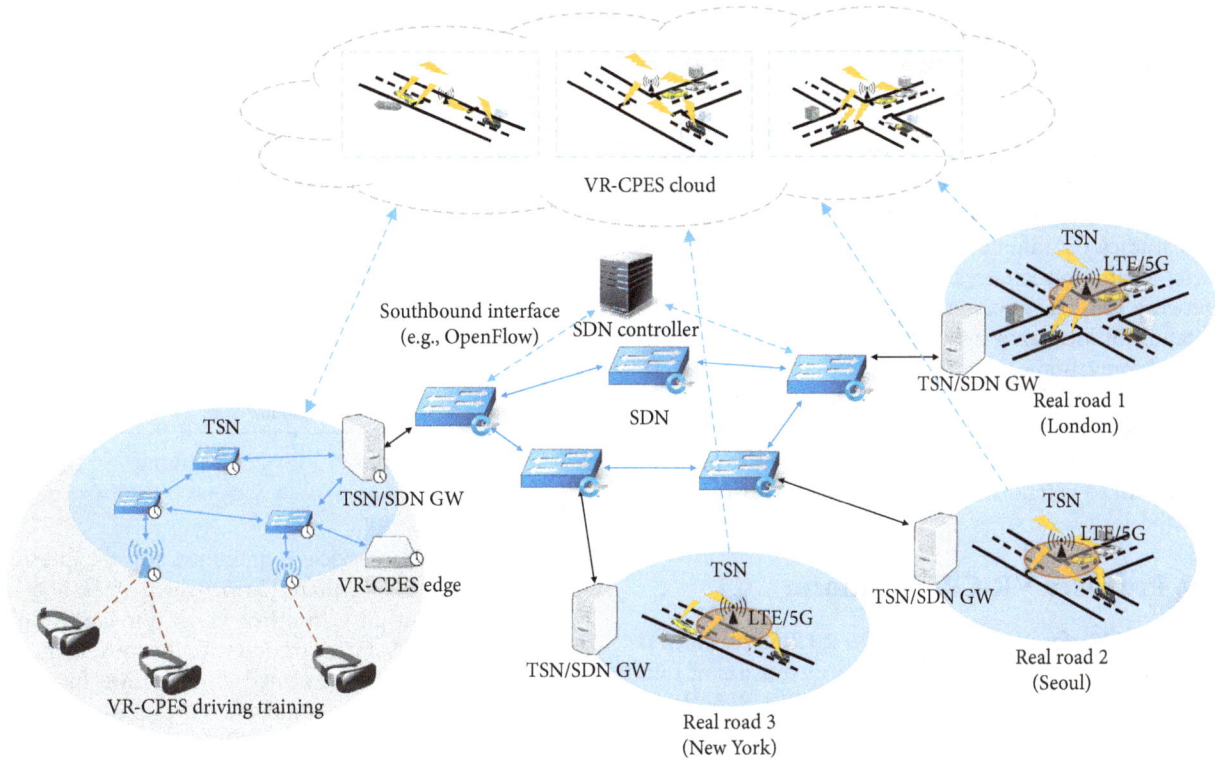

FIGURE 1: VR-CPES driver-training service.

networking (SDN) is a centralized network technology with a global network view that allows quick and dynamic configuration of the network, depending on the complex quality of service (QoS) of various network applications [21]. The SDN separates the control plane, which determines the network policy, from the data plane, which forwards the actual data. This decoupling of planes abstracts low-level network functionality and higher level service, which facilitates programmable network configuration. This feature is suitable for a large-scale IoT, which, for a network to be configured adaptively, depends on the different network requirements of many things [22]. In the SDN, the control plane receives network information of the data plane from a southbound interface, and the OpenFlow protocol is used dominantly for this operation [23]. The OpenFlow switch can identify all of the layer 2, layer 3, and layer 4 (L2–L4) protocols when configuring the path (i.e., the flow), so the SDN has the advantage of flow configuration in the reconfigurable and heterogeneous network condition [24]. However, it is difficult to calculate proper data transmission paths in systems where control data need to be transmitted with critical network requirements, such as CPS. Various studies have been conducted to extend SDN suitable for these systems [25]. In the

network for VR-CPES, it is necessary to transfer data depending on the QoS requirements of different applications, which are simulation for digital twin and VR rendering. In addition, time-based data transmission should be considered for seamless operation of a digital twin and physical thing.

2.3. Time-Sensitive Networking. Time-sensitive networking (TSN) is a set of standards under development by the Time-Sensitive Networking Task Group which is part of the IEEE 802.1 Working Group [26]. In the TSN standard, various technologies, such as time synchronization and frame preemption, are being developed to provide deterministic network service of low latency and low packet loss in ethernet. The predecessor of the TSN standard technology is the audio-video bridging (AVB), which started in the media industry to transmit high-quality audio and video data in real time using a standardized-ethernet network. Subsequently, there has been a requirement to transfer control/management data through the ethernet network in industry domains, such as automotive and factory automation. The standard is being extended to develop a reliable network based on time synchronization for each industry's

TABLE 2: QoS requirements for applications in VR-CPES.

Traffic class	Bandwidth	End-to-end delay	Jitter
Control data (vehicle)	Low	2.5 ms	Sub-microsecond
Safety data (vehicle)	Medium	45 ms	Sub-microsecond
Infotainment data (vehicle)	High	150 ms	100 ms
Video data (VR)	80 k~50 Mbps	<100~500 ms	<50~150 ms
Audio data (VR)	10~80 kbps	<100 ms	<100 ms

domain network. With the motivations developed to cover the requirements of various industries, TSN is utilized as a network technology in a system that requires transmission of critical control data and high-bandwidth data [27]. In addition, deterministic network technologies of TSN are being applied in the mobile network environment. In particular, the development of a standard for a fronthaul and various studies for mobile networks is in progress to meet the strict reliable low latency communication requirement of 5G network [28]. However, until now, TSN standards have a limitation of local area network due to problems such as time synchronization accuracy. In order to solve it, the IETF is in the process of extending the TSN to the wide-area network under the name DetNet [29], and the VR-CPES service also requires a dynamic and large-scale network environment, so extended technologies of the existing TSN are needed.

3. The Proposed VR-CPES Platform

3.1. VR-CPES Platform. The proposed VR-CPES platform is shown in Figure 2. The VR-CPES platform consists of four components: VR device, VR-CPES edge, VR-CPES cloud, and physical things. Each component transmits data through a TSN/SDN real-time network framework, which is described in Section 3.2. For the ease of explanation, we describe the operation of each component of the VR-CPES platform based on the driver-training service scenario in the introduction, as shown in Figure 3.

3.1.1. VR Device. The VR-CPES service should operate by interworking the corresponding vehicle according to the user's operation and the digital twins, which are reflected in the virtual environment from physical space. For this reason, in the VR device, it is very difficult to calculate the results for rendering—such as the perspective view associated with each digital twin according to the user's action—in order to immerse the user in the virtual environment. Therefore, an application that requires high-computing resources, such as a simulation, is operated in the VR-CPES edge and cloud. The VR device transmits only the user's behavior and eye tracking as sensor data and receives external-calculated data of digital twins and the virtual environment and then renders it.

3.1.2. VR-CPES Edge. The VR-CPES edge is a subsimulator for each VR device (i.e., each user). It estimates the commands of the user for the user's digital twin vehicle (DTV), based on the data received from the VR device. Since the

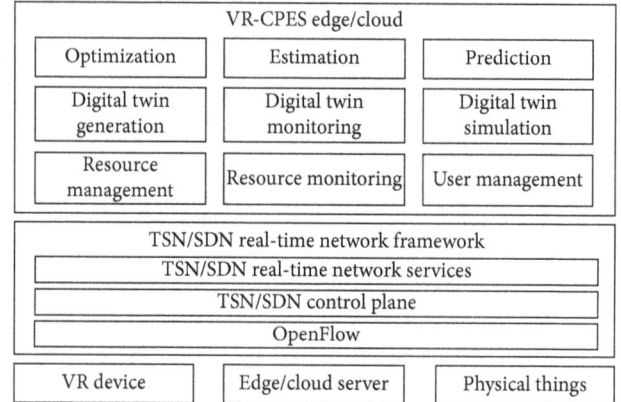

FIGURE 2: VR-CPES platform.

VR-CPES training service is intended for the person, it should operate at less than 13 ms, which does not affect human perception [30]. In other words, the control of the corresponding DTV interworking with the movement of the user, which should be executed in real time for the VR experience, needs to be simulated on the VR-CPES edge rather than the cloud. The VR-CPES edge performs simulation by receiving external environment data and previously executed simulation data for the scenario selected by the user in the VR-CPES cloud. The result data of simulation are transmitted back to the VR-CPES cloud and updated for analysis of DTV.

3.1.3. VR-CPES Cloud. The VR-CPES cloud is the main simulator of the VR-CPES platform and performs functions such as management and monitoring of VR-CPES things. In the cloud simulator, virtual environment models and digital twin models are generated based on data collected from physical space. In the VR-CPES cloud, unlike the VR-CPES edge, which considers only the user's DTV, the environment model and other digital twin models must interact with each other in the simulator. Therefore, simulation should be performed in a distributed cosimulation environment, rather than a single simulation environment [31]. Digital twin models are designed and interpreted based on the large amount of physical-space data, so that they can not only optimize the states of the corresponding things, but estimate and predict certain situations for appropriate control. This series of processes requires various analytical techniques, such as stochastic approach, machine learning, AI, and appropriate control theory. The VR-CPES cloud stores these simulated results of each situation, and the data are used as

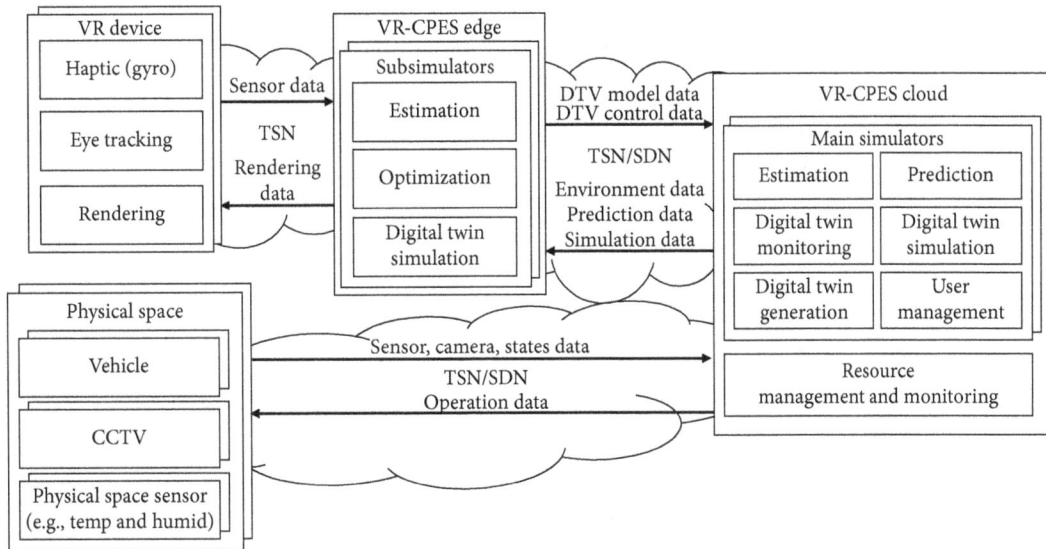

FIGURE 3: Structure of VR-CPES platform.

external environment input data of the DTV model in VR-CPES edge's subsimulator.

3.1.4. Physical Space. Physical space is a real physical space interworked with virtual reality environments. In a scenario, they are real road environments such as New York, Seoul, and London. Data of physical things that occur in the physical space are transmitted to the VR-CPES cloud and used to generate virtual space. Each physical thing must be utilized as digital twins, not only in the monitoring, but in the simulator. Therefore, unlike legacy IoT platform, real-time data transmission must be considered in VR-CPES platform, and this is supported by TSN/SDN real-time network framework.

3.2. TSN/SDN Real-Time Network Framework (TSRTnet). As mentioned above, the service of the VR-CPES platform requires real-time data transmission for VR experience of users. It can be achieved, not only by resource management according to the QoS of application level in each platform component, but by supporting data transmission according to the QoS of network level. TSN, a standard for real-time data transmission in ethernet networks, transmits data in frame units based on priority. High-priority frames are transmitted in further advance than low-priority frames through a TSN switch, and there are several algorithms to determine how ethernet frames are processed and how priority is assigned. The frames, which require hard real-time data transmission, such as physical states data in the VR-CPES service, are processed based on a time-aware shaper algorithm in TSN [32]. This time-based, scheduled-traffic transmission requires time synchronization between the nodes, which configure the network. The Precision Time Protocol (PTP) is used for time synchronization between nodes under sub-microsecond accuracy by means of hardware timestamps and compensating the time offset considering link delay [33]. In addition, the TSN

standards are extended for requirement of time-sensitive applications such as a multiple time domain [34].

The main application for the VR-CPES service runs on the VR-CPES cloud, not on the VR device, which is physically separated from the VR device runtime environment. It means the local network needs to extend to a large-scale network for the user's experience. TSN can guarantee time-sensitive data transmission, but it is limited to less than 7 hops switched network due to the inaccuracy of time synchronization. Although the TSN standard has recently been applied to the factory network considering the connection between TSN nodes based on the centralized network configuration concept in the industrial network, it is still not enough to satisfy the requirements of VR-CPES service [35]. Because it is difficult to practically configure the entire network with homogeneous TSN, we suggest TSN/SDN real-time network framework to plug multiple TSNs into SDN core network.

The structure of TSN/SDN real-time network framework (TSRTnet) for the VR-CPES platform is shown in Figure 4. The architecture consists of TSN/SDN gateways, an SDN controller, and OpenFlow switches. Each TSN is connected to an external SDN through the TSN/SDN gateway, which is managed by the SDN controller.

3.2.1. TSN/SDN Gateway. The TSN/SDN gateway is responsible for connecting the local TSN to the SDN backbone network. In this paper, we assume that the TSN/SDN gateway is capable of processing OpenFlow protocol to communicate with the SDN controller. The TSN nodes register and reserve time-sensitive streams for communication as talker and listener using Stream-Reservation Protocol (SRP). The SRP parameters used for TSN stream management requiring guaranteed QoS are shown in Table 3 [36]. The TSN nodes, such as physical things and the VR device, which need to be communicated with the VR-CPES cloud through SDN, reserve the TSN streams to the

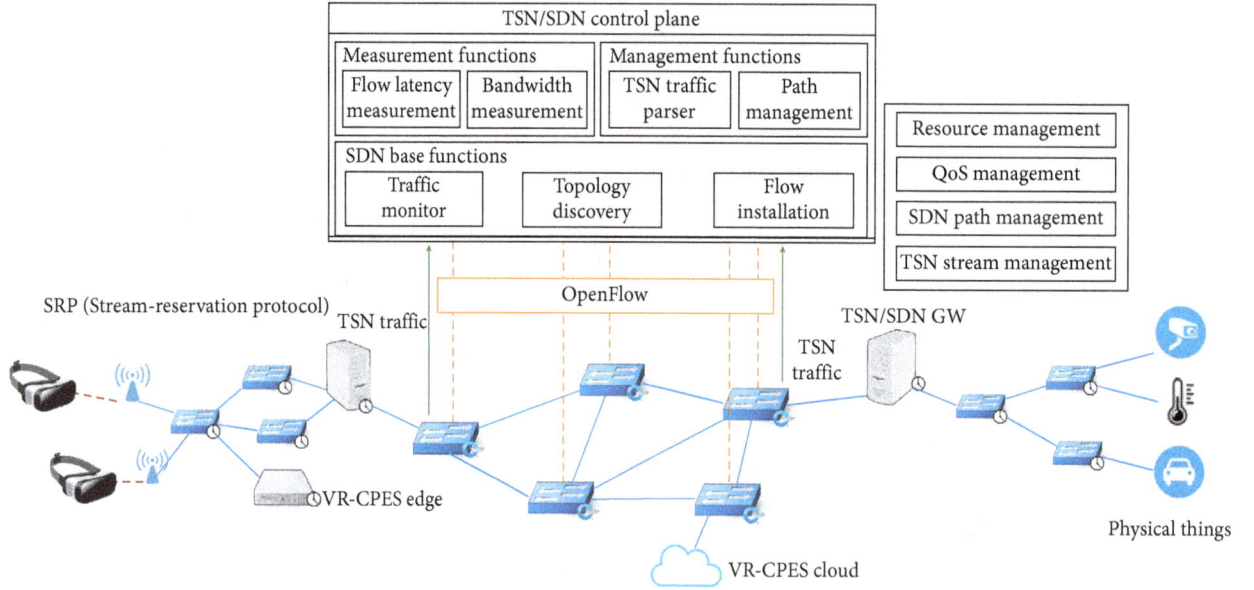

FIGURE 4: TSN/SDN real-time network framework.

TABLE 3: Stream-Reservation Protocol parameters.

StreamID	Stream identifier
Stream destination address	Destination address
Stream VLAN ID	VLAN identifier
MaxFrameSize	Traffic specification associated with a stream
MaxIntervalFrames	Traffic specification associated with a stream
Data frame priority	Priority value
AccumulatedLatency	Worst-case latency that a stream can encounter

TSN/SDN gateway by SRP. The TSN/SDN gateway encapsulates the SRP message of the connected node into a packet destined for the VR-CPES cloud. This packet is sent as a packet-in message to the SDN controller, and the SDN controller sets the flow of the SDN by parsing the SRP contained in this message.

3.2.2. SDN Controller. The SDN controller parses the packet-in message received from the TSN/SDN gateway and configures a path to communicate with VR-CPES cloud. The SDN controller has three base functions: traffic monitor, topology discovery, and flow installation. These functions set flows for configuring the path [37]. As the name implies, each function is responsible for monitoring traffic of the data plane, discovering network topology, and installing flow for packet forwarding of the data plane, respectively. Additionally, we designed four SDN functions to forward time-sensitive traffic to SDN core network:

(1) *Flow latency measurement.* This function measures the flow latency of each link to establish a network path according to a requirement of the application's end-to-end delay. The SDN controller has a global view of each flow latency through the latency-measurement function [38]. In this function, the SDN controller measures the latency using Internet Control Message Protocol (ICMP) every second.

(2) *Bandwidth measurement.* This function uses the bandwidth-measurement function to measure available bandwidth (ABW) of i-th link in SDN. Available bandwidth is calculated by the amount of packets forwarding to SDN flow in specific time. The amount of forwarded packets is sent to the controller through FlowStates message, and then the controller calculates available bandwidth using the period time T when the controller receives FlowStates message:

$$b_i(t) = \frac{count_i(t) - count_i(t - T)}{T},$$
$$ABW_i = c_i(1 - b_i). \quad (1)$$

(3) *TSN traffic parser.* This function is parsing the time-sensitive traffic, which is received from TSN/SDN gateway by OpenFlow packet-in message. The SDN controller parses the QoS of applications by checking the SRP parameters and VLAN header. These data are used as QoS information of applications to set a path in a path management function.

(4) *Path management.* Path management function sets path from source to destination based on the calculated latency and bandwidth of flows by measurement functions. Algorithm 1 shows the TSN/SDN path selection algorithm for selecting the candidate path from source to destination. Subsequently, the final path is selected from among the selected paths from Algorithm 1 using a path configuration flowchart. The TSN/SDN path configuration flowchart is shown in Figure 5.

```
       Input: N-Network Topology
              S-Source IP
              D-Destination IP
       Output: SP-Selected Path List
  (1) function PathSelection(N, S, D)
  (2)     for i = 1 to A_i do
  (3)         P ⟵ DelayCostShortest(N, S, D)
  (4)         DeleteList(P, 1)
  (5)         DeleteList(P, P_length)
  (6)         for i = 1 to P_length do
  (7)             RemoveNode(N, P[i])
  (8)         end for
  (9)         SP[i] ⟵ sort(P)
 (10)     end for
 (11)     return SP
 (12) end function
```

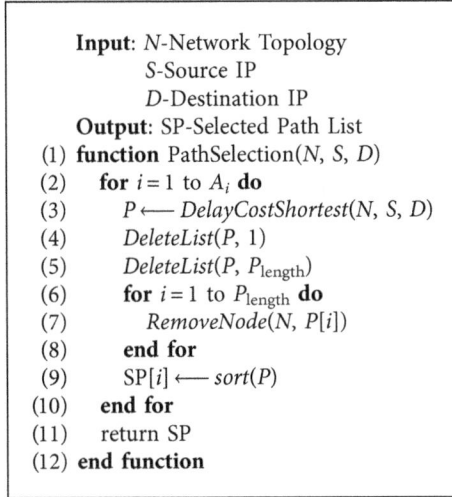

ALGORITHM 1: Pseudocode of the TSN/SDN path selection algorithm.

3.2.3. TSN/SDN Path Selection Algorithm.

The path selection algorithm we propose for TSRTnet is shown in Algorithm 1. In this algorithm, N is the network topology, S is the source IP, and D is the destination IP. The data-stream list to be connected according to application's QoS requirement is represented by A. Again, there are three types of data stream: physical states data, audio data, and video data. The DelayCostShortest function calculates the shortest path from S to D, and N assigns the cost to the link latency [39]. The selected path list as output of algorithm is SP.

3.2.4. TSN/SDN Path Configuration Flowchart.

The SDN controller configures the data-stream path as flow in a flow table of SDN core network, using the selected path (SP) list, which is the output of path selection algorithm. DS is a data-stream list, which is the same as A in the path selection algorithm, and it contains the application's QoS requirement. In the VR-CPES service, the TSN data streams are transmitted based on priority, which are identified by VLAN. In the TSN/SDN path configuration, the TSN data stream with highest priority is configured in advance. The SP and DS are ordered according to the highest-priority streams.

3.2.5. OpenFlow Switch Operation by SDN Controller.

The SDN controller processes the time-sensitive traffic and management path using OpenFlow protocol. The SDN controller sets the path based on the packet received in the OpenFlow packet-in message, and the path is divided into VLANs. As mentioned above, the QoS requirement of each application is classified by the SRP and VLAN of the TSN in the TSN traffic parser function. We assume that the SDN controller knows the QoS settings of each TSN for path configuration. The SDN controller sets the path, which is chosen from the TSN/SDN path configuration function, as flow using VLAN in the flow table of each OpenFlow switch. If the QoS of the application cannot be satisfied in the

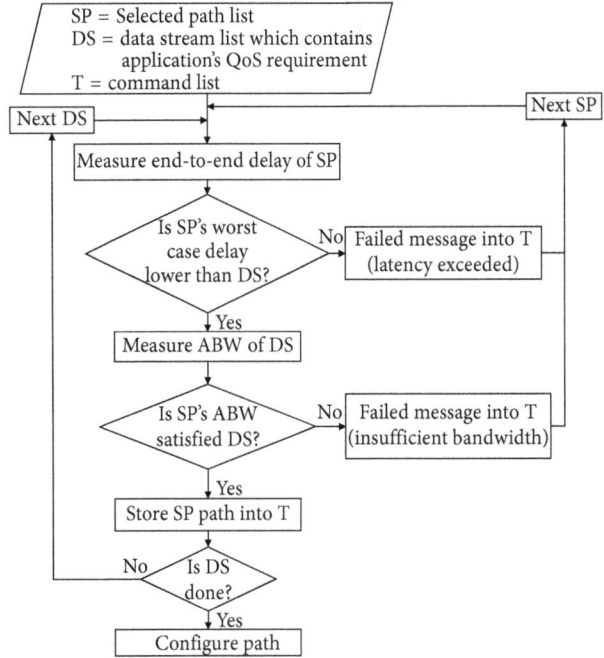

FIGURE 5: TSN/SDN path configuration.

network, the SDN controller sends the corresponding fail message to the TSN/SDN gateway.

4. Experiments and Evaluation

In VR-CPES driver-training service, physical states data, audio data, and video data require transmitting in real time, and all QoS requirements in network are mentioned above. To verify the real-time data transmission on the VR-CPES platform, we compared TSRTnet traffic with best-effort traffic. The experiment is performed under the Mininet emulator [40]. Figure 6 shows the network topology of the experiment. In the topology, each link bandwidth is 10 Mbps, and latency is 3 ms. Host 1 (H1) and Host 2 (H2) are the sources, and Host 3 (H3) is the receiver. We consider sources as the TSN/SDN gateways, and the receiver as the VR-CPES cloud of the VR-CPES driver-training service. Both H1 and H2 transmit 6 Mbps TSN streams to H3, which consists of 0.8 Mbps physical states data, 0.2 Mbps audio data, and 5 Mbps video data, respectively.

Figure 7 shows the packet loss of TSRTnet traffic and best-effort traffic. A packet loss that we measured is the sum of the loss packets in all three types of data delivered from the sources H1 and H2 to the receiver H3. In the proposed TSRTnet, the packet loss occurred almost 0% in both H1 (0.4%) and H2 (0.5%), but in the best-effort traffic case, a packet loss of 7% in H1 and 18% in H2 occurred. In the method of TSRTnet, packet loss scarcely occurred in all three data streams because the path is set by comparing the priority and reserving the bandwidth. On the contrary, in the case of best effort, all data are delivered with the same condition, which means they do not consider the QoS requirement. Therefore, the data with the highest bandwidth, such as video data, cause overhead on a certain link, and it incurs a large

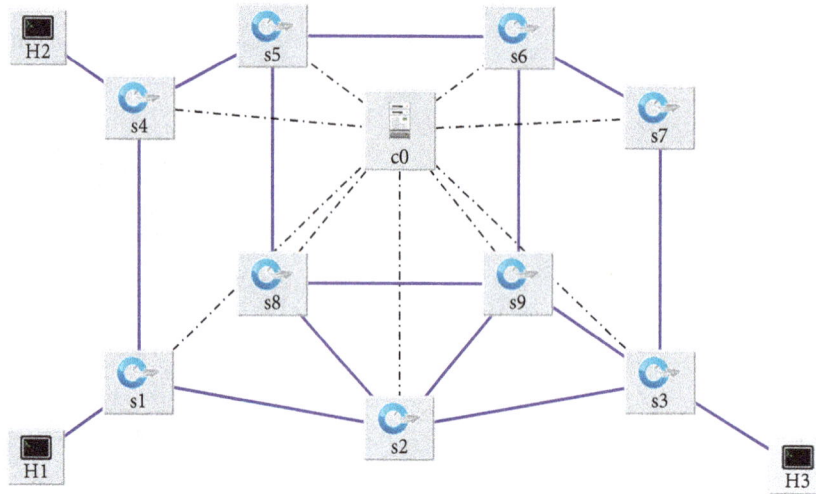

FIGURE 6: Network topology.

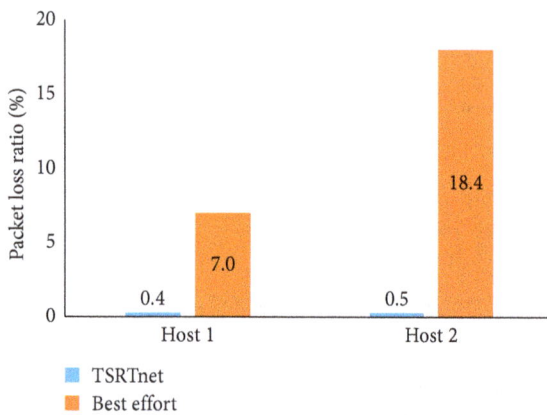

FIGURE 7: Packet loss of TSRTnet traffic and best-effort traffic.

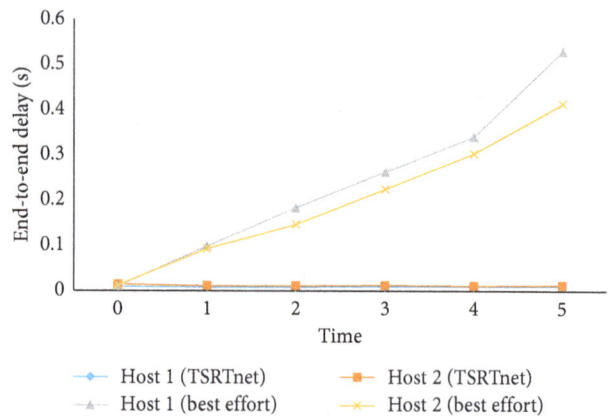

FIGURE 8: End-to-end delay of TSRTnet traffic and best-effort traffic.

amount of packet loss. The reason why the packet loss of H1 is lower than H2 is because the hop count of H1 is less than H2.

Figure 8 shows the comparison end-to-end delay between TSRTnet and best-effort traffic. In particular, we measured physical states traffic because it is highest priority traffic in our experiment. Both end-to-end delay variations of physical states traffic from H1 and H2 are constant in TSRTnet. Except for the delay due to the initial path configuration, it represents that the delay variation is less than 2 ms. On the contrary, the end-to-end delay of best-effort traffic increases continuously compared with the results from TSRTnet traffic. Until 0.2 seconds before packet loss, the end-to-end delay of best-effort was stable, as with TSRTnet traffic. Since then, however, the best-effort traffic delay for the three types of streams exceeded 100 ms in 1.1 seconds, which means it does not meet the requirements of the video stream with the highest end-to-end delay requirements. In addition, since H1 is close to the receiver, the traffic of H2 caused congestion in the same paths of H1. Consequently, the end-to-end delay of H1 is increased faster than H2.

The results show that the TSRTnet framework is more suitable for the VR-CPES service than the conventional network. As mentioned above, for seamless VR-CPES service, physical-state data, audio data, and video data are transmitted together, which means that data transmission according to complex QoS should be guaranteed in the network. TSRTnet shows almost zero packet loss compared to the conventional network, which transmits data with best effort. Packet loss is closely related to VR service interruption, which has a fatal impact on the user's VR experience. In addition, the constant latency variation of TSRTnet can satisfy the transmission of physical states data for time-critical simulation in VR-CPES.

5. Conclusion

In this paper, we proposed a VR-CPES platform that can provide novel VR-based educational services. VR-CPES platform solves the QoE problem of users caused by delay and loss of data transmission based on TSN/SDN real-time network, which makes time-critical data transmission between cyber things and physical things. Since VR-CPES services require strict time requirements, as well as bandwidth of multimedia and states data of physical things, the time-sensitive network services over the entire network, including core networks and local networks, are essential. The proposed TSN/SDN real-time network framework, called TSRTnet, effectively supports the requirements in terms of packet-loss ratio and end-to-end delay. TSRTnet includes a gateway function for the interworking between SDN core networks and TSN local networks, as well as a path selection algorithm for selecting candidate paths. As a result, TSRTnet showed stable performance of end-to-end delay variation within 2 ms and fewer packet losses compared to conventional best-effort network, which experienced very large delays and higher rates of packet loss.

Acknowledgments

This paper is partially supported by the Institute for Information and Communications Technology Promotion (IITP) funded by the Ministry of Science and ICT (MSIT) (2015-0-00816-004) and by the National Research Foundation of Korea (NRF) grant (no. 2017010875) from the Korea government.

References

[1] E. A. Lee, "CPS foundations," in *Proceedings of 47th ACM/IEEE Design Automation Conference on-DAC'10*, pp. 737–742, Anaheim, CA, USA, June 2010.

[2] M. Schluse, M. Priggemeyer, L. Atorf, and J. Rossmann, "Experimentable digital twins—streamlining simulation-based systems engineering for industry 4.0," *IEEE Transactions on Industrial Informatics*, vol. 14, no. 6, pp. 1722–1731, 2018.

[3] J. Lee, B. Bagheri, and H. A. Kao, "A cyber-physical systems architecture for industry 4.0-based manufacturing systems," *Manufacturing Letters*, vol. 3, pp. 18–23, 2015.

[4] F. Salim and U. Haque, "Urban computing in the wild: a survey on large scale participation and citizen engagement with ubiquitous computing, cyber physical systems, and internet of things," *International Journal of Human-Computer Studies*, vol. 81, pp. 31–48, 2015.

[5] X. Yu and Y. Xue, "Smart grids: a cyber–physical systems perspective," *Proceedings of the IEEE*, vol. 104, no. 5, pp. 1058–1070, 2016.

[6] N. Gavish, T. Gutiérrez, S. Webel et al., "Evaluating virtual reality and augmented reality training for industrial maintenance and assembly tasks," *Interactive Learning Environments*, vol. 23, no. 6, pp. 778–798, 2015.

[7] U. H. Govindarajan, A. J. C. Trappey, and C. V. Trappey, "Immersive technology for human-centric cyberphysical systems in complex manufacturing processes: a comprehensive overview of the global patent profile using collective intelligence," *Complexity*, vol. 2018, Article ID 4283634, 17 pages, 2018.

[8] T. H. J. Uhlemann, C. Lehmann, and R. Steinhilper, "The digital twin: realizing the cyber-physical production system for industry 4.0," *Procedia CIRP*, vol. 61, pp. 335–340, 2017.

[9] S. Weyer, T. Meyer, M. Ohmer, D. Gorecky, and D. Zühlke, "Future modeling and simulation of CPS-based factories: an example from the automotive industry," *IFAC-PapersOnLine*, vol. 49, no. 31, pp. 97–102, 2016.

[10] K. M. Alam and A. E. Saddik, "C2PS: A digital twin architecture reference model for the cloud-based cyber-physical systems," *IEEE Access*, vol. 5, pp. 2050–2062, 2017.

[11] L. Skorin-Kapov, D. Huljenic, D. Mikic, and D. Vilendecic, "Analysis of end-to-end QoS for networked virtual reality services in UMTS," *IEEE Communications Magazine*, vol. 42, no. 4, pp. 49–55, 2004.

[12] S. Tuohy, M. Glavin, C. Hughes, E. Jones, M. Trivedi, and L. Kilmartin, "Intra-vehicle networks: a review," *IEEE Transactions on Intelligent Transportation Systems*, vol. 16, no. 2, pp. 534–545, 2015.

[13] Y. Chen, T. Farley, and N. Ye, "QoS requirements of network applications on the Internet," *Information Knowledge Systems Management*, vol. 4, no. 1, pp. 55–76, 2004.

[14] G. Yang, L. Xie, M. Mäntysalo et al., "A health-IoT platform based on the integration of intelligent packaging, unobtrusive bio-sensor, and intelligent medicine box," *IEEE Transactions on Industrial Informatics*, vol. 10, no. 4, pp. 2180–2191, 2014.

[15] R. Lea and M. Blackstock, "City hub: a cloud-based IoT platform for smart cities," in *Proceedings of 2014 IEEE 6th International Conference on Cloud Computing Technology and Science (CloudCom)*, pp. 799–804, Hong Kong, China, December 2014.

[16] N. H. Motlagh, M. Bagaa, and T. Taleb, "UAV-based IoT platform: a crowd surveillance use case," *IEEE Communications Magazine*, vol. 55, pp. 128–134, 2017.

[17] oneM2M Standardisation Committee, *oneM2M Release 2 Specifications*, oneM2M, 2016, http://www.onem2m.org/technical/published-drafts.

[18] OIC Specification 1.1, 2016, https://openconnectivity.org/resources/specifications.

[19] GE Predix IoT Platform, 2017, https://www.ge.com/digital/predix.

[20] C. Mercer, *Internet of Things Platforms: Azure, AWS, IBM Watson and More-Which is the Best IoT Platform for Your Business*, Computer World, 2016, https://www.computerworlduk.com/galleries/data/best-internet-of-things-platforms-3635185/.

[21] S. Bera, S. Misra, and A. V. Vasilakos, "Software-defined networking for Internet of things: a survey," *IEEE Internet of Things Journal*, vol. 4, no. 6, pp. 1994–2008, 2017.

[22] A. Hakiri, P. Berthou, and A. Gokhale, "Publish/subscribe-enabled software defined networking for efficient and scalable IoT communications," *IEEE Communications Magazine*, vol. 53, no. 3, pp. 48–54, 2015.

[23] ONF, *OpenFlow Switch Specific Version 1.5.1*, Open Networking Foundation, 2015, https://www.opennetworking.org/wp-content/uploads/2014/10/openflow-switch-v1.5.1.pdf.

[24] S. Paris, A. Destounis, L. Maggi, G. S. Paschos, and J. Leguay, "Controlling flow reconfigurations in SDN," in *Proceedings of IEEE INFOCOM 2016-the 35th Annual IEEE International Conference on Computer Communications*, pp. 1–9, San Francisco, CA, USA, April 2016.

[25] R. W. Skowyra, A. Lapets, A. Bestavros, and A. Kfoury, "Verifiably-safe software-defined networks for CPS," in *Proceedings of the 2nd ACM International Conference on High Confidence Networked Systems*, pp. 101–110, Philedelphia, PA, USA, April 2013.

[26] Time-Sensitive Networking Task Group, 2018, http://www.ieee802.org/1/pages/tsn.html.

[27] M. Wollschlaeger, T. Sauter, and J. Jasperneite, "The future of industrial communication: automation networks in the era of

the Internet of things and industry 4.0," *IEEE Industrial Electronics Magazine*, vol. 11, no. 1, pp. 17–27, 2017.

[28] D. Chitimalla, K. Kondepu, L. Valcarenghi, M. Tornatore, and B. Mukherjee, "5G fronthaul-latency and jitter studies of CPRI over ethernet," *IEEE/OSA Journal of Optical Communications and Networking*, vol. 9, no. 2, pp. 172–182, 2017.

[29] Deterministic Networking, 2018, https://datatracker.ietf.org/wg/detnet/charter/.

[30] E. Bastug, M. Bennis, M. Médard, and M. Debbah, "Toward interconnected virtual reality: opportunities, challenges, and enablers," *IEEE Communications Magazine*, vol. 55, no. 6, pp. 110–117, 2017.

[31] S. Yun, J. H. Park, and W. Kim, "Data-centric middleware based digital twin platform for dependable cyber-physical systems," in *Proceedings of 2017 Ninth International Conference on Ubiquitous and Future Networks (ICUFN)*, Milan, Italy, July 2017.

[32] IEEE Std. 802.1Qbv, *Standard for Local and Metropolitan Area Networks-Media Access Control (MAC) Bridges and Virtual Bridged Local Area Networks Amendment: Enhancements for Scheduled Traffic*, IEEE, Piscataway, NJ, USA, 2016.

[33] IEEE Std. 1588-2008, *Standard for a Precision Clock Synchronization Protocol for Networked Measurement and Control Systems*, IEEE, Piscataway, NJ, USA, 2008.

[34] IEEE Std. 802.1AS, *IEEE Standard for Local and Metropolitan Area Networks-Timing and Synchronization for Time-Sensitive Applications in Bridged Local Area Networks*, IEEE, Piscataway, NJ, USA, 2010.

[35] IEEE Std. 802.1Qcc, *IEEE Approved Draft Standard for Local and metropolitan Area Networks-Media Access Control (MAC) Bridges and Virtual Bridged Local Area Networks Amendment: Stream Reservation Protocol (SRP) Enhancements and Performance Improvements*, IEEE, Piscataway, NJ, USA, 2018.

[36] IEEE. Std 802.1Qat, *IEEE Standard for Local and Metropolitan Area Networks, Virtual Bridged Local Area Networks, Amendment 14: Stream Reservation Protocol*, IEEE, Piscataway, NJ, USA, 2010.

[37] H. S. Kim, S. Yun, H. Kim, H. Shin, and W. Kim, "An efficient SDN multicast architecture for dynamic industrial IoT environments," *Mobile Information Systems*, vol. 2018, Article ID 8482467, 11 pages, 2018.

[38] D. Sinha, K. Haribabu, and S. Balasubramaniam, "Real-time monitoring of network latency in software defined networks," in *Proceedings of 2015 IEEE International Conference on Advanced Networks and Telecommunications Systems (ANTS)*, pp. 1–3, Kolkata, India, December 2015.

[39] F. Kuipers, T. Korkmaz, M. Krunz, and P. Van Mieghem, "Performance evaluation of constraint-based path selection algorithms," *IEEE Network*, vol. 18, no. 5, pp. 16–23, 2004.

[40] R. L. S. de Oilveira, A. A. Shinoda, C. M. Schweitzer, and L. R. Prete, "Using Mininet for emulation and prototyping software-defined networks," in *Proceedings of the 2014 IEEE Colombian Conference on Communications and Computing (COLCOM)*, pp. 1–6, Bogota, Colombia, June 2014.

13

Digi-Bags on the Go: Childminders' Expectations and Experiences of a Tablet-Based Mobile Learning Environment in Family Day Care

Kaisa Pihlainen [iD],[1] **Calkin Suero Montero,**[2] **and Eija Kärnä** [iD][1]

[1]*Department of Special Education, University of Eastern Finland, Joensuu, Finland*
[2]*Department of Computer Science, University of Eastern Finland, Joensuu, Finland*

Correspondence should be addressed to Kaisa Pihlainen; kaisa.pihlainen@uef.fi

Academic Editor: María D. Lozano

The use of mobile technologies is playing an increasingly important role in early childhood education (ECE) settings. However, although technologies are often integrated in ECE provided in day care centres, technology use in other ECE settings, such as in family day care, is rare. In this paper, we describe the Digi-bag, a tablet-based mobile learning environment deployed at several family day care homes, and present the expectations and first experiences of family day care personnel regarding the pedagogical use of Digi-bags together with 1- to 5-year-old children as well as their experiences of training to use the digital technology. The results of the pilot study indicate that the deployment of Digi-bags facilitates the pedagogical, creative, and regular use of digital technology with small children. The study also underlines the importance of providing opportunities to family day care personnel for peer support and peer learning in natural settings besides professional training in the use of digital technologies.

1. Introduction

Family child care, or family day care, has been defined as care given to a small group of children in the home of a child or a family day care practitioner [1]. In the USA, this is a common form of nonparental care for young children, with more than 1 in 5 children under five years of age receiving family child care [2]. In Finland, early childhood education (ECE) is provided in *day care centres* and *family day care homes* according to Finnish legislation [3] (when emphasizing care as a part of early childhood education, the abbreviation ECEC (early childhood education and care) is used in ECE literature. In this article, we use the abbreviation of ECE to emphasize the educational aspects of professionals working with young children). In total, 68% of 1- to 6-year-olds participate in ECE in Finland, of which 76% are in day care centres and 11% in family day care [4]. Although there are differences between day care centres and family day care [5], for example, family day care takes place at homes and the group size is limited to four children, and the same curriculum is used in both ECE settings.

Despite heavy debate, technology, in particular computer, use in early childhood has been shown to foster a number of positive traits in young children including self-confidence, increased spoken communication and cooperation skills as well as manual dexterity, nonverbal skills, and problem solving, among others [6]. Therefore, in recent years, technology has been used increasingly to support the development of children in ECE in Finland and elsewhere. Consequently, many day care centres are currently equipped with digital technologies for use in activities with children [7]. However, this is not often the case in many family day care homes in Finland [8]. Although the use of information technology devices and software is a requirement in the competence-based qualification of family day care personnel [9], day care homes seem to lack adequate technologies for pedagogical purposes, and furthermore, support and guidance is needed to keep the technical skills of day care personnel up to date [10]. This situation may place many young children and early childhood educators in an unequal position.

Furthermore, in Finland the requirements for family day care qualification include planning the ECE environment in a way that enables natural interactions for the children in terms of, for example, play, movement, exploration, artistic experiences, and expression, while at the same time providing a stimulating and safe and supportive environment for upbringing, care and learning [1]. It is also required to assess and adapt the learning environment regularly based on the children's needs and interests [1]. However, since good quality education and children's upbringing are ultimately based on the values of the personnel and their expectations for the future (e.g., [5]), it is crucial to study the current situation of family day care personnel regarding technology use, motivation, and experiences. Ways in which the current emphasis on technology use in ECE in Finland [11] could be integrated into the work of family day care personnel also needs to be studied.

To further our knowledge and understanding related to the abovementioned issues of technology use in family day care, three goals were set for this pilot study. Firstly, to determine the childminders' technology expectations and use prior to the Digi-bag trials. Secondly, to investigate the initial experiences of family day care personnel in using digital technologies during the digital backpack (Digi-bag, Section 2) trials with 1- to 5-year-old children in pedagogical settings. While employing the Digi-bags in their work, the family day care personnel had the possibility to use digital technologies such as tablets and Bee-Bot robots (https://www.bee-bot.us). We examine the changes in expectations and experiences of the participating family day care personnel during deployment of these digital technologies. The experiences of educational personnel are crucial to the realization of successful learning environments since being aware of and evaluating one's own beliefs and perceptions is regarded as a core component of educational know-how that supports the provision of high-quality ECE for small children [5]. Thirdly, since the Digi-bag trials were the first project to employ digital technologies in Finnish family day care [8], we were interested in collecting and analysing the training experiences of family day care personnel in the use of digital technologies in pedagogical settings. The findings will be used to modify the training of new childminders to use Digi-bags.

Our contribution puts forward both scientific and practical measures to develop ECE best practices to respond to the current technological changes in societies. Furthermore, our work advances this discussion by introducing the Digi-bag technology-enhanced mobile learning environment in family day care and describing the childminders' expectations and initial experiences of this learning environment. In doing so, we shed light on the technical requirements and pedagogical aspects of using mobile learning technologies in ECE as well as the need for provision of further education and training for ECE personnel.

2. Digi-Bags in Family Day Care

The Digi-bag concept was created out of the need to provide children and childminders (i.e., the family day care personnel that participated in the Digi-bag trials) in family day care with access to digital technologies comparable to those used in day care centres. We chose backpacks because they are easily adapted to the everyday lives of small (1- to 5-year-old) children (backpacks are used daily by Finnish school and preschool children, although 0- to 1-year-olds do not carry them on their backs), and through them, we can combine familiar and new digital and nondigital tools to create hands-on, flexible, approachable, and playful learning experiences. As such, the Digi-bags aim at fulfilling the role of granting children equal opportunities for learning through digital technologies regardless of the ECE setting.

Each Digi-bag consists of a bag equipped with a tablet (Lenovo TAB 3, Android 6.0, 32 GT) with a protective cover, Internet connectivity, and headphones, as well as instructions and tips for using the tablet. Furthermore, each childminder had a personal memory card to be used for saving data, including photos and videos. Tablets were chosen as the main mobile technology device for family day care since with them all of the advantages of a modern-day computer can be obtained, and they are light, portable, and easy to use [12] even for very small children (e.g., two-year-olds) [13]. Furthermore, the characteristics of interaction between childminder and child in a family day care environment also support the use of tablets. That is, the distance between the adult and children is markedly smaller in family day care compared to day care centres [14]. Furthermore, compared to day care centres, children move more in family day care [14], which emphasises the advantages of mobile technology in the family day care settings.

For the trials, five Digi-bags were prepared, and the contents were customized based on one of five core content areas of the Finnish National Core Curriculum for Pre-primary Education [1]. The core content areas are (see also Table 1) as follows:

(1) *Self-expression*: learning self-expression through music, arts, handicrafts, and oral and physical expression,

(2) *Language*: learning language skills,

(3) *Me and my community*: learning to understand the variety of people's backgrounds, history, ethics, and religion in society and communities and to participate in the community,

(4) *I explore and act*: learning mathematics, technologies, and environmental education,

(5) *I grow and develop*: learning to look after oneself, managing daily activities from the viewpoints of physical exercise, nutrition, consumer skills, health, and safety.

We chose to equip each backpack for each core content area with tools to strengthen the practical implementation of the new curriculum [1] as well as to support the pedagogical grounding for the use of mobile technologies. Table 1 gives a brief description of each core pedagogical area, the practical methods of teaching and learning, and the tools and materials provided with the Digi-bag to support the pedagogical approach.

Each Digi-bag differed in terms of the kind of software installed on the tablet and nondigital devices, and the toys

TABLE 1: Core content areas of the Finnish National Core Curriculum for Pre-primary Education; tools and materials in the Digi-bags; practical examples of pedagogical ways to use them.

Core area of the Finnish curriculum	Pedagogical function of the Digi-bag	Tools and materials (digital & nondigital) in the Digi-bag
(1) *Self-expression*: emphasis on music, arts, handicrafts, and oral and physical expression	Use of imagination and creativity, experimenting and expressing own ideas	Tablet and software; chicken shakes; drum; finger puppets; story book; playdough
(2) *Language*: emphasis on developing language skills	Naming and classifying objects, practicing social, interactional, and cultural skills, self-expression, and a combination of these skills	Tablet and software; card game: emotions; nursery rhymes; games supporting storytelling and singing
(3) *Me and my community*: emphasis on history, society, ethics, religion	Getting to know different people, languages, and cultures in the local community, habits, playing, festivals	Tablet and software; maxi domino game; card game; book on games; activities for children
(4) *I explore and act*: emphasis on mathematics, ICT, and environmental education	Reading pictures, numbers, media, and text; naming objects, learning concepts, using and producing various contents, using digital documentation, exploring the world	Tablet and software; Bee-Bot robot; magnifying glass; foam shapes
(5) *I grow and develop*: emphasis on physical exercise, nutrition, consumer skills, health, safety	Practicing dressing, eating alone, taking care of personal hygiene and own property, moving safely inside and outside, recognizing emotions, planning, implementing, and evaluating activities together, learning common rules and trust	Tablet and software; mini bean bags; gross motor exercise cards; book: theory and tips on children's physical exercise in ECE

and games chosen to support the core content areas of the curriculum (Figure 1). For example, a Digi-bag for supporting children's language skills (core area 2) included materials for storytelling, playing with concepts, and listening to nursery rhymes, while a Digi-bag for multiliteracy and ICT competence development (core area 4) included a magnifying glass and a Bee-Bot robot (Table 1). The Digi-bags were compiled in backpack form to enable easy mobility between childminders. The participating childminders were also encouraged to write down their experiences of using the Digi-bags as well as their ideas and tips to share with other users in order to foster the circulation and adoption of new of pedagogical ideas. We believe that circulation of materials and ideas among the childminders fosters motivation and curiosity in both the children and childminders.

The Digi-bags combined tablets with zero- or low-technology tools and materials based on four pedagogical aspects. Firstly, use of a tablet alongside more traditional and familiar tools used in ECE supports the notion that mobile learning technologies are a natural part of the everyday life of children [15], as is the case with zero- or low-technology devices and materials. Secondly, providing a variety of tools together allows the children and personnel to freely choose and express their preferences for tools for learning, which is widely connected with motivation (e.g., [6]). Thirdly, providing a tablet with other tools and materials may trigger the imagination of the children and personnel to combine various tools for playing and learning (see [16] for more details about hybrid play). Finally, we avoided overemphasis on technology over pedagogy and activity content by providing a mixture of digital technology and nondigital traditional tools. Our goal was to strengthen pedagogically oriented use of technologies by combining them with pedagogical and content knowledge (see [17]). In practice, use of the Digi-bags depends on each childminder and the group of children and the degree of adult-driven and child-driven activity.

For the Digi-bag pilot study, childminders received five Digi-bags and a two-hour training session on use of the tablets in April 2017. As the Digi-bags were ready to use, the participating personnel introduced the bags to the children in their care immediately. The first experiences from this pilot group are presented in this study and will be used to modify the training and the contents of the following ten Digi-bags to be prepared in Autumn 2017 based on the five core areas of the new curriculum (see [11]). The bags will then be deployed for a 14-week period with all the remaining family care personnel in the municipality in which the trials were conducted.

3. Methodology

This pilot study of the Digi-bag trials was conducted as a case study [18]. A case study is an inquiry into a specific phenomenon (the case) within its real-world context [19]. In this study, the goal was to understand five childminders' expectations and initial experiences of the Digi-bags as part of their daily work with small children, and their experiences of the training in digital technology use. In addition, the results of the pilot case study were expected to provide information on how mobile technology-enhanced learning environment such as the Digi-bag could enrich the education of small children in family care and how childminder training in the use of technology in their daily work could be provided.

3.1. Participants. The local supervisor of the childminders introduced the Digi-bag concept and a description of the pilot study to all childminders in the municipality, and we used voluntary sampling to recruit five female childminders to participate in the trials. The number of participants was limited to five to allow one Digi-bag per childminder. The age of the childminders varied from 49 to 59 years ($M = 56$ years, $SD = 4.2$). Their educational background reflected the variety

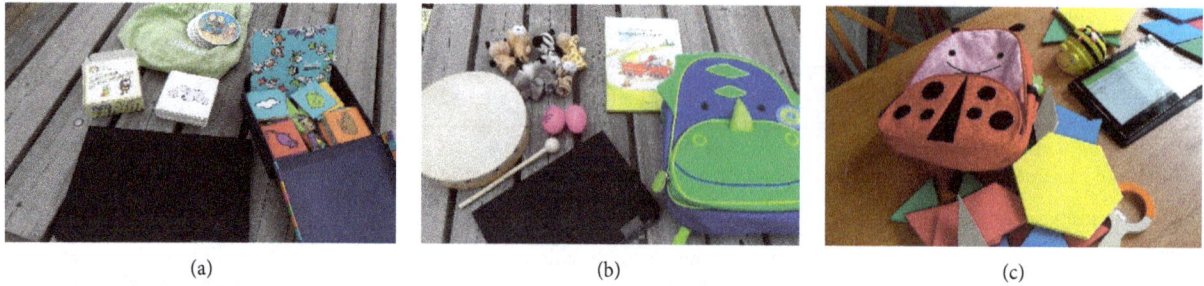

FIGURE 1: Examples of the Digi-bags and their contents. (a) *Languages* (bag not shown); (b) *Self-expression* (playdough not shown); (c) *I explore and act.*

of backgrounds of the childminders, which is also observed at the national level: two childminders had official family day care qualification, one was a day care assistant and one had a comprehensive school education. One childminder did not mention her educational background. It is notable that one of the qualified family day care childminders also had a degree in information technology and two childminders were currently undergoing further education in ECE.

All of the participating childminders were highly experienced, having worked in the field from 15 to more than 30 years ($M = 23$ years, $SD = 6.7$). The composition of the child groups reflected the typical, nation-wide situation in family day care. That is, each childminder took care of four children, both boys and girls. The children's age varied from one to five years. None of the children received special support.

3.2. Data Collection and Analysis. The data collection was conducted using short questionnaires with both closed and open-ended questions and unstructured interviews. Prior to the data collection, the researchers met all participating childminders as well as their local supervisor in order to negotiate the practicalities of the pilot and the best ways for data collection. For example, the childminders requested flexible data collection methods, and thus, questionnaires were provided in both online and paper formats.

The data were collected in three phases. Firstly, the childminders filled in a questionnaire to record their thoughts and expectations regarding the Digi-bag trials prior to the pilot launch. These data were collected in April 2017. Two childminders used an online form and three preferred to fill in a paper version of the questionnaire. The questionnaire included questions on the childminder's background; their use of digital technologies in educational contexts and leisure time; their expectations regarding the use of digital technology with children in ECE; their feelings regarding the use of digital technology in childminding; their thoughts on how other people (children, parents, other childminders, etc.) might perceive the use of digital technology in childminding; and their expectations of the Digi-bag pilot.

Secondly, the participating childminders reported their initial experiences of the Digi-bag use after the trials. The data were collected in July-August 2017. Three childminders responded using the online questionnaire. Thirdly, the written data were supplemented with unstructured interviews. One of the researchers visited three childminders in August 2017 to discuss their initial experiences of the Digi-bags. These qualitative data were analysed using classical content analysis [20, 21]. The closed questions were analysed using descriptive statistical methods such as frequencies and percentage. The study practices were based on generally accepted ethical principles for scientific research. As the research was carried out in natural settings, the rights and protection of the research participants were the first priority. Participation was voluntary, and all participating childminders and the guardians of the children in care gave their written consent prior to the trials.

4. Results

4.1. Technology Expectations and Use Prior to the Trials. The childminders had some experience of using digital technologies prior to the Digi-bag trials, mostly from pastime activities. However, digital technologies were less used for pedagogical activities with children in family day care (Table 2). Nevertheless, three childminders described using digital technologies sometimes (varying from weekly to less than monthly use) in different pedagogical ways, such as finding craftwork instructions or searching for information about animals. They also used digital devices for reading fairy tales, taking photos, and listening to music. Two childminders, however, had no prior experience of using digital technologies in a pedagogical context with children.

When asked about the different ways they could use digital technology with children in ECE, the childminders mostly mentioned uses that are already familiar to them, such as searching for information online, taking photos, listening to music, and watching videos. The childminders' responses also included examples of taking photographs not only to capture images and memories but also to collect children's feedback or as means to react to children's current interests. For example, a child could take photos from his or her day or of their favourite objects.

In addition, the childminders mentioned age-appropriate digital games that could support the children's development and learning (Figure 2). For example, one childminder wrote that "now the children are interested in letters (*so there is a need for*), a game with which we could listen to and write letters and words."

Overall, all of the childminders positively anticipated the use of digital technologies. They described the children's excitement to use digital devices not only at home but also in family day care. From their personal viewpoint, the

TABLE 2: Childminders' digital technology use prior to the Digi-bag trials (F).

Childminder	How often do you use the following digital technologies in your free time? (F)				How often do you use the following digital technologies with children in family day care? (F)			
	Mobile phone	Computer	Tablet	Other[1]	Mobile phone	Computer	Tablet	Other
#1	5	4	1	1	3	3	2	1
#2	5	5	—	—	2	2	—	—
#3	5	4	5	3	4	2	4	—
#4	5	2	2	—	1	1	1	1
#5	5	5	—	—	1	1	1	1

[1]Video/photo cameras and Blu-ray players.
F = frequency; 5 = daily; 4 = weekly; 3 = couple of times a month; 2 = less than monthly; 1 = not at all; — = data missing.

(a) (b)

FIGURE 2: Children using a tablet in family day care.

childminders related technology use to easing and enriching their work tasks with the children and as a way of learning something new. Learning new activities seemed to be exciting for the childminders as one respondent wrote, "I have not been a huge mobile phone user before, so it will be interesting to see how the Digi-bag trial goes with us, because I will also be learning almost from scratch." Besides positive reactions, some respondents also shared some concerns towards technology use in family day care. One childminder expressed that "the challenge is how I can offer anything new. It feels like the children are cleverer than I am." Some childminders also worried about a possible decrease in playtime but simultaneously recognized the possibility for new play and games to emerge.

4.2. First Experiences with the Digi-Bags. During the trials, the childminders used the Digi-bags once or twice a week with the children. Tablets were regarded as versatile tools in family day care and were used both indoors and outdoors. Two interviewed childminders emphasized that tablets were very beneficial for taking photos and videos, which were useful for recalling earlier activities. As one childminder noted, "I think it is great that you can take photos of trips. The kids love to watch them afterwards and wonder at what they did." Besides these familiar activities with tablets, the childminders also discovered new, creative ways to use tablets in daily life with the children. For example, one childminder described a situation before a day nap as follows: "*(tablets are handy because now)* I can read fairy tales from the tablet also in a dark room."

Although one childminder reported that their children were not really interested in using a tablet in family day care as it was an ordinary device at home, most of the childminders had overall positive initial experiences of using

the tablets provided with the Digi-bags. They explained how the children each took turns using the tablet while the other children observed the user at a close distance. Some childminders reported that they were very surprised to see how smoothly and effortlessly turn-taking and rule-following took place among the children. Childminders also reported that they had found ways to communicate and negotiate the rules of tablet use. Children were taught, for example, to store the tablet in a safe place and not to leave it on the floor.

When the children used the tablet, the childminders observed them from a close distance and were able to provide help if needed. Although in most cases the children used the tablets in groups, the childminders emphasized the importance of noticing each child as an individual, whose unique habits and needs in using the technology had to be taken into account. In this way, the childminders could, for instance, intervene when they observed a child's excessive use of the tablet.

Furthermore, the childminders referred many times to the children's acquisition of skills while using the tablet. For example, one childminder explained that even a one-year-old child was able to play with a tablet by herself when a childminder helped her to navigate the tablet screen and "turn off different pages." Learning was connected with playful activities enabled by the different software and technologies. The childminders mentioned, for example, that the children learned letters by playing games on the tablets, practiced musical skills outdoors using QR codes that were read with the tablets, and used Bee-Bot robots to learn traffic rules. As one childminder described: "the children had fun steering the Bee-Bots with the arrow keys. They drove forward, left, right, and backward along streets on their traffic play mat. At the same time we discussed, for example, how a pedestrian should behave in traffic, what do the red, yellow, and green colours mean on traffic lights, and so on." Another childminder described how children who were taking photos and videos with a tablet of what they could see, hear, and do during a forest walk learned also turn-taking, listening to instructions, and following them. Moreover, the childminders were pleased to see that, in practice, the use of digital technologies did not diminish the children's free playtime. On the contrary, as one childminder mentioned, "I haven't seen other play give way (*to tablet use*), far from it, the kids are just playing (*without tablets*) as well."

The respondents also mentioned some critical issues related to the technical characteristics of the digital devices. For example, in bright daylight, it was difficult to see pictures on the tablet screen; the tablet was also too big and heavy for some children to carry. The need for a waterproof cover for the tablet and a softer bag to put the tablet in was also expressed. Apart from these few critical comments, the childminders' initial experiences of Digi-bag deployment in family day care were overall positive.

4.3. Experiences of Training in Digital Technology Use. The childminders expressed their eagerness to learn new perspectives regarding their work in family day care. As one childminder puts it, "it is not only the kids that are learning, also the childminders are getting tips. I really like that there is always something new, especially things that relate to my work (*so*) things do not get stuck in a rut. I do not want to give up what is practical, but I can always take on new ideas."

The childminders also indicated their need for training in the use of digital technologies in family day care. After a two-hour technical training session, some of the childminders remained unclear regarding their own role in learning to use the technologies. Being accustomed to participating in officially organized training, they did not know whether they should investigate further uses of the tablets on their own. As one respondent wrote, "sometimes we wonder what to do. (*Wait for*) training or experiment by ourselves?" Some childminders seemed to have had unsatisfactory prior experiences of ICT training, which may have affected their expectations of training in Digi-bag use.

Nevertheless, for those who implemented self-regulated learning, the learned new uses seemed to be rewarding. For example, one childminder who had used digital technologies prior to the training wrote that "(*it was*) nice to try QRs (*for day care activities*) when I understood how easy it is to do them." Her activities and experiments also motivated other childminders who noticed the concrete benefits and easiness of preparing QR codes in practice. One childminder who observed the other childminders' work with QRs described, "I also got interested in that immediately because you can do such cool things together with the kids. And it is something I want to learn." Furthermore, the childminders explained the need for simple training programmes in the introductory phase of the new devices as well as access to written easy instructions, especially about digital games' use.

5. Discussion

The childminders' expectations of tablet use prior to the trials equated with their actual use of tablets at work, including

information searching, music listening and, photo and video shooting. Even though these ways of using tablets are pedagogically justifiable (e.g., [7]), they represent only a small proportion of their use potential. We interpret this as reflecting a need to provide training for childminders about varied pedagogical possibilities of using these devices. Furthermore, training should focus on the core areas of the curriculum to support the work of the childminders [9] as well as children's upbringing, care, and learning in the most suitable ways [1].

Some childminders reported having fears prior to the Digi-bag trials of being unable to create motivating activities for their children through digital technology. These fears disappeared during the trials and, instead, the childminders described both their desire and practical ways to use tablets as pedagogical tools to support children's individual upbringing and learning. Examples such as combining the use of Bee-Bot robots with teaching traffic rules, applying QRs in forest trips, and using a tablet for reading fairy tales at nap time in a dark room reflected the motivation of the childminders to actively employ creative ways of using tablets in family day care. Motivation was spread by both sharing own practical experiences as well as listening to the experiences of peers.

The results from the first experiences of the Digi-bag trials revealed that childminders perceived tablets as a natural part of children's current and future life. The childminders reported that the children have continued using the technologies on a weekly basis in family day care after deployment of the Digi-bags. This indicates that the tablets have increased the pedagogical use of technology in the ECE settings under study; as prior to the trials, their use varied from weekly to less than monthly. Simultaneously, the Digi-bags also increased the diversity of technology use, such as using devices outside, in dark spaces, and with small children. This diversity reflects the benefits of using tablets for pedagogical activities as have also been noted elsewhere (e.g., [12, 22]). After the first experiments with tablets, the childminders contested their own preconceptions about the decrease in children's playtime. This is crucial, since play is regarded as a central part of supporting children's development, learning, and well-being in ECE settings [1].

The concerns that the childminders had about the children's use of digital technologies during the trials were mostly linked with the technical characteristics of the tablets, such as size, weight, and the functionality of the screen in outdoor use. Even though technical characteristics, such as lightness, portability, and ease of use, were the main reason for choosing tablets for family day care use, we noticed their limitations when the users are small children. Therefore, we hope to provide more child-friendly digital technologies, including tablets and smart phones, even for small children.

The childminders themselves also asked for more training in technology use. This request is relevant since the use of technologies in ECE was included in the latest core curriculum for pre-primary education launched in 2016 in Finland. As Friedrichs-Liesenkötter [23] states, ICT training for ECE personnel "can be regarded as part of the response to the debate around the quality and the professionalization of ECE." Based on the experiences of the childminders, both training organized by professionals and possibilities for peer support are needed to further educate childminders in the use of digital technologies. Peer collaboration with tablets has been recorded also elsewhere as supporting learning in further education [22]. Our study shows that peer support and working together with peers may motivate less experienced childminders to experiment with digital technologies in natural settings.

When considering the expectations and first experiences of the childminders, it should be taken into account that only five childminders participated in the Digi-bag preliminary trials. However, the findings of usability research indicate that five individuals can reveal around 80% of key issues during technology testing [24]. Moreover, five participants were specifically chosen in order to pilot five Digi-bags that were based on the five core content areas of the Finnish National Core Curriculum for Pre-Primary Education. Although the number of participants was limited in our trials, their experiences corresponded well with previous studies. However, further Digi-bag trials should be undertaken with larger participant numbers and more varied cultural and/or educational settings.

6. Conclusion

This paper introduced the Digi-bag mobile learning environment launched for use in family day care with small children and their childminders. The preliminary study indicated that technologies were used in family day care weekly in versatile and creative ways. In order to support childminders in employing digital mobile technologies in family day care, further training is needed. Training should reflect the current situation of the childminders as technology users with attention paid to the attitudes, fears, and frustrations of the childminders. Alongside official training, childminders should be provided opportunities for peer support and peer learning in authentic settings. To this end, the Digi-bag trials provided a fruitful starting point for implementing a mobile learning environment in family day care.

Finnish ECE is holistic in nature, emphasizing the interrelatedness of upbringing, teaching, and care [5]. Mobile learning environments seem to fit well with these goals. In order to strengthen children's equal rights to learning and use of digital technologies [25], we hope that in future the Digi-bag mobile learning environment will be deployed and studied in various day care settings. It would be interesting to study in detail how technology initialization happens: Do children take the initiative or are tablets used based on the childminder's prompting? How do these actions vary from each other? Ways of providing further education for childminders, such as peer training, also need further research.

Acknowledgments

Our sincere thanks to the ECE managers and childminders in the Liperi municipality for their collaboration in this study.

References

[1] National Board of Education, "Varhaiskasvatussuunnitelman perusteet 2016, (Fundaments of early childhood education)," *Määräykset ja ohjeet*, vol. 17, 2016, http://www.oph.fi/download/179349c_varhaiskasvatussuunnitelman_perusteet_2016.pdf.

[2] E. M. Kryzer, N. Kovan, D. A. Phillips, L. A. Domagall, and M. R. Gunnar, "Toddlers' and preschoolers' experience in family day care: age differences and behavioral correlates," *Early Childhood Research Quarterly*, vol. 22, no. 4, pp. 451–466, 2007.

[3] Finlex, *Varhaiskasvatuslaki (Early Childhood Education Law)*, Ministry of Social Affairs and Health 36/1973, Helsinki, Finland, 2015.

[4] S. Säkkinen and T. Kuoppala, "Varhaiskasvatus 2015, (Early childhood education 2015)," Statistical report 21/2016, National Institute for Health and Welfare, Helsinki, Finland, 2016.

[5] K. Karila, T. Kosonen, and S. Järvenkallas, *Varhaiskasvatuksen Kehittämisen Tiekartta Vuosille 2017–2030, (The Road Map for Developing Early Childhood Education for 2017–2030)*, vol. 30, Publications of Ministry of Education and Culture, Helsinki, Finland, 2017.

[6] K. McCarrick and Xiaoming, "Buried treasure: the impact of computer use on young children's social, cognitive, language development and motivation," *AACE Journal*, vol. 15, no. 1, pp. 73–95, 2007.

[7] N. Vernadakis, A. Avgerinos, E. Tsitskari, and E. Zachopoulou, "The use of computer-assisted instruction in preschool education: making teaching meaningful," *Early Childhood Education Journal*, vol. 33, no. 2, pp. 99–104, 2005.

[8] U. Puustinen, "Konstan digiloikka, (Konsta's digital leap)," *Motiivi*, vol. 6, pp. 14–17, 2017.

[9] Finnish National Agency for Education, *Perhepäivähoitajan Ammattitutkinto 2013. Näyttötutkinnon Perusteet, (Further Vocational Qualification of Childminders 2013. Basics for Competence-Based Qualification), Decree 15/011/2013, Decrees and Instructions*, vol. 24, Finnish National Agency for Education, Helsinki, Finland, 2013.

[10] P. Lindberg, "Tieto- ja viestintätekniikan mahdollisuudet perhepäivähoidossa (The possibilities of information and communications technology in family day care)," in *Villistä Valvottuun, Valvotusta Ohjattuun. Perhepäivähoidon Ohjauksen Historia ja Nykytilan Haasteet (From Wild to Controlled, from Controlled to Supervised. The History of Family Day Care and Today's Challenges)*, S. Parrila, Ed., vol. 4, pp. 100–111, Sosiaali- ja Terveysministeriön Julkaisuja, Helsinki, Finland, 2005.

[11] National Board of Education, "Esiopetuksen opetussuunnitelman perusteet, (National core curriculum for pre-primary education 2014)," 2014, http://www.oph.fi/download/163781_esiopetuksen_opetussuunnitelman_perusteet_2014.pdf.

[12] E. B. Miller and M. Warschauer, "Young children and e-reading: research to date and questions for the future," *Learning, Media and Technology*, vol. 39, no. 3, pp. 283–305, 2014.

[13] Common Sense Media, *Zero to Eight: Children's Media Use in America 2013*, Common Sense Media, San Francisco, CA, USA, 2013.

[14] J. Reunamo, "Mitä perhepäivähoidossa todella tapahtuu?, (What really happens in family day care?)," 2012, http://blogs.helsinki.fi/reunamo/files/2012/06/pph.pdf.

[15] A. Suoninen, "Lasten mediabarometri 2013: 0-8-vuotiaiden mediankäyttö ja sen muutokset vuodesta 2010, (Children's media

barometer 2013: the media use of 0–8 years and its changes from 2010)," Nuorisotutkimusverkosto/Nuorisotutkimusseura, Verkkojulkaisuja 75, 2013.

[16] H. Tyni, A. Kultima, T. Nummenmaa, K. Alha, V. Kankainen, and F. Mäyrä, "Hybrid playful experiences: playing between material and digital," TRIM Research Reports, no. 19, 2016, http://tampub.uta.fi/handle/10024/98900.

[17] P. Mishra and M. J. Koehler, "Technological pedagogical content knowledge: a framework for teacher knowledge," Teachers College Record, vol. 108, no. 6, pp. 1017–1054, 2006.

[18] R. K. Yin, Case Study Research: Design and Methods, SAGE, London, UK, 5th edition, 2013.

[19] R. K. Yin, "Validity and generalization in future case study evaluations," Evaluation, vol. 19, no. 3, pp. 321–332, 2013.

[20] M. W. Bauer, "Classical content analysis: a review," in Qualitative Researching with Text, Image and Sound, M. W. Bauer and G. Gaskell, Eds., pp. 132–152, SAGE, London, UK, 2000.

[21] K. Krippendorff, Content Analysis: An Introduction to Its Methodology, SAGE, London, UK, 2004.

[22] J. Butcher, "Can tablet computers enhance learning in further education?," Journal of Further and Higher Education, vol. 40, no. 2, pp. 207–226, 2016.

[23] H. Friedrichs-Liesenkötter, "Mediaeducational habitus of future educators in the context of education in day-care centers," Journal of Media Literacy Education, vol. 7, no. 1, pp. 18–34, 2015.

[24] J. Nielsen and T. K. Landauer, "A mathematical model of the finding of usability problems," in Proceedings of the ACM INTERCHI'93 Conference, Amsterdam, The Netherlands, April 1993.

[25] A. Third, D. Bellerose, U. Dawkins, E. Keltie, and K. Pihl, Children's Rights in the Digital Age: A Download from Children around the World, Young and Well Cooperative Research Center, Melbourne, VIC, Australia, 2nd edition, 2014, https://www.unicef.org/publications/files/Childrens_Rights_in_the_Digital_Age_A_Download_from_Children_Around_the_World_FINAL.pdf.

14

The Urban Nexus Approach for Analyzing Mobility in the Smart City: Towards the Identification of City Users Networking

Federica Burini, Nicola Cortesi, Kevin Gotti, and Giuseppe Psaila ⓘ

University of Bergamo, Bergamo, Italy

Correspondence should be addressed to Giuseppe Psaila; giuseppe.psaila@unibg.it

Academic Editor: Joaquin Huerta

We present an interdisciplinary approach that makes possible to learn how citizens live in the city by the means of mobile social media data, that is, volunteered geographical information provided by the inhabitants through social media and mobile apps, by adopting a new reticular approach to spatial analysis. In particular, we present the general notions as background of our work, an investigation methodology to apply whenever such an analysis task must be performed, and a digital environment of tools and frameworks to support the methodology.

1. Introduction

One of the main long-term goals of city administrators is to make their city attractive for residents and city users (tourists, commuters, migrants, etc.). Nowadays, the buzzword of *smart city* is used, to characterize ideal cities of the future. However, since real cities are far away from the utopian smart cities of the future, in order to understand how to improve and make the city *smarter* than it currently is, administrators need information about how the city is lived now, what are the critical issues, and what are the positive aspects to develop.

Nowadays, city users can significantly yet unconsciously contribute to the production of data useful for administrators, by the use of social media. In fact, city users often post geolocated messages that describe their activities throughout the city, by means of their smartphones. Such a big quantity of information could provide unexpected insights concerning how people live their city or the city they are visiting, and could be exploited to improve services, public transportations, viability, and so on.

However, to achieve this ambitious goal, many contributions are needed: the first technical contribution is given by the ability of gathering mobile data; the second is the ability of transforming and querying gathered mobile data; and the third is a flexible tool that provides analysts with the

ability of studying *Big Geo-Data* by means of a simply-to-read visualization system. From a methodological point of view, there is a need to find an approach helping researchers in the investigation and analysis of the user-generated contents, in order to provide unexpected knowledge to administrators for understanding urban-mobility issues. In other words, an interdisciplinary approach is needed in order to make data scientists and computer scientists be aware of the main elements that are useful to geographers and urban space analysts to understand mobility and spatiality of individuals. Then, an interdisciplinary work can produce a methodology as well as software tools and frameworks to conduct on such an investigation.

In this paper, we present the *Urban Nexus Approach*, developed within the *Urban Nexus* project, an *Excellence Initiative* of the *University of Bergamo* (Italy) that involves *Anglia Ruskin University* (Cambridge, UK) and *EPFL* (Lausanne, Switzerland). In short, the goal of this project is to develop a methodology and tools to study how city users move within the city, use the urban space, and share their experience in places. The *Urban Nexus Approach* is a modular methodology that relies on a set of software tools and frameworks for gathering data sets coming from various sources of information (in particular, mobile social media), transforming and querying possibly geotagged data, visualizing them on maps, and analyzing them in order to reconstruct networking of individuals within the city (i.e., spaces and connections they produce during their movement).

In this paper, first of all we introduce the general conceptual framework that provides basic concepts as background of our work. On these concepts, we define our approach and define the investigation methodology that constitutes the first main contribution of the paper. Then, the *Urban Nexus digital environment* is extensively presented: it is a pool of software tools and frameworks specifically devised to support the investigation methodology; the digital environment is the second main contribution of this paper. In fact, we remark that we do not want to present specific results concerning specific cities; we present a general approach that is both methodological and technical.

The paper is organized as follows. Section 2 introduces the multidisciplinary background of our research project. Section 3 introduces the investigation methodology that is driving our work. Section 4 presents the *Urban Nexus digital environment*, that is, the pool of tools and frameworks that together make possible to perform the analysis. Finally, Section 5 addresses concluding remarks.

2. Background

In this section, we present the background of our research. It is a multidisciplinary background, since the research project is multidisciplinary.

2.1. A New Perspective about the Contemporary City. A great question geographers are trying to give an answer is "what is the contemporary city?" [1]. An approach could be to interpret the "city" as a node of a globalized network, where a local dimension and a global dimension no longer exist separately; rather, these dimensions interact, by reconfiguring urban contexts with their centralities, their axis, their full and empty spaces, and their internal and external connection [2].

In other words, if the base of contemporary life is movement, the elements from which to explore the city are its people (stakeholders of mobility) and the places they visit, viewed as nodes of a network that creates unity and cohesions [3].

2.1.1. The Concept of Rhizome. In the context of globalization, the "contemporary city" assumes a polycentric and reticular configuration: it is no longer subdivided into center and periphery; rather, it is viewed as an "osmotic-centered system" of mobility. In fact, it is inserted into a globalized network, where local scale and global scale interact by reconfiguring centrality, axes, and internal and external connections of the city [1].

The creation of networks among the multiple places of the contemporary city is one of the processes that characterize the mobility of the inhabitants in the era of globalization: a new reticular dimension emerges, based on connections activated among places, exploited by individuals in their life experience; connections could be real (transportation infrastructures) and virtual (information about places published on the web or on social media, possibly produced by citizens).

Such a reticularity, produced by experience of individuals in urban space, is termed *rhizomatic*, resuming a concept born in the field of botany and then reelaborated in the philosophical field: "Compared to centric (even polycentric) systems, hierarchical communication, and predetermined connections, a *rhizome* is an acentric, nonhierarchical, and not meaningful system" ([4], page 33). The concept of *rhizome* is refined in spatial terms by Jacques Lévy: "a rhizome is the space of individual action in mobility, but also in the multiform relationship with other individuals" ([5], page 19). A further definition could be the following: "a rhizome is a family of networks, characterized by the absence of identifiable boundaries and a meeting between topological metric inside and topographic metrics outside" ([5], pages 18-19). In other words, a rhizome belongs to the topology metric, that is, to a discontinuous space, based on nodes and connections that produce a network without beginning, without end, and without well-defined boundaries because it is the result of the experience in space of individuals. From a very pragmatical point of view, we can guess that "a rhizome is a set of places frequently lived by a single user and by many users, on the basis of material and virtual connections among them."

Based on this perspective, understanding the contemporary city means understanding city users, that in turn means understanding their rhizomes.

2.2. The Role of Social Media and Big Data. It is now clear that the identification of rhizomes is the crucial point of our research, but this is not an easy task. The reason is that each person has his/her way to experience the city, conditioned by the places where he/she lives, the places where he/she works, his/her interests, and so on.

An unexpected help comes from the popularity of social networks that stimulated new behaviors by people. In fact, many social network users continuously post messages about their day-by-day life. Through mobile apps for smartphones, they can post georeferenced messages that, if gathered, could reveal their movements and their habits. Many social networks provide an API (application programming interface) for getting posted messages; the result is that a huge amount of data can be easily gathered about habits of single users (people become sensors of themselves [6, 7]).

Consequently, in order to study rhizomes, we could rely on the potential of Big Data [8–11], by exploiting techniques and Big Data mapping systems. Big Data could foster the analysis of function and use of urban spaces based on the needs of inhabitants and city users. Data sets coming from an incredible variety of sources, such as social networks, mobile phone companies, public authorities, and national statistical institutes, could be cross-analyzed, aiming at understanding habits, flows, and relationships of inhabitants and city users (residents, commuters, tourists, migrants, etc.); the goal is to reveal different ways of experiencing the world and of managing the distances within it [12].

Such a kind of research has been already introduced in Europe by the *European Statistical System* and accepted by the Italian National Statistics (named ISTAT), particularly from the "Commission to study and guide the choices of ISTAT on Big Data." Public administrations could get great advantages in taking decisions about mobility infrastructures and many other issues, if they were provided with tools to exploit to improve their knowledge about how city users live in the city (a smart city should learn about itself).

2.3. The Urban Nexus Project: An Interdisciplinary Approach to Study the Smart City. The *Urban Nexus* project (see the seminal paper [13] for more information) aims at developing a scientific and educational cooperation among *University of Bergamo* (Italy), *Ecole Polytechnique Fédérale de Lausanne* (Switzerland), and *Anglia Ruskin University* in Cambridge (UK) (the research project is an initiative of the University of Bergamo in Italy, under the scientific coordination of Emanuela Casti and Federica Burini at the CST-DiatheisLab; see: http://urbannexus.unibg.it). The overall goal is to *develop a methodology* to analyze contemporary cities through an integrated and interdisciplinary approach, in order to foster renovation and improvement of accessibility to the city. The three cities hosting the involved universities, that is, Bergamo (Italy), Lausanne (Switzerland), and Cambridge (UK) will be three real case studies.

The Urban Nexus project aims at involving researchers with different competences, mainly, geographers, spatial analysts, and computer scientists. In fact, in order to collect and cross-analyze large amounts of data, skilled computer scientists are necessary; they have to provide flexible tools, able to manage many and possibly huge data sets, each of them possibly having different formats and heterogeneous structures. Nevertheless, the choice and analysis of data must be carried on, and the results are interpreted by geographers and spatial analysts.

The activities of the project are devoted to provide two main contributions:

(i) The definition of an *investigation methodology* that, relying on the concept of rhizome, is able to drive researchers to analyze data coming from various Big Data sources in order to understand the contemporary city.

(ii) The development of a novel *digital environment* that provides tools and frameworks able to effectively support the investigation methodology; in fact, nowadays, powerful software tools are a necessary condition to effectively perform social and geographical analysis based on Big Data.

2.4. Technological Context. The term *Big Data* is a buzzword very poplar nowadays. Behind it, many aspects are hidden, so that many researchers have been working on the topic.

In particular, a famous paper is the study by Kitchin [14], where the 3 Vs model is proposed, in order to characterize the topic: Big Data means *Volume, Velocity,* and *Variety.*

This observation is important, because people usually think about Big Data more or less only in terms of volume; however, data change very fast and can come very fast; furthermore, they can be really various, that is, a large number of different data sets coming from a large number of sources should be integrated together, in order to get the desired results. Our context is characterized by volume and variety.

Currently, technological approaches to data exchange and diffusion has converged on JSON (JavaScript Object Notation) as de facto standard for information interchange. In fact, social network APIs use JSON to represent data sets they can export. Many web services adopt the same approach, and on open data portals, JSON data sets are becoming more and more popular, in place of CSV and XML. In particular, as far as XML is concerned, JSON is substantially playing the role for which XML was designed: the reason is that an XML document is hard to manage within programs; on the contrary, object-oriented data structures of programming languages can be easily serialized into JSON, and JSON files can be easily deserialized into object-oriented main memory data structures.

On the same track, the GIS community as well has adopted JSON: the recent standard named *GeoJSON* [15] is a JSON-based format to describe geographical information layers.

The consequence of the diffusion of JSON is that DBMSs able to store and manage large collections of JSON data sets have become necessary. On this technology track, MongoDB [16] is certainly the most famous NoSQL database system (see [17] for an overview about NoSQL databases): in fact, it is able to store and query heterogeneous collections of JSON objects (i.e., objects with different structures) in the same collection. In spite of its popularity, the query language is specifically taught for programmers, being based on JavaScript; consequently, it is not suitable for geographers and analysis that are not familiar with hard procedural programming.

The reader can guess that large data sets, as well as a large number of data sets, are difficult to analyze by hands. Useful techniques that could greatly help in analyzing such a mess of data could come from the Information Retrieval area: in particular, the well known *Page-Rank* algorithm [18] could be a good starting point. The first version of Google search engine was based on it; its strength is the reticular view of linked web pages; consequently, we expect that the approach could be adopted for and adapted to the analysis of reticularity of contemporary cities.

More than twenty years ago, the area of data mining has provided very interesting techniques to analyze frequent cooccurrences of items in market-basket analysis, known as *itemset mining* and *association rule mining* [19–21]. Originally developed for the retail industry and applied to relational databases [22], they are very useful for a variety of problems. In our context, they could be applied to address the problem of identifying rhizomes of city users, moving from traces of people: in fact, a rhizome could be seen as the frequent cooccurrence of places in tours/trips performed by city users. Furthermore, other techniques developed on the side of itemset mining and association rules mining could provide inspiration for developing novel techniques and tools. In particular, the need for integrating several data sets may recall the techniques that make effective the integration of association rules extracted from within multiple transaction databases (see [23–26]).

3. Methodological Approach

The first contribution provided by the project is the definition of an investigation methodology, toward the identification of rhizomes from within data describing mobile users of social media. Independently of the specific data set and independently of the specific goal of the analysis, the analysis process should proceed performing the following steps:

(1) *Identification of sources*: it is necessary to identify sources of potentially useful data sets that should be as meaningful as possible.

(2) *Selection, transformation, and analysis* of data produced by mobile social media users and their *cartographic representation* to understand mobility in urban spaces.

(3) *(Towards the) reconstruction of rhizomes* from mobile data produced by social media users.

All these steps will be separately described in the next subsections.

3.1. Identification of Sources.
The first step to perform is the collection of data sets, from relevant data sources. However, it is not easy to understand what data sources are relevant: different data sources provide different data sets that could give different points of view of the problem.

In particular, in order to analyze the mobility of inhabitants, the following sources could be considered.

(i) *Statistical sources.* National and European Statistics Institutions continuously provide data sets about citizenship in general and about mobility in particular. These data sets are *certified* in the sense that their quality is controlled, and they can be considered *reliable*. Of course, since they are the result of a rigorous selection and analysis process, they are not up to date. Furthermore, they do not describe habits of single people.

(ii) *Collaborative sources* rely on the wish of people to participate to an investigation, typically through web applications, social media, and mobile apps. The participatory process can generate very useful crowd-sourced information and Volunteered Geographic Information (VGI) concerning single-city users [27–30].

(iii) *Private sources* could effectively integrate previous data sets. In particular, as far as detection of mobility is concerned, mobile phone companies own very detailed data sets about movements of phone users. The data sets they could provide are very large and very rich, in terms of information about the sequence of visited places and how long a person stays in a place. However, these data sets are usually difficult to be obtained because phone companies are jealous of their data.

Thus, depending on the selected data sets, more or less complex and more or less long acquisition processes must be carried on to get the desired data sets.

3.2. Selection, Transformation, Analysis, and Cartographic Representation.
Once gathered, data sets must be put in a form suitable for getting useful information. Several activities could be done.

(i) Selection. Among all data sets, it is necessary to select only the relevant data. Many considerations can affect this task, for example, the typology of places, the typology of city users, the time period of interest, and so on.

(ii) Transformation. To analyze selected data (or, better, to cross-analyze selected data), a process of transformation and integration is necessary. For example, if places visited by users are characterized by the latitude and longitude, perhaps it could be necessary to perform a geo-coding activity, that is, substitute punctual coordinates with the name of places.

(iii) Analysis. Transformed data sets should be analyzed by analysis, in order to get some useful hints about what interesting information could it be possible to extract.

(iv) Cartographic Representation. The analysis could be effectively supported by a cartographic representation of selected and transformed data, in order to conduct a visual-analysis process.

3.3. (Towards the) Reconstruction of Rhizomes.
Following a reflective approach [31], the goal of this step is to analyze the rhizomatic spatiality that emerges from data sets describing city users. We identified a multiparadigm approach, that is, multiple paradigms can be jointly applied to analyze data from different perspectives.

3.3.1. Site Analysis.
Places that emerge from data sets (e.g., visited by single users) constitute a reticular view of the space, where they are the nodes of a network. Several dimensions characterize the analysis.

(a) Localization: by analyzing images and text, locations could be identified or better characterized with respect to simple coordinates.

(b) Time analysis: time elapsed on a node and distance in time or in space between two nodes could reveal interesting information (e.g., about accessibility of nodes).

(c) Categories of nodes: nodes (places) can be categorized, in order to perform a more specific analysis; for example, a place can be categorized as *public spaces* (e.g., open spaces such as squares and roads and mobility places such as stations, airports, and public urban transport stops), *semipublic spaces* (such as monuments, religious, and historical places), or *semiprivate places* (such as hotels, restaurants, and shops), as identified by Jaques Lévy ([1], page 57).

(d) Opinion of users: the opinion of users in relation to places could give a sentiment polarity about it.

3.3.2. Connection Analysis (Networks). The reticular view of cities demands for analyzing connections between nodes. Aspects to discover are hereafter reported.

(a) *Accessibility* (cycle-pedestrian, automobile, public transport on wheel or on iron, air).

(b) *Quantitative relationships between nodes*: a ranking method could reveal the strength of relationships between nodes, on the basis of the reticular perspective. In this respect, we defined an algorithm named *Node Rank*, described in Section 4.5.1.

On the same line, data mining tools, such as frequent itemset mining [19] are necessary, in order to reconstruct rhizomes based on a quantitative approach.

3.3.3. Cartographic Representation. This is a necessary support to analysts. However, traditional cartography could not be satisfactory, in order to visualize relationships on the reticular space; for this reasons, we are going to perform research work by experimenting representations that possibly meet the reflexive approach (inspired by [31]).

The investigation methodology so far presented cannot be deployed without software tools that assist analysts during the overall process.

4. The Urban Nexus Digital Environment

The investigation methodology described in Section 3 must be supported by several software tools. These tools constitute the *Urban Nexus Digital Environment*. From the technical point of view, this environment constitutes the main result of the project, with tools and frameworks specifically developed moving from needs merged during the definition of the investigation methodology.

Figure 1 depicts the digital environment, showing the components that we describe hereafter.

The *Urban Nexus digital environment* is divided in four sections, depending on the task performed by tools and analysts:

(i) On the left-hand side, we find the *Data Acquisition* section that covers tasks devoted to acquiring data sets. In particular, we consider, as data source, *Open Data Portals*, that have become valuable and precious data sources, and *Twitter*. We also generically consider any kind of data source that provides useful data.

(ii) On the top of the figure, we find the Integration and Transformation section, devoted to integrate and transform the collected data. Usually, integration

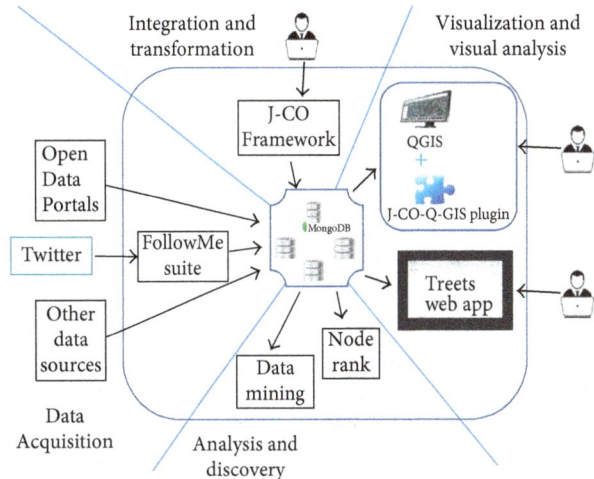

FIGURE 1: The *Urban Nexus digital environment*.

and transformation tasks are complex and, often, tedious activities, but they are crucial for conducting investigation activities.

(iii) On the right-hand side, we find the Visualization and Visual Analysis section that copes with those tasks performed by analysts based on a visual analysis of the data, as well as with mapping of data.

(iv) On the bottom of the figure, we find the Analysis and Discovery section that copes with tasks requiring massive analysis of data, based on possibly novel data mining and knowledge discovery algorithm.

In the center of the digital environment, we find data sets: in fact, source data, intermediate results, and final results are the value of investigation activities. Nevertheless, data must be gathered, integrated, transformed, analyzed, and visualized. Such tasks are performed by the tools that constitute the *Urban Nexus digital environment*.

The investigation methodology asks to collect data coming from many sources, such as social media like *Twitter*, open data portals, and many other public sources of information. Such data sets are usually provided as JSON collections, due to its capability of describing complex and heterogeneous data. Nowadays, JSON has become the de facto standard representation adopted to share social media data and open data sets. Consequently, the storage service in the *Urban Nexus digital environment* is provided by *MongoDB*, the very popular NoSQL DBMS that natively stores collections of JSON objects.

Let us briefly describe the components of the *Urban Nexus digital environment*.

(i) The FollowMe Suite gathers data from Twitter, in order to discover traveling users and track them. It monitors a pool of airports to detect users that posted geolocated posts in the airport area and tracks these users for 8 days, in order to find out traveling users.

(ii) The J-CO Framework is devoted to transform and query collections of possibly geotagged JSON data sets. It provides a query language, named J-CO-QL, that allows analysts to specify complex high-level queries by means of declarative operators that do not require programming skills; operators of the language deal with spatial representation in a native way.

(iii) QGIS is the very famous free GIS software that can be used to visualize geolocated data so far gathered and stored within MongoDB databases. A plug-in, named J-CO-QGIS, was developed to allow easy visualization of JSON geotagged data sets obtained by means of the J-CO Framework.

(iv) The Treets web app is a visual exploratory tool that allows analysts to explore traces of single users, by analyzing their posts (texts and pictures), by means of a cartographic representation of posts and traces.

(v) Node Rank is an algorithm developed to study the centrality of places in travelers' trajectories. Furthermore, other data mining tools, such as frequent itemset mining algorithms, are going to be used, in order to discover frequent patterns of places visited by tourists and, more in general, city users.

In the rest of this section, we introduce the components of the *Urban Nexus digital environment*.

4.1. The FollowMe Suite. The *FollowMe* suite [32] is an open and interoperable pool of tools [33], developed to discover and track social network users through their geotagged posts. The suite originates from the idea of trying to track *Twitter* users, but the suite is easily extensible and its architecture is open, so that new external components can be easily added (in fact, for a period, we also tracked *Instagram* users [34]). The goal is to gather nonauthoritative information about tourists that visit a given city. In particular, we are interested in tracking flying tourists, that post when they are waiting for their plane in the origin airport, or when they are waiting for their luggages in the destination airport.

We now illustrate the general overview of the approach and of the suite, illustrated in Figure 2.

The general vision is that *Twitter* users post a geotagged message when they are either in the origin or destination airport, at the beginning of their trips; then, during their trips, they capture pictures or write comments about a place, possibly by means of a geotagged post as well. All these posts are sent by the *Twitter* mobile app to the *Twitter Farm* that hosts servers and stores data. At this point, it is the turn of the *FollowMe* suite.

(i) The Hang Post Finder is responsible to query the Twitter API to look for messages posted in the area of a monitored airport (Twitter API provides a service to search for geolocated tweets, given the coordinates of the center and the radius of an area of interest). These messages are called hang posts, because they are the hang to discover users to follow.

(ii) The Timeline Tracker follows each user identified by means of hang posts, by inspecting his/her timeline,

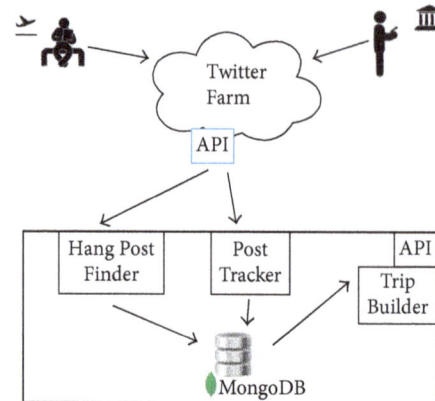

FIGURE 2: The *FollowMe* suite.

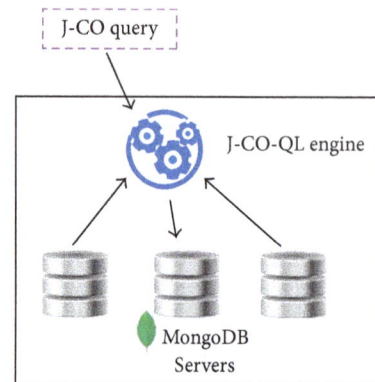

FIGURE 3: The *J-CO* framework.

that is, the history of messages posted by the user to find out geolocalized messages posted in the next 8 days after the date of the hang post.

(iii) Posts are stored within a MongoDB database as JSON objects.

(iv) The Trip Builder is the component invoked to query the database of gathered messages and reconstruct trips, in order to later study them. The *Trip Builder* can be invoked through the API that provides an external access to this query service. Again, JSON is adopted to generate output data sets, as well as classical CSV files (loosing, in this case, some information and flexibility of representation).

4.2. The J-CO Framework. The framework we conceived for cross-analyzing multisource georeferenced data is named *J-CO* (*JSON COllections*) and is constituted by the following main components depicted in Figure 3:

(i) One or more *NoSQL* databases, namely *MongoDB*, to support JSON objects storage from multiple sources

(ii) A *Data Model* that makes explicit the role of geotagging (when present) in collected JSON objects, as well as a clear *Execution Model* on top of which to build a novel query language

(iii) A novel query language, named *J-CO-QL*, that provides high-level operators for transforming heterogeneous collections of JSON objects, able to deal in a native and simple way with geotagging and spatial relationships

(iv) The *J-CO-QL* Engine that evaluates queries by reading/writing data in a *MongoDB* database (it receives queries and executes them on top of the NoSQL servers)

4.2.1. Data Model. The basic concept on which we rely to define the *J-CO-QL* language is the JSON object. Fields (object properties) can be simple (numbers or strings), complex (i.e., nested objects), and vectors (of numbers, strings, and objects). As far as the georeference field is concerned, we rely on the *GeoJSON* standard [15, 35]. In particular, we assume an object's geometry field named ~geometry, defined as a *GeometryCollection* type for the *GeoJSON* standard. The absence of this top-level field means that the object does not have an explicit geotag.

As an example, consider the JSON objects reported in Figure 4, describing *Points of Interest* (POIs), in the city of Bergamo. In this example, two shops are represented with a point-like georeference: notice the ~geometry field, that describes the longitude and latitude (resp.) of the centroids of the shops' locations on Earth, as defined by *WGS* (*World Geodetic System*) *84* (this standard is our default *CRS–Coordinate Reference System*).

In a *NoSQL* environment such as *MongoDB*, a *Database* is defined as a set of collections c, while a *Collection* is represented by a name c.name and its instance, that is, a vector of JSON objects. To manipulate JSON collections and to store their results in new collections, we need operators (Section 4.2.2) that satisfy the *closure property*, that is, they get collections and generate collections. This is a first characteristic of the *J-CO-QL* language.

J-CO-QL queries transform collections stored in databases and generate new collections which will be stored in turn into a possibly different database, to achieve data persistence. For simplicity, we call such databases as *Persistent Databases*.

4.2.2. Query Process. A *state s* of a query process is a pair *s* = (tc, IR), where tc is a collection named *Temporary Collection*, while IR is a database named *Intermediate Results database*.

Each operator application starts from a given query process state and generates a new query process state. When a query is executed by the *J-CO-QL* Engine, the resulting tc (*Temporary Collection*) will be optionally stored to an IR (*Intermediate Results*) database, that could be taken as input by a subsequent operator application.

The query process starts from the empty temporary collection $s_0 \cdot tc = \varnothing$ and the empty intermediate results database $s_0 \cdot IR = \varnothing$. Thus, *J-CO-QL* provides operators (named *start operators*) able to start the computation, taking collections from the persistent databases, while other

```
[
    {
        "id": 321,
        "category": "shop",
        "name": "Shop 321",
        "~geometry":
        {
            "coordinates": [
                9.186973,
                45.467843
            ],
            "type": "Point"
        }
    },
    {
        "id": 456,
        "category": "shop",
        "name": "Shop 456",
        "~geometry":
        {
            "coordinates": [
                9.205654,
                45.477872
            ],
            "type": "Point"
        }
    }, ...
]
```

FIGURE 4: Excerpt of collection POIs.

operators (named *carry-on operators*) carry on the process continuously transforming the temporary collection and possibly saving it into the persistent databases or, for temporary results, into the intermediate results database IR. This way, new subtasks can be started, by taking collections both from persistent databases and from IR.

Innovative Features. The need to support complex transformation processes that typically pass through the generation of several intermediate results motivates the intermediate results database IR. Intermediate collections are stored explicitly into IR, to be later used for creating the target collection. In fact, it would be inappropriate to store them into the persistent databases that should store source and target collections. Collections stored in IR are clearly intermediate, and will disappear from the system at the end of the process.

4.2.3. J-CO-QL: Language and Execution Engine. The key components of the *J-CO* Framework are the query language, named *J-CO-QL*, and its execution engine. With respect to other query languages for *JSON* objects, the main innovations provided by *J-CO-QL* are the following:

(i) Typically, other query languages for JSON collections are unable to deal with heterogeneous objects in the same collection at the same time (in general, several queries must be written and then their results united together). J-CO-QL provides operators

[key]	date	postId	time	userId	~geometry		
0	2015-04-01	58313545517835690	06:10:20	John	[-] Object, 2 properties		
					coordinates	[-] Array, 2 items	
						0	45.66561062
						1	9.69930355
					type	Point	
1	2015-04-06	58313905695025460	06:54:01	Alketa	[-] Object, 2 properties		
					coordinates	[-] Array, 2 items	
						0	45.66561062
						1	9.69930355
					type	Point	

(a)

[key]	category	id	~geometry		
0	shop	7547	[-] Object, 2 properties		
			coordinates	[-] Array, 2 items	
				0	9.44838445411
				1	45.7409545188
			type	Point	
1	shop	7562	[-] Object, 2 properties		
			coordinates	[-] Array, 2 items	
				0	9.696728220119999
				1	45.6921785928
			type	Point	

(b)

FIGURE 5: (a) Objects from collection TrackedPosts, describing tracked posts. (b) Objects from collection POIs.

specifically designed to deal with objects with different structure within the same operator application.

(ii) Second, other languages for querying JSON data are conceived for programmers, or for people having programming skills. In contrast, operators provided by J-CO-QL are high-level operators that allow analysts to think directly to objects structure; they do not have to write low-level procedures.

(iii) Finally, but not less important, J-CO-QL directly deals with georeference possibly contained in JSON objects because the data model explicitly deals with them through the ~geometry field.

In the rest of the paper, we will present relevant operators of *J-CO-QL* by explaining their use in transformation tasks. The reader can refer to [36], where the language is introduced in more details.

4.2.4. J-CO-QL by Example: Cross-Analyzing Tweets and POIs.
In this section, we will show the power of *J-CO-QL*, by showing its application to a real case concerning smart cities. Meanwhile, a sketch of fundamental operators is given; for a detailed introduction, refer to our internal report [37].

The first collection to consider is named TrackedPosts; it contains georeferenced posts gathered from *Twitter* by the *FollowMe* suite (Section 4.1) since May 1, 2015. The second collection is named POIs and contains georeferenced Points of Interest. We have two specific goals: (a) we want to discover the most attractive POIs in the area of the city of Bergamo; (b) we want to discover tourist traces that mostly visited POIs.

Each post in collection TrackedPosts is described by a JSON object having several fields, that is, postId, userId, date, and time. An excerpt of this collection is reported in Figure 5(a), where objects are pretty printed in a graphical way. Notice the ~geometry field: it was not present in the original JSON representation of posts; it has been added by performing some preprocessing activities by means of *J-CO-QL* (they are not reported here because they could be tedious for the reader).

As far as Points of Interest (POIs) are concerned, we started from the 4 collections we gathered from Open Data Portals, one for each POI category: *hotel*, *shop*, *museum*, and *architectural*. Similarly to the preprocessing task performed for posts, by means of the *J-CO-QL* we preprocessed them in order to obtain a unique and homogeneous collection of JSON objects, containing all the POIs, named POIs. Fields of objects are id, category (which denotes the category of the POI, e.g., hotel, shop, museum, and architectural), and ~geometry. An excerpt of collection POIs is reported in Figure 5(b).

Moving from collections TrackedPosts and POIs, we can perform transformations suitable to reach our goals, that is, (a) to rank POIs on the basis of the number of posts nearby and (b) rank user traces with respect to the number of POIs of a given type their posts are nearby. Both goals rely on the ability of discovering which posts are nearby which POI. To this end, the *J-CO-QL* language provides the SPATIAL JOIN operator, able to couple objects coming from two collections on the basis of a (metric or topological) spatial condition on geometries. This operator is the key of the process that is reported in Figures 6 and 7. Hereafter, we will describe the process in details, by introducing the *J-CO-QL* operators.

4.2.5. Ranking POIs and Posts.
The general goal of the analysis process is to compute a score for POIs and a score for traces. The score for POIs is defined by

$$\text{Sc}(p) = \left| \{\text{tp} \in \text{TrackedPosts} | \text{dist}(p, \text{tp}) \leq 100\} \right|. \quad (1)$$

The score $\text{Sc}(p)$ of a POI p is the number of posts with distance less than 100 meters from the POI's coordinates (function dist gives the distance in meters).

As far as users traces are concerned, we rank them with respect to the four categories of POIs we considered, separately. The score is defined in

$$\text{Sc}(t, c) = \sum_{\text{tp}_j \in t, p_i \in \text{POIs100}(c, \text{tp}_j)} \frac{1}{1 + \text{dist}(p_i, \text{tp}_j)}. \quad (2)$$

For a trace t and a POI category c, the score $\text{Sc}(t, c)$ is the sum of inverse of the distance (in meters) between

```
SPATIAL JOIN OF COLLECTIONS
  TrackedPosts@SmartData as post,
  POIs@SmartData as poi
ON DISTANCE(M) <= 100 SET GEOMETRY RIGHT
CASE WHERE
      WITH .poi.category, .poi.id,
           .post.postId, .post.userId
      GENERATE
        {category: .poi.category,
         poiId: .poi.id
         postId: .post.postId,
         userId: .post.userId,
         dist: DISTANCE(M),
         inv_dist: 1 / (1 + DISTANCE(M))}
      KEEPING GEOMETRY
  DROP OTHERS;
SET INTERMEDIATE AS PostsAndPOIs;

GROUP
  PARTITION WITH STRING .poiId
   BY .poiId, .category
   INTO posts
   GENERATE {.category, .poiId, .posts,
            count: SIZE(.posts)}
     SETTING GEOMETRY .posts[1].(~geometry)
  DROP OTHERS;
SAVE AS WeightedPOIs@SmartData;
```

FIGURE 6: Query that computes the scores for POIs.

```
GET COLLECTION PostsAndPOIs;
GROUP
 PARTITION WITH STRING .poiId
  BY .userId, .category
  INTO posts
  GENERATE {.category, .userId, .posts,
            score: SUM(.posts, .inv_dist)}
     DROPPING GEOMETRY
DROP OTHERS;

FILTER
 CASE WHERE score>= 0.03
  GENERATE {.userId, .category, .score}
DROP OTHERS;
SET INTERMEDIATE AS RankedTraces;

JOIN OF COLLECTIONS
    TrackedPosts@SmartCities AS post,
    RankedTraces AS RT
 CASE WHERE .post.userId = .RT.userId
 GENERATE {score: .post.score,
           postId: .post.postId,
           userId: .post.userId,
           category: .RT.category}
    SETTING GEOMETRY AS .Post.~geometry
DROP OTHERS;

GROUP
 PARTITION WITH .score, .postid,
                .userId, .~category
  BY .score, .userId, .category
  INTO posts
  GENERATE {.score, .userId, .category}
    SETTING GEOMETRY AS POLYLINE(.posts)
 DROP OTHERS;
SAVE AS RankedTracks@SmartCities;
```

FIGURE 7: Query that computes the scores for tourists' traces.

a point tp_j in trace t and a POI p_i of category c at distance no greater than 100 meters from tp_j. (POIs $100(c, tp_j)$ is the set of POIs of category c at distance no greater than 100 meters from tp_j), plus 1 meter. This way, each term cannot be infinite, and it is always less than or equal 1 (it is 1 when tp_j and p_i have the same coordinates). Note that we have four scores for each trace, one for each POI category.

So, the concrete problem of the analyst is to compute the scores for POIs and traces. In the following, we will describe how a *J-CO* user could perform the analysis described above, by executing two simple *J-CO-QL* queries.

(1) Computing Scores for POIs. The first query (Figure 6) has the goal of creating a new collection named WeightedPOIs, where each object describes a POI and has a score field which is the number of posts nearby to the POI. We can split the query in two parts: the first one executes a SPATIAL JOIN between collections TrackedPosts and POIs; the second part performs a GROUP operation, in order to group together all posts nearby to the same POI and derive the score field (i.e., the number of nearby posts).

The key operator for this query is the SPATIAL JOIN operator (introduced in our language in [38]). It takes two collections (either stored in a *MongoDB* persistent database or in the intermediate results database IR) and builds pairs of objects coming from the two input collections, on the basis of the relationship existing between their geometries. In this case, we join objects l_i from collection TrackedPosts, with objects r_j from collection POIs, creating an object $t_{i,j}$ for every i, j. Both source collections are stored in the persistent database SmartData. The two collections are aliased, within the operator, as post and poi, respectively. As far as the

geometry of $t_{i,j}$ is concerned, in our case, we specified SET GEOMETRY RIGHT, meaning that we keep the geometry of the right object r_j (in this case, the POI); alternatively, we can specify LEFT, INTERSECTION, ALL (which merges the two geometries). Notice that the output object $t_{i,j}$ has three fields: one has the name of the left collection alias (i.e., post) and its value is the left object l_i; one has the name of the right collection alias (i.e., poi) and its value is the right object r_j; the third one is the ~geometry field. Finally, the object $t_{i,j}$ is generated (i.e., the left and right objects are joined) only if the spatial join condition specified after the ON keyword is met: in this case, the distance between the geometries of the two objects must be no greater than 100 meters.

The WHERE condition after the CASE keyword selects output objects having the desired fields (WITH predicate): these objects are restructured by the GENERATE action, that also adds two fields, that is, the distance value and its inverse, respectively, named dist and inv_dist.

We then store those results in a new collection named PostsAndPOIs into the intermediate results database *IR*.

After that, the GROUP operation takes the before-generated collection and, by grouping the objects in the collection by values of fields, poiId, and category, creates a collection of distinct POIs. This collection is similar to the original POIs collection, but each object has a count field, that is, the number of posts with distance equal or less than 100 meters from the POI. Notice that field posts is the array of all posts grouped together; since in the previous spatial join, the geometry of POIs was kept, all grouped objects in array posts have the same geometry; thus, we take the ~geometry field of the first object in the array as ~geometry field of the overall object (SETTING GEOMETRY option in the GENERATE action of the GROUP operator).

The collection so far obtained is saved into the persistent database SmartData with name WeightedPOIs.

(2) Computing Scores for Traces. The second query (Figure 7), computes the scores, for each POI category, of tourists' traces. The query is composed by four parts:

(1) First of all, moving from the intermediate collection PostsAndPOIs, the GROUP operator computes the score of each trace for each category.

(2) The FILTER operator selects traces having score of at least 0.03 for at least one category, meaning that we want only traces with at least 3 posts in the surrounding of at least one of the given category.

(3) Tourists' traces are cloned, in order to associate them with all the different scores (one for each category), by the JOIN operator.

(4) Finally, the GROUP operator generates one object for each trace for each score category, in order to generate a geometry representing the trace by means of a polyline, to show it on the map.

In details, the first GROUP operator takes posts in collection PostsAndPOIs and groups objects having the same value for fields userId and category, generating the array field named posts, containing all posts grouped together. The GENERATE action adds the score field, by summing the values of field inv_dist of objects in the array posts.

The FILTER operator takes the temporary collection and keeps only objects that satisfy the WHERE condition; then, it generates a restructured version of them. In this case, we want traces (identified by userId and labeled by field category) with a score of at least 0.3. By using the GENERATE action, we obtain objects having only fields userId, category, and score. The output collection is stored into the intermediate results database with name RankedTraces.

In the third part of the query, the JOIN operator joins the intermediate collection RankedTraces (aliased as RT) with collection TrackedPosts (aliased as post) in the SmartCities persistent database. The WHERE condition specifies that we join objects of the two input collections if they have the same value for fields userId in both objects. The behavior is similar to the SPATIAL JOIN, apart from the ~geometry field, that is not automatically dealt with. In practice, we extend posts with the category and the score of the trace for that category. Generated objects have fields score, userId, and category; the ~geometry field is

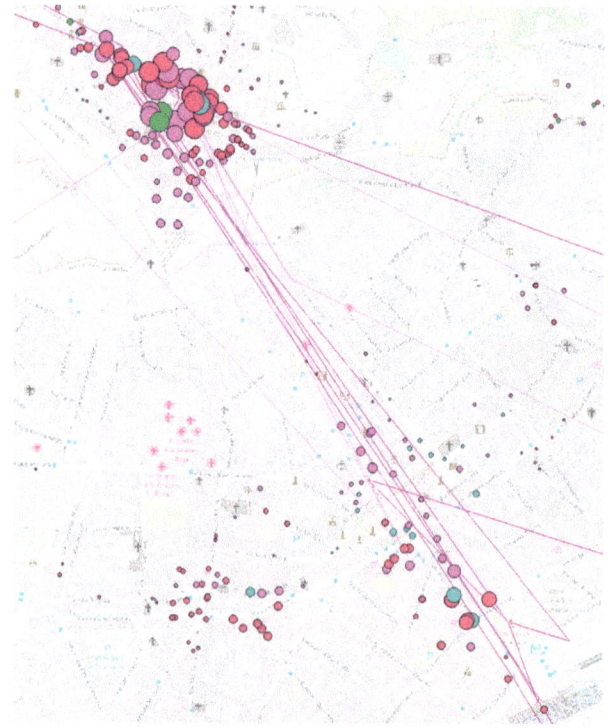

FIGURE 8: Visualization of POIs (circles), having more than one posts in its surroundings, with size proportional to their score (colored points) and tourists' traces (lines) having the best score for architectural POIs (the greater the score is, the higher the score is).

taken form the ~geometry of the post, as specified by the SETTING GEOMETRY clause.

Finally, we can create a collection of objects that describe a trace labeled with the category and the score, where ~geometry is a *polyline* representing the track of all posts in the trace. This is done by the last GROUP operator that groups objects (posts) on the basis of fields score, userId, and category, deriving the final ~geometry by specifying SETTING GEOMETRY AS POLYLINE, which derives a polyline from each point grouped into array posts. We finally store the output collection as RankedTraces in the SmartCities persistent database.

Actually, we clustered tourists' traces by the category of their favored POIs. In particular, if a user mostly visits architectural POIs (and posted georeferenced content nearby them), his/her track will be classified as an architectural trace. Otherwise, if a user only posts social georeferenced content when near to a shop, the *shop* category will be assigned to his/her trace.

In Figure 8, we can see a visualization for the POIs in distinct colors depending on their category (red, green, blue, and purple circles represent distinct categories of POIs), where each circle size represents the score (proportional to the number of posts nearby). Tourists' traces (lines) are depicted with the color of the POIs' category: it can be seen that the violet user traces come closer to many architectural POIs, than to other categories of POIs, since the violet color is associated with architectural POIs and the stronger is the color of the trace, the higher is the score of the trace.

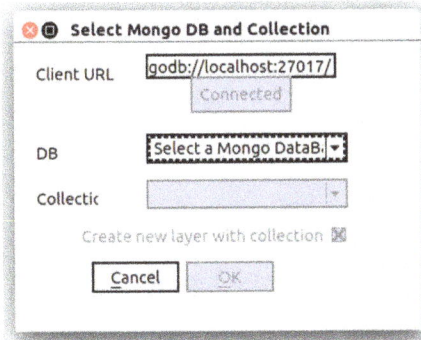

FIGURE 9: *J-CO-QGIS*: Db browser.

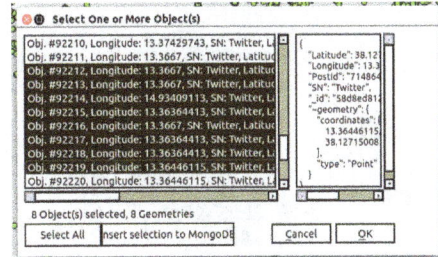

FIGURE 10: *J-CO-QGIS*: collection viewer.

FIGURE 11: Visualization of selected georeferenced posts (red stars) over points of interests in Bergamo (colored points).

4.3. The J-CO-QGIS Plug-In for QGIS. In order to provide analysts with a powerful tool for querying and visualizing georeferenced JSON data within classical *GIS* software, we developed a plug-in for *QGIS* (currently the most popular open-source free *GIS* software) named *J-CO-QGIS*.

J-CO-QGIS provides several features, in particular

(i) a DB Browser, which allows users to connect to MongoDB persistent databases and to select the collections to show;

(ii) a Collection Viewer, which takes collections from MongoDB persistent databases and loads those objects with geometrical representation (having the ~geometry field) into QGIS, so as to map them;

(iii) a Query Issuer, which is a text editor that allows users to write J-CO-QL queries and to send them to the *J-CO-QL Engine*.

When the plug-in is launched, it shows the *DB Browser* window (Figure 9): first of all, the user must specify the connection string to the *MongoDB* server (the connection string has the form mongodb://user:password@example.com/ the_database? authMechanism = SCRAM-SHA-1). By clicking on the *Connect* button, the plug-in actually connects to the specified *MongoDB* server and shows the list of available databases. By browsing this list, shown in the *DB* list-box, the user chooses the database from which to get the desired collection. The content of the *Collection* list-box is updated, with the list of collections available in the chosen database.

At this point, the *Collection Viewer* window is open (Figure 10). In this window, it is possible to select objects of interest to show as a *QGIS* information layer; the right-hand side area shows the full structure of selected *JSON* objects. The *Insert selection to MongoDB* button reexports the selected objects into a new collection within the persistent database, whose name will be asked in a prompt. The *OK* button shows the layer containing the selected objects on the map.

As an example, suppose we have two collections, named POIs and TrackedPosts (deeply described in Section 4.2.4). The former describes Points of Interest and the latter the location from which tourists wrote geotagged posts to a social network. Figure 11 depicts the layers created by selecting objects in the two collections: POIs are depicted as colored circles, while posts' locations are depicted as red stars (the user can choose how to visualize geotags).

4.4. The Treets Web Application. *Treets* is a web application developed to provide analysts with a visual-analysis tool for geolocated tweets gathered by the *FollowMe* suite (Section 4.1). It has been designed to enable people, that do not have coding or GIS skills to explore data, find out information from trajectories, texts, and pictures in tweets, to help them in the analysis process. To achieve this goal, *Treets* provides features such as filtering, visualization, and data exporting.

In this section, we present functionalities of *Treets*, to show how this visual-analysis tool can help analysts in their exploratory activity.

In the left-hand side of the home page (Figure 12), a map dynamically shows traces and tweets. The panel in the right-hand side of the home page shows some information and provides filters (later discussed and shown in Figure 13).

On the map, tweets are represented as points. Clicking on a point, a pop-up window appears, containing details of the chosen tweet, including text, images, and external links (Figure 14). By clicking on the pop-up window, a new browser tab is open, and the user is redirected to *Twitter* web site, namely to the page showing the tweet itself.

On the map in the home page of *Treets*, tweets posted by the same user are linked in the chronological order by colored lines forming a *trace*; each trace has a different color. By clicking on a trace, a pop-up window appears with the user name, the number of tweets in the trace and a button to export tweets in the trace as a CSV (Comma Separated Value) file.

As far as filtering of tweets shown on the map is concerned, *Treets* provides five different alternatives:

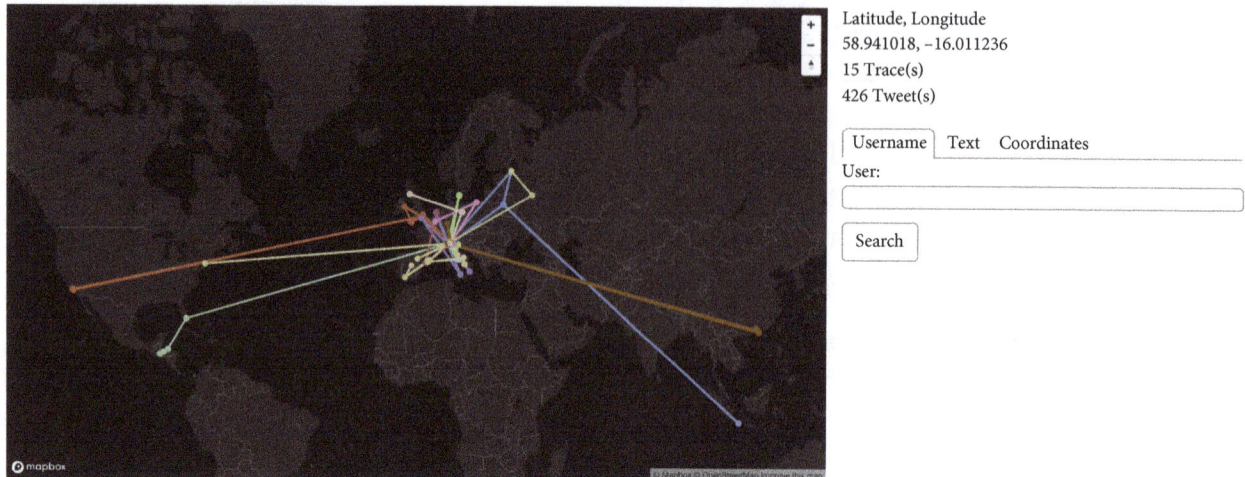

FIGURE 12: *Treets* visual-analysis tool: home page.

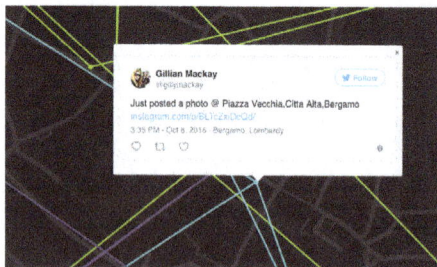

FIGURE 13: *Treets*: search form.

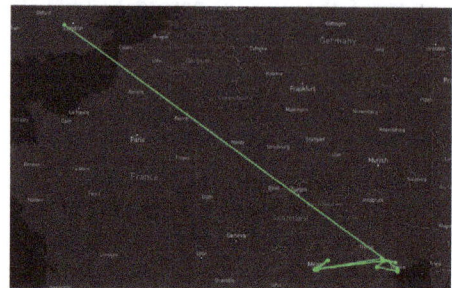

FIGURE 15: *Treets*: trace of a Californian tourist.

FIGURE 14: *Treets*: visual inspection of a tweet on the map.

FIGURE 16: *Treets*: traces with at least a tweet in the search (blue) area.

(i) *By Username*: to retrieve the trace of a desired user (see, e.g., the trace in Figure 15, that shows the trip of a Californian tourist).

(ii) *By Text*: to find tweets containing the desired text and display the full traces they are part of on the map. Double quoted texts permit to find an entire sentence within texts e.g., to look for Accademia Carrara, the ancient art museum in Bergamo, the search string is "Accademia Car-rara," which is different from "Carrara Academy" (anyway, the search is performed in a case-insensitive way).

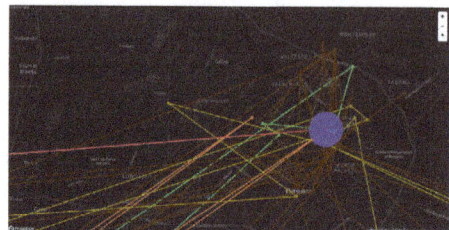

(iii) *By Location*: by double clicking on the map, the Latitude and Longitude fields in the search form will be automatically filled (they can also be manually edited); thus, the desired radius of the circumference can be specified; the *Show Circle* button shows the search circle on the map; the *Search* button will filter tweets falling in the search circle and all traces with at least one selected tweet will be shown on the map (Figure 16).

(iv) *By Location and Text*: this search option allows analysts to combine location-based and text-based filtering.

(v) *By J-CO-QL*: J-CO-QL queries can be specified; furthermore, it is also possible to select a *MongoDB* server and a database, from which to select a collection of JSON objects to show (only geotagged objects) on the map; this way, more skilled analysts can make use of one single web tool for both performing integration and transformation of data sets and to visualize them on the map.

As an example, we investigated tweets gathered by the *FollowMe* suite in order to discover unusual tourists that visited the city of Bergamo, with respect to our expectation of *usual tourist*. By performing a text search, looking for tweets talking about *Accademia Carrara*, that is, the ancient art museum in Bergamo, we found out the tourist whose trace is shown in Figure 15: this is a Californian tourist, that landed in Milan Linate airport, visited some museums in Milan, then went to Bergamo to spend some days, and then moved to Venice making some intermediate stages. Looking at texts, it is clear that this tourist is an art passionate. Notice that even though this tourist is not statistically relevant alone, an analysis could reveal interesting outcomes, suggesting improvement of synergy among museums and cities, building a network in northern Italy to attract such kind of tourists. Furthermore, from a technical point of view, it suggests to develop tools to perform analysis and discovery tasks on an automated basis that would never be possible to conduct by hand.

4.5. Analysis and Discovery Tools.

In the *Urban Nexus digital environment*, this section is the newest and less developed. The idea is to develop tools that perform knowledge discovery tasks. The reader can think about data mining tools, but also other kind of analysis are acceptable; it depends on what the methodology asks for.

Currently, we identified the following needs:

(i) Analysis of paths followed by users, in terms of visited places and visited POIs.

(ii) Discovering the most frequent patterns of personal spatial dimensions, by means of frequent itemset mining techniques.

(iii) Trajectory analysis techniques could be developed, in order to better understand how city users move in the city.

Currently, we have worked on data collected by the *FollowMe* suites, that is, geolocated tweets of possibly traveling users. The analysis of such a data set inspired a novel technique to study the *centrality* of places, that is, rank places to let the central places in tours emerge; this technique is described in Section 4.5.1.

As far as discovering the most frequent patterns of personal spatial dimensions is concerned, we are going to address this problem. In our opinion, having the set of places visited by one single person, the well-known *frequent itemset mining* technique [19–21] could be very helpful in revealing common patterns of space usage. In particular, we could find out common patterns of one single person, as well as common patterns of groups of people; we also foresee that unexpected communities could emerge. Certainly, we are going to apply this approach to collected tweets, but we plan to adopt a volunteered approach to gather more detailed data.

Finally, better insights could be provided by trajectory analysis techniques. In previous works [39, 40], some algorithms for clustering trajectories detected from tweets gathered by the *FollowMe* suite were developed and tested. However, we think that the results are not satisfactory for the investigation methodology developed within the *Urban Nexus* project; consequently, they are not part of the *Urban Nexus digital environment*.

4.5.1. Node Rank.

We now present the *Node Rank* technique, developed to discover centrality of places visited by tourists.

Consider a corpus of *geolocated* tweets, gathered by the *FollowMe* suite (Section 4.1), containing the traces of tourists that visited a given city. A tweet t has a *date_time* property, that is, the date and time the tweet was posted, as well as a *user_id*, denoting that the user that posted the tweet. For each user, it is possible to sort his/her tweets on the basis of property *date_time*, thus obtaining the *trip* of user *user_id*:

$$\text{Trip (user\_id)} = \langle t_1, \ldots, t_n \rangle, \tag{3}$$

where $n \geq 1$ and such that, for each pair t_i and $t_{(i+1)}$, it is $t_i \cdot \text{date\_time} < t_{(i+1)} \cdot \text{date\_time}$.

Being geolocated, that is, with associated latitude and longitude of a point on the earth, a tweet t corresponds either to the place from which the user posted the tweet, or to the place the user selected while geotagging the tweet. The consequence is that a trip describes in which order a user moved and which places he/she visited, or places he/she was referring to.

Thus, we can represent a trip on a graph, where nodes correspond to places (on the basis of their coordinates), while edges represent a move of a user from a place denoted by a tweet t_i to a place denoted by a tweet $t_{(i+1)}$. Obviously, in our analysis, it is not relevant to study a single trip, but all trips together, so in the same graph, we represent all collected trips.

As an example, Figure 17 shows a sample graph that represents 13 places (nodes) and 12 moves (edges) that are obtained from several trips of different users described by their tweets. As far as nodes are concerned, the bold number in the top left corner is the node identifier, while the number reported in the center of nodes is the number of tweets posted in that place. As far as edges are concerned, the number reported beside an edge denotes the number of moves emerged from tweets.

At this point, we can formulate the problem: *we want to define a ranking measure for places, depending on the movement of citizens, able to measure if a place is central, on the basis of incoming and outgoing moves (edges).*

Hereafter, we present the formal definition of the *Node Rank* approach. A complete description can be found in [41].

The Node Rank Method. The *Node Rank* method was inspired by the *Page-Rank* method [18, 42], due to the similarity between hypertextual links and moves of tourists in a network modeling a city. The idea is that when a tourist is visiting a city, his/her decision to visit a place could be influenced not only by the popularity of places from which he/she come from, but also by the popularity of places he/she

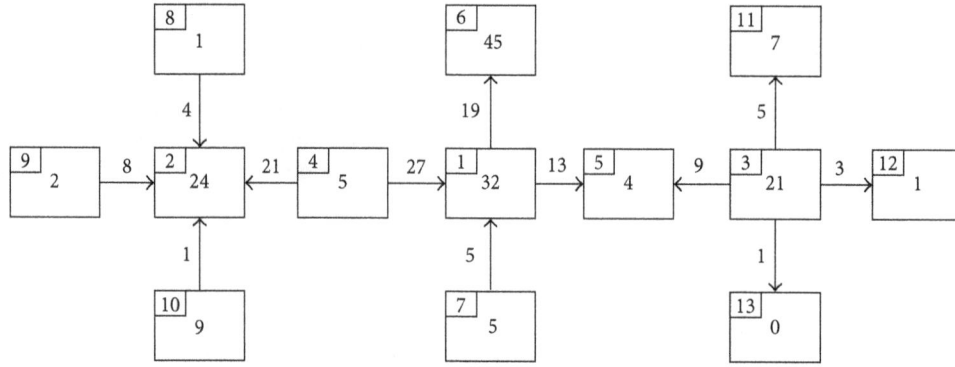

FIGURE 17: Sample graph of visited places.

is going to visit. Then, it is necessary to consider different kinds of relevance:

(i) *Inbound Relevance.* A place (a node in the graph) receives an inbound relevance, that is, a relevance influenced by the popularity of nodes from which inbound edges start.

(ii) *Outbound Relevance.* A place receives an outbound relevance, that is, a relevance influenced by the popularity of nodes to which outbound edges point to.

(iii) *Compound Relevance.* The relevance of each place is obtained by composing inbound relevance and outbound relevance.

To illustrate the idea, consider the sample graph depicted in Figure 17. If we consider the inbound relevance, node 2 should be intuitively the most relevant, since it has four incoming edges; this means that the associated place is of interest for many tourists and is the target where to end the trip. If we consider node 3, we can see the opposite situation, that is, it has four outgoing edges, so it represents a place that is important as starting point for tourist trips. Therefore, we argue that the rank of node 2 and node 3 should be the same.

But what about node 1? It has two incoming nodes and two outgoing nodes, so its role is different with respect to the role played by node 2 (a favorite target) and node 3 (a favorite starting point): it is crucial as intermediate place through many trips, that is, it is the place where many tourist passed, during their trips, in the middle of their paths. Since the goal of our research is to discover such crucial places, we need to combine both inbound relevance and outbound relevance.

The basic set of equations computes either the inbound relevance or the outbound relevance of each node u, denoted as $NR_{in}(u)$ and $NR_{out}(u)$, respectively.

$$NR_{in}(u) = (1-d) + d \sum_{v \in IB(u)} \frac{NR_{in}(v)}{N_{in}(v)},$$

$$(4)$$

$$NR_{out}(u) = (1-d) + d \sum_{v \in OB(u)} \frac{NR_{out}(v)}{N_{out}(v)},$$

where $N_{in}(u)$ is the number of incoming edges, while $N_{out}(u)$ is the number of outgoing edges. $IB(u)$ and $OB(u)$

denotes, respectively, nodes connected to u by an inbound edge and an outbound edge for node u.

Note that we obtain two systems of equations: one for inbound ranks and one for outbound ranks.

For each node, it is necessary to combine inbound and outbound ranks, by means of the following formula.

$$NR(u) = (NR_{in}(u) + NR_{out}(u)) \times adj(u). \quad (5)$$

The sum of the inbound and outbound ranks is corrected by means of an adjustment factor:

$$adj(u) = 1 - \frac{1}{1 + \log_k (1 + N_s(u) + N_{in}(u) + N_{out}(u))},$$

$$(6)$$

where $N_s(u)$ is the sum of number of single tweets, and self-loops for node u. The adj parameter has been introduced to deal with single tweet trips, that is, with users that posted only one tweet in their trips, as well as self-loops, that is, users that posted two or more consecutive tweets from the same place. The base of the logarithm k is based on the maximum value of

$$N_{tot} = \max_{u \in B} (N_s(u) + N_{in}(u) + N_{out}(u)), \quad (7)$$

that is, the place with maximum number of tweets (B is the total set of nodes in the graph). So, the base of the logarithm k is defined as

$$k = \sqrt[2]{N_{tot}}. \quad (8)$$

The final rank of node is the sum of the inbound rank and the outbound rank. However, we could obtain ranks that still precisely do not represent the actual importance of place: from our perspective, a place must be valuable for both tourism and mobility. For these reasons, given the same number of links, a node with variety of incoming and outgoing links is more important than uniform ones. Thus, the final ranks $\overline{NR}(u)$ are given by the following equations:

$$\overline{NR}(u) = (NR_{in}(u) + NR_{out}(u)) \times \left(1 + \frac{(N_{in}(u))/(N_{out}(u))}{perc}\right),$$

$$(9)$$

$$\overline{NR}(u) = (NR_{in}(u) + NR_{out}(u)) \times \left(1 + \frac{N_{out}(u)/N_{in}(u)}{perc}\right).$$

$$(10)$$

If $N_{in}(u) < N_{out}(u)$, (9) is used to compute the final score for node u, otherwise (10) is used. The correction factor (by which $(NR_{in}(u) + NR_{out}(u))$ is multiplied) is introduced, in order to benefit nodes with more variety, for instance nodes with the same number of incoming and outgoing edges receive the maximum boost, meanwhile nodes with only one of the two types of edges do not receive any boost. The factor *perc* must not be chosen equal to zero: if its value is positive and near to zero, it will produce a great boost, instead if its value is a large number it will nullify the possible increase; moreover, we can also choose a negative value if we want to give a negative boost.

For our purpose, perc = 5 is suitable, since it gives a maximum boost of 20%, with respect to the noncorrected ranks $(NR_{in}(u) + NR_{out}(u))$.

The algorithm we developed to solve the equations so far defined is outside the scope of this paper (see [41]). However, we report a sample result obtained by analyzing a pool of tweets concerning the city of Bergamo. Table 1 shows how places are ranked by the *Node Rank* approach, on the basis of analyzed tweets, while Table 2 shows the same places ordered on the basis of the simple number of tweets posted in that places. The reader can see how the order changes: the places that gain the top-most positions emerge as the key places through which most of tourists pass before visiting other places during their trip.

The potential applications of this approach are manifold, at various orders of magnitude: changing the scale, we can analyze local areas (such as cities), regions, countries, and so on.

5. Conclusions

The analysis of mobility by exploiting user-generated data through mobile devices is an ambitious task. In fact, it demands for an integrated multidisciplinary approach: geographers and analysts must define a suitable methodology to gather mobile-users data sets and investigate them; computer scientists must develop novel and powerful tools and frameworks to perform very complex tasks on data, such as data gathering, transformation, integration, analysis, and visualization. This is the ambition of the *Urban Nexus* project.

Based on this premise, we can summarize the contributions of this paper. First of all, we introduced a novel perspective toward smart cities, where a city is smart since it learns about how city users live in the city, in order to improve infrastructures and services on the basis of needs in terms of mobility; the concept of rhizome, born in other scientific areas, is providing a novel and crucial approach, supporting the reticular view of city-life experience.

Second, the paper provides an investigation methodology that could be used to drive any kind of analysis process of mobile-user data, towards the reconstruction of rhizomes.

Third of all, we show the large variety of tools necessary to perform such an investigation, moving from a real-life data set gathered from *Twitter*; this data set contains traces of traveling people through Europe, reconstructed by following georeferenced messages they posted during their trips. Then,

TABLE 1: Rank places of Bergamo.

Place	Score
Aeroporto, Orio al Serio	121,60,882
Oriocenter, Orio al Serio	1,915046
Piazza vecchia, Città Alta	1,203989
Stazione Inferiore Funicolare	1,187952
Sistema Piacentiniano	0.852875
Porta Nuova	0.698300
Piazza Mercato delle Scarpe, Città Alta	0.521350
Fiera di Bergamo	0.490758
Stazione FS	0.475586
Piazza Giacomo Matteotti	0.459371

TABLE 2: Number of tweets for place in Bergamo.

Place	Number of tweets
Aeroporto, Orio al Serio	1448
Stazione Inferiore Funicolare	103
Oriocenter, Orio al Serio	100
Piazza Vecchia, Città Alta	70
Piazza Giacomo Matteotti	46
Piazza Pontida	28
Da Mimmo	23
Sistema Piacentiniano	23
Castello di San Vigilio	20
Stazione FS	16

the *J-CO* framework supports the integration and transformation of data sets, while the *Treets* web application allows analysts to deeply investigate traces and habits of mobile users. Finally, towards the reconstruction of collective rhizomes, we presented the *Node Rank* method, that is able to give a quantitative evaluation of visited places on the basis of the reticular approach. In practice, we could say that a large variety of Big Data requires a large variety of tools to analyze them.

Obviously, we cannot consider our research to be at the end. Paradoxically, we are at the beginning of a long research activity. This is true in relation to several aspects. First of all, traces of traveling users detected through *Twitter* are not the only possible ones. Many other sources could be used, such as the one produced by *Google Timeline*, that users could voluntarily provide; they could give a different perspective about habits of users, but the investigation methodology does not change.

Second, novel strategies to transform and integrate data could be developed, on the basis of automatic algorithms that could cluster data and characterize them based on texts and images; furthermore, sentiment analysis techniques could be applied to detect sentiment polarity of messages.

Finally, as sketched throughout the paper, data mining techniques based on the concept of frequent itemset mining could be developed and applied, so as to actually reconstruct rhizomes from global perspectives.

We are confident in the development of the three areas mentioned above. Thus, many research activities could be and will be performed, definitely enlarging the *Urban Nexus digital environment* and, at the end, the possibility for smart cities to learn from city users' experience. Nevertheless, the overall integrated approach, which is the main contribution of this paper, is still suitable and perfectly adequate to support our future work.

References

[1] J. Lévy, *L'Invention du Monde: Une Géographie de la Mondialisation*, Presses de Sciences Po Paris, Paris, France, 2008.

[2] E. W. Soja, *Postmetropolis Critical Studies of Cities and Regions*, Nlackwell, London, UK, 2000.

[3] E. Casti and F. Burini, *Centrality of Territories: Verso la Rigenerazione di Bergamo in un Network Europeo*, Bergamo University Press, Bergamo, Italy, 2015.

[4] G. Deleuze, F. Guattari, J. V. Pérez, and U. Larraceleta, *Rizoma: (Introducción)*, Pre-Textos, València, Spain, 2003.

[5] J. Lévy, T. P. L. Romany, and O. P. Maitre, "Rebattre les cartes. topographie et topologie dans la cartographie contemporaine," *Réseaux*, vol. 34, pp. 17–52, 2016.

[6] F. Girardin, F. Calabrese, F. Dal Fiore, C. Ratti, and J. Blat, "Digital footprinting: uncovering tourists with user-generated content," *IEEE Pervasive Computing*, vol. 7, no. 4, 2008.

[7] M. F. Goodchild, "Citizens as sensors: the world of volunteered geography," *GeoJournal*, vol. 69, no. 4, pp. 211–221, 2007.

[8] M. Batty, "Big data, smart cities and city planning," *Dialogues in Human Geography*, vol. 3, no. 3, pp. 274–279, 2013.

[9] M. Batty, "Big data and the city," *Built Environment*, vol. 42, no. 3, pp. 321–337, 2016.

[10] M. Graham and T. Shelton, "Geography and the future of big data, big data and the future of geography," *Dialogues in Human Geography*, vol. 3, no. 3, pp. 255–261, 2013.

[11] R. Kitchin, "Big data and human geography: opportunities, challenges and risks," *Dialogues in Human Geography*, vol. 3, no. 3, pp. 262–267, 2013.

[12] M. Lussault, *L'Homme Spatial: la Construction Sociale de l'Espace Humain*, Vol. 363, Seuil Paris, Paris, France, 2007.

[13] F. Burini, D. E. Ciriello, A. Ghisalberti, and G. Psaila, "The urban nexus project: when urban mobility analysis, vgi and data science meet together," in *Mobile Information Systems Leveraging Volunteered Geographic Information for Earth Observation*, pp. 111–130, Springer, Cham, Switzerland, 2018.

[14] R. Kitchin, *What Does Big Data Mean for Official Statistics?*, Discover Society, Bristol, England, 2015.

[15] H. Butler, M. Daly, A. Doyle, S. Gillies, S. Hagen, and T. Schaub, "The geojson format," Tech. Rep., Internet Engineering Task Force, Fremont, CA, USA, 2016.

[16] K. Banker, *MongoDB in Action*, Manning Publications Co., Shelter Island, NY, USA, 2011.

[17] C. Strauch, *Nosql Databases*, 2011, http://www.christof-strauch.de/nosqldbs.pdf.

[18] S. Brin and L. Page, "Reprint of: the anatomy of a large-scale hypertextual web search engine," *Computer Networks*, vol. 56, no. 18, pp. 3825–3833, 2012.

[19] R. Agrawal, T. Imieliński, and A. Swami, "Mining association rules between sets of items in large databases," *ACM SIGMOD Record*, vol. 22, no. 2, pp. 207–216, 1993.

[20] R. Agrawal and R. Srikant, "Fast algorithms for mining association rules," in *Proceedings of 20th International Conference on Very Large Data Bases, VLDB*, vol. 1215, pp. 487–499, San Francisco, CA, USA, September 1994.

[21] G. Grahne and J. Zhu, "Fast algorithms for frequent itemset mining using fp-trees," *IEEE Transactions on Knowledge and Data Engineering*, vol. 17, no. 10, pp. 1347–1362, 2005.

[22] R. Meo, G. Psaila, and S. Ceri, "A new SQL-like operator for mining association rules," in *Proceedings of the 22st VLDB Conference*, Bombay, India, September 1996.

[23] X. Wu, C. Zhang, and S. Zhang, "Database classification for multi-database mining," *Information Systems*, vol. 30, no. 1, pp. 71–88, 2005.

[24] X. Wu and S. Zhang, "Synthesizing high-frequency rules from different data sources," *IEEE Transactions on Knowledge and Data Engineering*, vol. 15, no. 2, pp. 353–367, 2003.

[25] S. Zhang, Q. Chen, and Q. Yang, "Acquiring knowledge from inconsistent data sources through weighting," *Data & Knowledge Engineering*, vol. 69, no. 8, pp. 779–799, 2010.

[26] S. Zhang, X. Wu, and C. Zhang, "Multi-database mining," *IEEE Computational Intelligence Bulletin*, vol. 2, no. 1, pp. 5–13, 2003.

[27] M. Haklay, "Citizen science and volunteered geographic information: overview and typology of participation," in *Crowdsourcing Geographic Knowledge*, pp. 105–122, Springer, Berlin, Germany, 2013.

[28] B. Hecht and M. Stephens, "A tale of cities: urban biases in volunteered geographic information," *ICWSM*, vol. 14, pp. 197–205, 2014.

[29] D. Z. Sui, S. Elwood, and M. Goodchild, *Crowdsourcing Geographic Knowledge: Volunteered Geographic Information (VGI) in Theory and Practice*, Springer Science & Business Media, Berlin, Germany, 2012.

[30] M. Zook, M. Graham, T. Shelton, and S. Gorman, "Volunteered geographic information and crowdsourcing disaster relief: a case study of the Haitian earthquake," *World Medical & Health Policy*, vol. 2, no. 2, pp. 7–33, 2010.

[31] E. Casti, *Reflexive Cartography: A New Perspective in Mapping*, Vol. 6, Elsevier, New York, NY, USA, 2015.

[32] A. Cuzzocrea, G. Psaila, and M. Toccu, "Knowledge discovery from geo-located tweets for supporting advanced big data analytics: a real-life experience," in *Model and Data Engineering*, pp. 285–294, Springer International Publishing, Berlin, Germany, 2015.

[33] G. Bordogna, A. Cuzzocrea, L. Frigerio, G. Psaila, and M. Toccu, "An interoperable open data framework for discovering popular tours based on geo-tagged tweets," *International Journal of Intelligent Information and Database Systems*, vol. 10, no. 3-4, pp. 246–268, 2017.

[34] A. Cuzzocrea, G. Psaila, and M. Toccu, "An innovative framework for effectively and efficiently supporting big data analytics over geo-located mobile social media," in *Proceedings of the 20th International Database Engineering & Applications Symposium*, pp. 62–69, ACM, Montreal, QC, Canada, ACM, July 2016.

[35] T. E. Chow, "Geography 2.0: a mashup perspective," in *Advances in Web-Based GIS, Mapping Services and Applications*, S. Li, S. Dragićević, and B. Veenendaal, Eds., pp. 15–36, Taylor & Francis Group, London, ISBN 978-0-415-80483-7, 2011.

[36] G. Bordogna, S. Capelli, D. E. Ciriello, and G. Psaila, "A cross-analysis framework for multi-source volunteered, crowd-sourced, and authoritative geographic information: the case study of volunteered personal traces analysis against transport network data," *Geo-spatial Information Science*, pp. 1–15, 2017.

[37] S. Capelli, P. Fosci, F. Marini, and G. Psaila, "J-co-ql: A flexible query language for complex geographical analysis of heterogeneous geo-tagged json data sets," Tech. Rep., University of Bergamo, Bergamo, Dalmine, Italy, 2017.

[38] G. Bordogna, S. Capelli, and G. Psaila, "A big geo data query

framework to correlate open data with social network geo-tagged posts," in *Proceedings in AGILE 2017 International Conference*, Paris, France, July 2017.

[39] G. Bordogna, L. Frigerio, A. Cuzzocrea, and G. Psaila, "Clustering geo-tagged tweets for advanced big data analytics," in *Proceedings of Big Data (BigData Congress), 2016 IEEE International Congress*, pp. 42–51, IEEE, San Francisco, CA, USA, June 2016.

[40] G. Bordogna, L. Frigerio, A. Cuzzocrea, and G. Psaila, "An effective and efficient similarity-matrix-based algorithm for clustering big mobile social data," in *Proceedings of Machine Learning and Applications (ICMLA), 2016 15th IEEE International Conference*, pp. 514–521, IEEE, Anaheim, CA, USA, December 2016.

[41] N. Cortesi, K. Gotti, G. Psaila, F. Burini, K. T. Lwin, and M. Hossain, "A network-based approach to discover rank places visited by tourists from geo-located tweets," in *Proceedings of Software, Knowledge, Information Management and Applications (SKIMA 2017), 11th International Congress*, IEEE, Chengdu, China, 2017.

[42] L. Page, S. Brin, R. Motwani, and T. Winograd, "The pagerank citation ranking: bringing order to the web," Tech. Rep., Stanford InfoLab, Stanford, CA, USA, 1999.

15

An F-Score-Weighted Indoor Positioning Algorithm Integrating WiFi and Magnetic Field Fingerprints

Sinem Bozkurt Keser (ID),[1] **Ahmet Yazici,**[1] and **Serkan Gunal**[2]

[1]*Department of Computer Engineering, Eskisehir Osmangazi University, Eskisehir, Turkey*
[2]*Department of Computer Engineering, Anadolu University, Eskisehir, Turkey*

Correspondence should be addressed to Sinem Bozkurt Keser; sbozkurt@ogu.edu.tr

Academic Editor: Carlos T. Calafate

Indoor positioning systems have attracted much attention with the recent development of location-based services. Although global positioning system (GPS) is a widely accepted and accurate outdoor localization system, there is no such a solution for indoor areas. Therefore, various systems are proposed for the indoor positioning problem. Fingerprint-based positioning is one of the widely used methods in this area. WiFi-received signal strength (RSS) is a frequently used signal type for the fingerprint-based positioning system. Since WiFi signal distribution is nonstationary, accuracy is insufficient. Therefore, the performance of indoor positioning systems can be enhanced using multiple signal types. However, the positioning performance of each signal type varies depending on the characteristics of the environment. Considering the variability of the performances of different signal types, an F-score-weighted indoor positioning algorithm, which integrates WiFi-RSS and MF fingerprints, is proposed in this study. In the proposed approach, the positioning is first performed by maximum likelihood estimation for both WiFi-RSS and magnetic field signal values to calculate the F-score of each signal type. Then, each signal type is combined using F-score values as a weight to estimate a position. The experiments are performed using a publicly available dataset that contains real-world data. Experimental results reveal that the proposed algorithm is efficient in achieving accurate indoor positioning and consolidates the system performance compared to using a single type of signal.

1. Introduction

Location-based services (LBSs) have become more popular with the recent advancements in mobile computing technology. One of the critical components of the LBS is positioning systems that can be divided into two categories, namely, outdoor and indoor positioning. Global positioning system (GPS) is a well-known solution in the outdoor environment, whereas it gives poor accuracy in the indoor area since it usually requires line-of-sight (LOS) propagation to obtain acceptable accuracy [1]. Therefore, an indoor positioning method is required.

Various technologies are developed for solving indoor positioning problem such as global system for mobile communications (GSM) [2], radio-frequency identification (RFID) [3], ultrasonic [4], Bluetooth (BT) [5], wireless fidelity (WiFi) [6], and magnetic field (MF) [7]. These technologies have both advantages and disadvantages depending on their nature and applications. For example, the GSM-based system uses existing infrastructure, but it does not offer reasonable accuracy for indoor areas. RFID-based and ultrasonic-based indoor positioning systems (IPSs) have reasonable accuracies, but they need the installation of additional signals. BT-based IPS has a short operating range and poor predictability. Therefore, it is recommended that BT is used as a supplementary technology in an IPS. WiFi-based IPS is the most widely deployed system when compared to other systems. Besides, WiFi-based indoor positioning is preferred to other technologies due to both massive deployments in the indoor area and extensive usage of WiFi-enabled devices. WiFi-based indoor positioning methods have some drawbacks, such that WiFi signals deteriorate over time, which leads to inaccurate position estimates. Therefore, the accuracy of only WiFi signals may not be adequate for some applications and can be enhanced using multiple signals. In recent researches, MF-based positioning is also considered as a complementary technique to WiFi-based positioning technology to improve the positioning accuracy. The MF-aided positioning is very popular nowadays since all smartphones include a magnetometer to collect the magnetic data freely when needed. Overall, the positioning performance of each signal type varies depending on the characteristics of the environment.

Considering the variability of the performances of different signal types, an F-score-weighted indoor positioning algorithm, which integrates WiFi-RSS and MF fingerprints, is proposed in this study. In the proposed approach, firstly, positioning is performed by maximum likelihood estimation for both WiFi-RSS and MF signal values to calculate the F-scores of each signal type. Then, each signal type is combined using F-score values as a weight to estimate a position. The experiments are performed using the publicly available dataset which contains real-world data [8]. The results show that the proposed algorithm is effective and efficient in achieving good indoor positioning accuracy and consolidates the system performance compared to that of just one signal type. The main contributions of this study are that the proposed algorithm considers the environment structure when estimating the position and takes the advantages of two signal types into account at the same time to enhance the positioning performance.

The rest of the paper is organized as follows. Related works for indoor positioning technologies are given in Section 2. Section 3 introduces the preliminaries for the fingerprint-based positioning. The proposed algorithm is given in Section 4. Section 5 provides the experimental work and the related results. Finally, Section 6 includes the conclusions of the work.

2. Related Works

Several methods have been proposed for positioning in the indoor environment. These methods can be classified as triangulation, proximity, pedestrian dead reckoning (PDR), vision analysis, and fingerprinting. Triangulation is a geometric-based method that uses signal parameters to calculate the mobile unit (MU) position [6]. The triangulation method suffers from non-line-of-sight (NLOS) conditions; therefore, it gives erroneous results for indoor positioning. In proximity, the MU location is estimated as the antenna position that receives the strongest signal from the MU [9]. Therefore, a dense grid of antennas with known positions is used. This method is generally used in RFID technology. It requires additional hardware, has low resolution, and has poor accuracy. So, it is impractical for indoor positioning. In PDR, the position of the MU is calculated using the previously calculated position, speed, and the direction of the MU [10]. Since the current position is relative to the previous position, the errors are cumulative. Moreover, the sensors in most smartphones do not provide very accurate data. Therefore, PDR is not solely adequate for an IPS. Vision analysis has a high complexity since it requires the setup of the large image database and the real-time communication between the server and the MU [11]. Therefore, it produces an undesirable solution for indoor positioning.

WiFi-received signal strength- (WiFi-RSS-) based fingerprinting method is a widely adopted approach due to its relatively high accuracy and modest cost [12]. It utilizes the existing wireless local area network (WLAN) infrastructure. However, it has some challenges such that WiFi-RSS values suffer from the multipath effect, which leads to erroneous position estimate. Therefore, it can be enhanced using supplementary technologies such as BT or MF. But, using BT with WiFi-RSS-based fingerprinting method is not a good choice. Since both of the technologies operate at the same frequency (2.4 GHz), the signal inference is inevitable. On the other hand, MF has some advantages such that MF does not suffer from NLOS conditions or multipath effects in indoors and it is easy to obtain MF measurements using today's smartphones. Fingerprinting method can be applied to constitute an MF-based fingerprinting method for indoor positioning [13]. However, it has sensitivity to certain materials. Besides, MF strength diminishes rapidly with distance. Therefore, an MF-based fingerprinting method is best utilized as a supplementary method to WiFi-RSS-based fingerprinting method for indoor positioning. The positioning accuracy of WiFi-based positioning can be enhanced using MF data. A hybrid fingerprint-based positioning, which uses WiFi-RSS and MF values, is proposed in [14, 15]. These studies use both signal types to construct a single hybrid fingerprint map for indoor positioning without any weighting. But, the positioning performances of these signal types vary depending on the characteristics of the environment. In this study, concerning the regarding works, two separated fingerprint maps are utilized and integrated with a novel F-score-based weighting approach. The weights are selected depending on the performance of the individual signal type for a given environment. The details of the proposed works are given in Section 4.

3. Preliminaries

Fingerprint-based indoor positioning is a widely used method due to several factors such as its easy implementation, low cost, and the promising positioning accuracy. It includes two phases: fingerprint-based mapping (training) and positioning [12]. In the fingerprint-based mapping phase, the location dependent characteristics of a signal, which is recorded at reference points (RPs), are stored in a database to form a fingerprint map. In the positioning phase, the mobile unit fingerprint is compared to the fingerprints in the database to estimate the position of the mobile unit. RADAR [16] and HORUS [17] are typical IPS based upon WiFi-RSS-based fingerprinting method. Fingerprint-based indoor positioning is illustrated in Figure 1.

The processes of fingerprint-based mapping and the algorithms used for training and positioning are given in the following subsections.

3.1. Fingerprint-Based Mapping. Fingerprint-based mapping is started with dividing the test area into equal-sized grids; then, signal values are recorded from the center of each grid as a fingerprint. Various studies adopt fingerprint-based indoor positioning in the literature. These studies store the

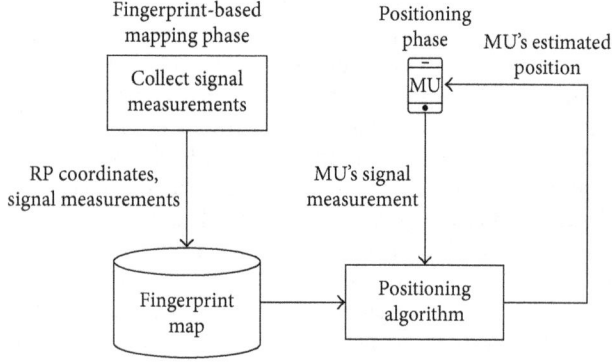

FIGURE 1: Fingerprint-based indoor positioning.

WiFi-RSS values for their database [18, 19]. Recent studies recognize the efficiency of the MF and store the samples from the magnetometer to setup their database [13, 20]. Among these databases, the publicly available databases are quite limited [18–20]. Besides, these databases contain one type of signal measurement. In the previous studies, a multi-signal fingerprint database that includes WiFi-RSS and MF for indoor positioning was proposed [8]. In this study, this database is used. The processes of obtaining the fingerprint-based maps are given below.

3.1.1. Radio Map. Radio map is constructed by dividing the experimental area into equal-sized grids [16]. The center of each grid represents the reference points (RPs) where WiFi-RSS values of the radio signals transmitted by APs are collected. These WiFi-RSS values are stored into the radio map as a fingerprint with the known coordinates of RPs. The fingerprint at the ith RP in the radio map is stored as

$$\text{FP}_i = \left\{ \text{lb}_i, x\text{Coord}_i, y\text{Coord}_i, \left(\text{MAC}_{i,1}, \text{RSS}_{i,1}\right), \left(\text{MAC}_{i,2}, \text{RSS}_{i,2}\right), \ldots, \left(\text{MAC}_{i,k_i}, \text{RSS}_{i,k_i}\right)\right\}, \tag{1}$$

where FP_i is the fingerprint information at RP_i, lb_i is the label of the RP_i, $x\text{Coord}_i$ and $y\text{Coord}_i$ are the x and y coordinates of RP_i, $\text{MAC}_{i,j}$ and $\text{RSS}_{i,j}$ are the MAC address and WiFi-RSS values of the jth AP received at RP_i, and k_i is the number of available APs at RP_i.

3.1.2. Magnetic Map. Magnetic map is established using the same procedure as the radio map construction. Each fingerprint in the magnetic map contains the x, y, and z values of MF strength values that are obtained by a magnetometer sensor on a mobile device [20]. Magnetometer sensor returns the x, y, and z values of the MF strength values in relation to the device orientation; therefore, the magnitude of each axis may differ as the device's orientation changes, even when it stays in the same position. So, they must be converted to the world coordinates before being stored in the magnetic map. The accelerometer and gyroscope included on the mobile phone can be used to convert the device orientation to world coordinates. Yaw (ϕ) and pitch (θ) angles obtained from the accelerometer are integrated with heading angle (ψ) from the MF and gyroscope sensors using Kalman filter to obtain the orientation of the mobile device. The orientation angles are used to construct rotation matrix as seen in the following equation:

$$R_x(\phi) = \begin{bmatrix} 1 & 0 & 0 \\ 0 & \cos\phi & -\sin\phi \\ 0 & \sin\phi & \cos\phi \end{bmatrix},$$

$$R_y(\Theta) = \begin{bmatrix} \cos\Theta & 0 & \sin\Theta \\ 0 & 1 & 0 \\ -\sin\Theta & 0 & \cos\Theta \end{bmatrix}, \tag{2}$$

$$R_z(\psi) = \begin{bmatrix} \cos\psi & -\sin\psi & 0 \\ \sin\psi & \cos\psi & 0 \\ 0 & 0 & 1 \end{bmatrix}.$$

Then, the local MF strength measurements are multiplied by the rotation matrix which includes orientation angles to calculate the global MF strength measurements as follows:

$$B_p = R_x(\phi)R_y(\Theta)R_z(\psi)B_e, \tag{3}$$

where B_e and B_p are the MF strength vectors in the device and world coordinate orientations, respectively. Then, using the global values of the MF strength vector (B_p), the magnetic fingerprint at the ith RP in the magnetic map is represented as

$$\text{MFP}_i = \left\{\text{lb}_i, x\text{Coord}_i, y\text{Coord}_i, \text{global}_{i,x}, \text{global}_{i,y}, \text{global}_{i,z}\right\}, \tag{4}$$

where MFP_i is the magnetic fingerprint information at RP_i, lb_i is the label of the RP_i, $x\text{Coord}_i$ and $y\text{Coord}_i$ are the x and y coordinates of RP_i, and $\text{global}_{i,x}$, $\text{global}_{i,y}$, and $\text{global}_{i,z}$ are the global x, y, and z values of MF strength in relation to the world coordinates at RP_i.

3.2. Training and Positioning. The fingerprint-based positioning system usually includes two stages: training and positioning. Given the radio map or magnetic map, a model is constructed in the training phase. In the positioning phase, the model is used to calculate the position of the mobile unit by comparing measurements of the newly obtained data with the recorded fingerprints from the training data. Depending on the application, the fingerprint map can be divided into training and test data. The training data are used to train the model by pairing the input with actual output. Also, the test data can be used for either positioning or evaluating the wellness of the model.

In the literature, various algorithms are used to construct a model for the positioning. In [16], K-nearest neighbour (KNN) is applied to estimate the position. In another study, a decision tree is applied for positioning

[21]. The target location is predicted using Naïve Bayes (NB) estimate in [17]. Maximum likelihood estimation (MLE) is one of the most popular algorithms among them which take the standard deviation of the measurements into account [22]. It provides higher accuracy when compared to the other algorithms [23]. Additionally, the calculated likelihood values for different signal types are useful for constructing hybrid solutions for the indoor positioning problem. In the MLE algorithm, the probability of obtaining the mobile device fingerprint F' at the ith RP whose fingerprint F_i is given by

$$p(F'|F_i) = \frac{1}{(\prod s_i)(2\pi)^{n/2}} e^{-(1/2)\sum(x_i' - \overline{x}_i/s_i)^2},$$

$$F_i = (\overline{x}_i, s_i), \tag{5}$$

$$F' = (x_1', x_2', \ldots, x_n'),$$

where $\overline{x}_i$ and s_i are the mean and the standard deviation of the signal measurements at the ith RP, x_i' is the signal measurement at an unknown location, and n is the dimension of the fingerprint map. The mean and the standard deviation are calculated by

$$\overline{x}_i = \frac{1}{m} \sum_{j=1}^{m} x_i^j,$$

$$s_i = \sqrt{\frac{1}{m-1} \sum_{j=1}^{m} \left(x_i^j - \overline{x}_i\right)^2}, \tag{6}$$

where m is the number of collected measurements for the RP_i and x_i^j is the jth measurement from AP_i for the radio map. For the magnetic map, x_i^j is MF strength values as given in (4).

The computational cost of (5) can be reduced by taking its natural log:

$$p(F'|F_i) = -\ln(\prod s_i) - \frac{n}{2}\ln 2\pi - \frac{1}{2}\sum\left(\frac{x_i' - \overline{x}_i}{s_i}\right)^2. \tag{7}$$

Since the second term of (7) is constant, it can be ignored. Equation (7) is rewritten by multiplying -1 and defining $c = \ln(\prod s_i)$ as

$$p(F'|F_i) = c + \frac{1}{2}\sum\left(\frac{x_i' - \overline{x}_i}{s_i}\right)^2. \tag{8}$$

Now, (5) is converted to (8) [24]. For the mobile device fingerprint F', (8) is calculated for each RP, and then the label of the RP is returned by calculating

$$\arg\min_i p(F'|F_i). \tag{9}$$

Additionally, the wellness of the training model can be evaluated using a separated test data. Model evaluation is carried out as follows: after calculating the estimated position of each test data using (9), the confusion matrix is generated with the estimated position labels and actual position labels. The confusion matrix is a basis for calculating the terms such as true positive (TP), false negative (FN), true negative (TN),

TABLE 1: Confusion matrix for binary case representation.

	Positive (estimated)	Negative (estimated)
Positive (actual)	TP	FN
Negative (actual)	FP	TN

TP: when the user is in RP "i" and is classified as RP "i"; TN: when the user is not in RP "i" and is not classified as RP "i"; FP: when the user is not in RP "i" and is classified as RP "i"; FN: when the user is in RP "i" and is not classified as RP "i".

and false positive (FP) [25]. The binary case representation of the confusion matrix is given in Table 1.

Then, F-score can be calculated using the confusion matrix as follows:

$$\text{F-score} = \frac{2 * \text{TP}}{2 * \text{TP} + \text{FP} + \text{FN}}. \tag{10}$$

In the literature, the F-score values are generally used to evaluate the training model performance using the test data. In this study, the F-score values are used to evaluate the model (9) performance for each signal type. Besides, these F-score values are used as the weights of each signal type in the positioning phase to enhance the performance of the indoor positioning system.

4. The Proposed Algorithm

Traditional fingerprint-based positioning algorithms are initialized with constructing database by collecting measurements from the experimental area. The measurements usually contain only WiFi-RSS values obtained from the APs in the region of interest. RFKON database [8], on the other hand, contains both WiFi-RSS and MF strength values for each RP. The measurements in the RFKON database are used for constructing radio map and magnetic map, which are explained in the first subsection of Section 2. The radio map and the magnetic map are utilized as inputs for the proposed F-score-weighted indoor positioning algorithm. The pseudocode of the algorithm is given in Algorithm 1.

As seen in the Algorithm 1, the process of the proposed algorithm is started with normalization of data in each map. For this process, max-min normalization method is employed. Then, each map is split into training data (60%) and test data (40%) in Step 2 of the training-testing phase. The test data are used for obtaining F-score weight values of each signal type. In Step 3 of the training-testing phase, the mean and the standard deviation of each signal type are calculated by (6) to obtain training model parameters. In the next step, (9) is applied to both signal types separately for obtaining estimated RP labels. Then, F-score values of each signal type for each RP are calculated using (10). In the last step of training-testing phase, WiFi-RSS-based model (MLE_{WiFi}) and MF-based model (MLE_{MF}) are generated for the F-score-weighted indoor positioning.

In the positioning phase, newly obtained WiFi test data, MF test data, and the models are utilized as the inputs. In Steps 1 and 2 of the positioning phase, likelihood values are calculated at each RP for both signal types. Then, the final position is estimated using the formula in step 4 of the

Training-testing phase:
Inputs: Radio Map, Magnetic Map.
Outputs: WiFi-RSS-based model (MLE_{WiFi}), MF-based model (MLE_{MF}).
(1) Normalize each instance in radio map and magnetic map using min-max normalization procedure.
(2) Split the radio map and the magnetic map as training data (60%) and test data (40%).
(3) Use (6) to obtain $\bar{x}_i$ and s_i for the training data of each signal type.
(4) Calculate the RP labels for both test data type using (9) separately.
(5) Apply (10) to calculate the F-score values of each signal type per RP using the calculated RP labels and the actual RP labels.
    The F-score values are stored as the weight of each signal type as $\text{weight}_{\text{WiFi},i}$ and $\text{weight}_{\text{MF},i}$.
(6) Construct WiFi-RSS-based model (MLE_{WiFi}) and MF-based model (MLE_{MF}) as follows:
    $\text{MLE}_{\text{WiFi},i} = (\text{lb}_i, x\text{Coord}_i, y\text{Coord}_i, \bar{x}_i, s_i, \text{weight}_{\text{WiFi},i})$,
    $\text{MLE}_{\text{MF},i} = (\text{lb}_i, x\text{Coord}_i, y\text{Coord}_i, \bar{x}_i, s_i, \text{weight}_{\text{MF},i})$.

Positioning phase:
Inputs: MLE_{WiFi}, MLE_{MF}, WiFi New Test Data, MF New Test Data.
Outputs: estimated position.
(1) Apply MLE_{WiFi} with (8) using WiFi New Test Data to obtain likelihood values of each RP ($p_{\text{WiFi}}(F'|F_i)$, $i = 1, \ldots, k$, where k is the number of RPs).
(2) Apply MLE_{MF} with (8) using MF New Test Data to obtain likelihood values of each RP ($p_{\text{MF}}(F'|F_i)$, $i = 1, \ldots, k$, where k is the number of RPs).
(3) Normalize likelihood values using max-min normalization method to obtain $p_{\text{WiFi}}(F'|F_i)$ and $p_{\text{MF}}(F'|F_i)$.
(4) Use the following equation to calculate the final position:
    $\arg\min_i \; p_{\text{WiFi}}(F'|F_i)' * \text{weight}_{\text{WiFi},i} + p_{\text{MF}}(F'|F_i)' * \text{weight}_{\text{MF},i}$.

ALGORITHM 1: Pseudocode of the proposed method.

positioning phase in Algorithm 1. Note that $\text{weight}_{\text{WiFi},i}$ and $\text{weight}_{\text{MF},i}$ are the WiFi-RSS and MF F-score weights for ith RP, respectively. Depending on the region of the indoor environment, WiFi signals or MF signals may result in better accuracy for the positioning calculation. Then, the corresponding F-score value of the signal will be higher and would be dominant for positioning.

5. Experimental Work

The proposed method is compared to existing methods in the literature for the test area of RFKON database. The system architecture of our IPS is shown in Figure 2. It contains two major units named as Gezkon and Konsens. Gezkon is a mobile application and it is responsible for collecting WiFi-RSS and MF strength values from the test area. Konsens is a server, which is used to estimate the position of the mobile device and also responsible for updating and calibrating RFKON database. The communication between Konsens and Gezkon is achieved by sensor nodes through the data distribution service layer.

The details of the test area and experimental results are provided in the following subsections.

5.1. Test Area. The layout of the test area for RFKON database is shown in Figure 3. In this area, real-world indoor localization experiments are conducted to evaluate the performance of the proposed algorithm. The test area is at the Technopark building in Eskisehir Osmangazi University. The area is divided into grid squares (2.4 m × 2.4 m) and the center of each grid is reported as RP. There are 20 RPs in the experimental area. The measurements are collected from

the 1st floor of this area at each reference point. 40 WiFi and 40 magnetic field measurements are collected at each RP for the training and test phases. Hence, 800 measurements are collected for constructing the training set and 800 measurements for the test set.

As seen in Figure 3, there are five sensor nodes, which are represented with small red squares deployed at the locations on the first floor. The red stars represent the RPs used to collect signal measurements from the area.

5.2. Experimental Results. WiFi-RSS values or MF values may provide better positioning accuracy depending on the structure of the indoor area. If there are different fingerprint maps for each signal type, the F-score values can be used to understand accurateness for each map. In our experimental area, F-score values are obtained for radio map and magnetic map as in Figure 4. Those heat maps illustrate the accurateness of positioning in the given RPs.

One can observe from Figure 4(a) that F-scores are relatively high in the upper right regions. That is, WiFi-based positioning offers high performance for these regions. However, F-scores are getting lower in the lower-left regions. In other words, WiFi-based positioning offers lower performance for those regions. In the meantime, the heatmap for MF-based positioning, which is shown in Figure 4(b), offers almost a complementary performance. That is, F-scores are relatively low in the upper right regions and relatively high in the lower-left regions. In the middle regions of the map, both WiFi and MF yield similar performance. Therefore, utilizing each signal type with F-score-weighted values concurrently in the proposed positioning algorithm can enhance the performance of the IPS.

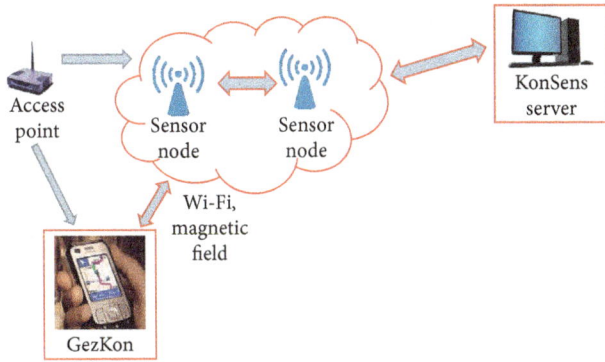

FIGURE 2: System architecture of our IPS.

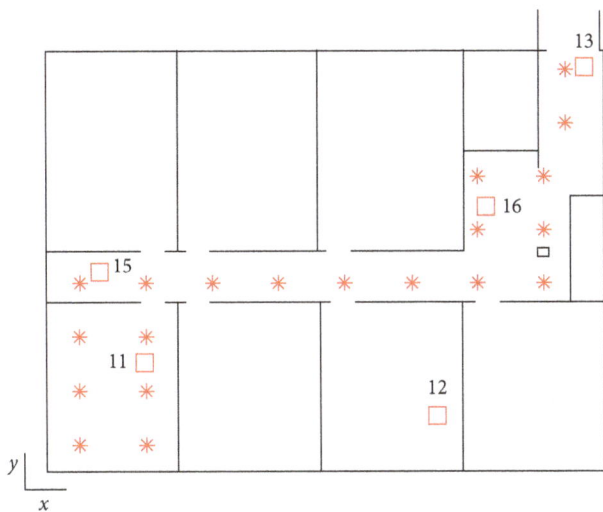

FIGURE 3: The map of the floor where the experiments are conducted.

FIGURE 4: Heatmap of F-scores for (a) WiFi and (b) MF data using MLE algorithm.

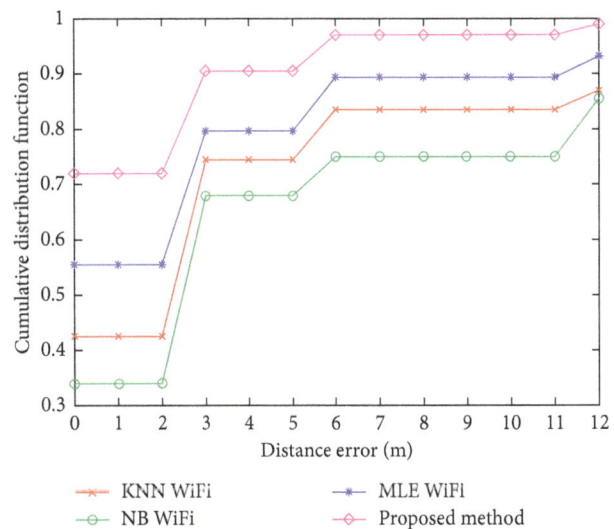

The proposed positioning algorithm is compared to KNN algorithm that is used in RADAR [16] and NB algorithm that is used in HORUS [17]. The experiments are conducted by comparing KNN, NB, and MLE algorithms using only one signal measurement. Therefore, firstly, the proposed algorithm is compared to KNN, NB, and MLE algorithms with WiFi-RSS data. Then, the comparison is carried out when applying KNN, NB, and MLE algorithms with MF data. Figure 5 illustrates the distribution of localization error comparatively.

As shown in Figure 6, the performance of the proposed method is higher than those of the other algorithms when MF strength measurements are used. It is clear from Figures 5 and 6 that the proposed algorithm outperforms the other algorithms for each type of signal measurement. In Table 2, the positioning performances of the respective algorithms are consolidated for each signal type considering a positioning error less than 3 m and 6 m.

The proposed algorithm offers a positioning error less than 3 m for 91% of test data and less than 6 m for 97% of the test data. This performance is far higher than those of the other algorithms. Moreover, the proposed algorithm can

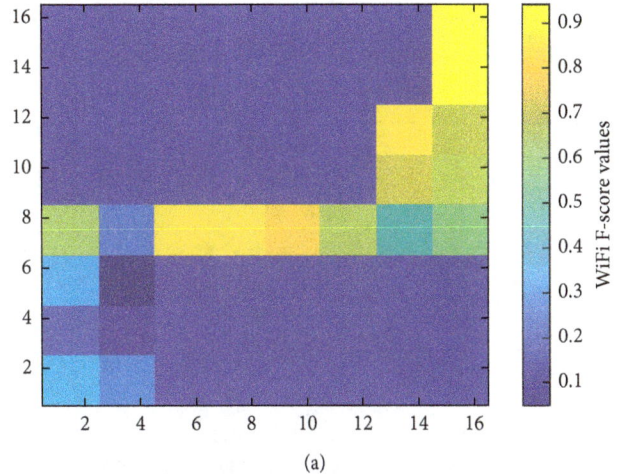

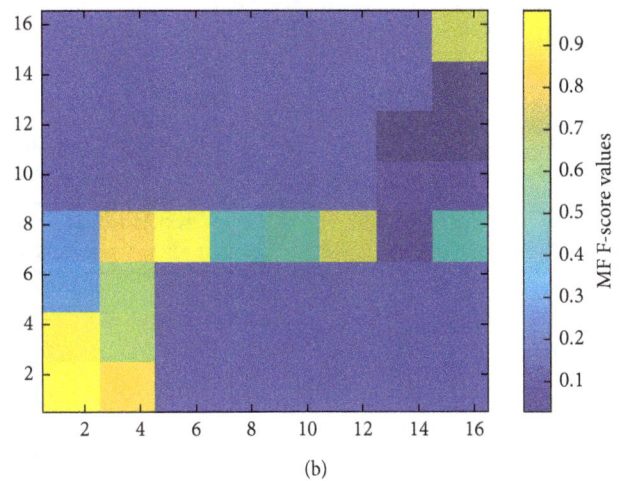

FIGURE 5: Performances of the indoor positioning algorithms using only WiFi-RSS data. One can note that the proposed algorithm is apparently superior to the other algorithms when WiFi-RSS signal measurements are utilized.

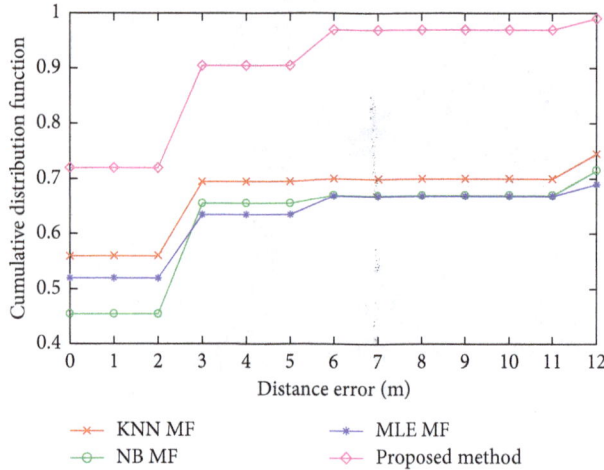

FIGURE 6: Performances of the indoor positioning algorithms using only MF data.

TABLE 2: Consolidated performances of the indoor positioning algorithms using single signal type.

Method	Precision (%)	
	<3 m	<6 m
Proposed algorithm	**91%**	**97%**
MLE WiFi	84%	92%
KNN WiFi	75%	84%
NB WiFi	64%	75%
MLE MF	64%	67%
KNN MF	69%	70%
NB MF	66%	67%

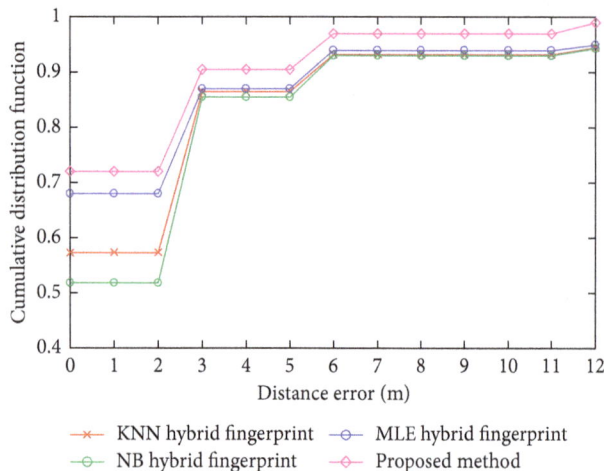

FIGURE 7: Performances of the indoor positioning algorithms using hybrid fingerprint data.

effectively integrate MF and WiFi signals for more accurate positioning.

The studies [14, 15] use WiFi and MF signal types to construct a hybrid fingerprint map for fingerprint-based indoor positioning to enhance the accuracy of their systems. Therefore, the proposed algorithm is also compared to

TABLE 3: Consolidated performances of the indoor positioning algorithms using hybrid data.

Method	Precision (%)	
	<3 m	<6 m
Proposed algorithm	**91%**	**97%**
MLE hybrid fingerprint	87%	94%
KNN hybrid fingerprint	85%	93%
NB hybrid fingerprint	83%	92%

MLE, KNN, and NB that use hybrid fingerprint map for positioning. Comparative results are given in Figure 7.

As seen in the figure, all algorithms offer improved performance, thanks to the hybrid fingerprint map. However, the proposed algorithm is still the best when compared to the other algorithms. In Table 3, the positioning performances of the respective algorithms are listed for hybrid data considering a positioning error less than 3 m and 6 m.

Though the performances of the other algorithms are improved with the help of hybrid data, the proposed method still surpasses the others algorithms.

6. Conclusion

In this study, an F-score-weighted indoor positioning algorithm that combines WiFi and MF fingerprints is proposed. In the first phase of the proposed method, MLE algorithm is utilized for positioning using WiFi and MF signal values separately to calculate the F-score values of each signal type. Then, each signal type is integrated by utilizing F-score values as weights to estimate the final position. The experiments are conducted using the publicly available dataset containing real-world data.

As a result of the experiments, it is revealed that the positioning performance of each signal type varies depending on the characteristics of the environment. As given in Test Results section, WiFi and MF signals offer their best performances in different parts of the experimental area. Considering the variability of the performances of different signal types, the proposed F-score-weighted indoor positioning algorithm, which integrates WiFi-RSS fingerprints and MF fingerprints, offers the best performance with respect to the other hybrid approaches in the literature.

Acknowledgments

This work is supported by the Scientific and Technological Research Council of Turkey (TUBITAK) under Grant no. 1130024.

References

[1] I. Getting, "The global positioning system," *IEEE Spectrum*, vol. 30, no. 12, pp. 36–47, 1993.

[2] R. Górak, M. Luckner, M. Okulewicz, J. Porter-Sobieraj, and P. Wawrzyniak, "Indoor localisation based on GSM signals: multistorey building study," *Mobile Information Systems*, vol. 2016, Article ID 2719576, 17 pages, 2016.

[3] B. Kyoungsoo and J. Yoo, "RFID based indoor positioning system using event filtering," *Journal of Electrical Engineering & Technology*, vol. 12, no. 1, pp. 335–345, 2017.

[4] U. Yayan, H. Yucel, and A. Yazıcı, "A low cost ultrasonic based positioning system for the indoor navigation of mobile robots," *Journal of Intelligent & Robotic Systems*, vol. 78, no. 3-4, pp. 541–552, 2015.

[5] Q. Wang, Y. Feng, X. Zhang, Y. Sun, and X. Lu, "IWKNN: an effective Bluetooth positioning method based on isomap and WKNN," *Mobile Information Systems*, vol. 2016, Article ID 8765874, 11 pages, 2016.

[6] H. Liu, H. Darabi, P. Banerjee, and J. Liu, "Survey of wireless indoor positioning techniques and systems," *IEEE Transactions on Systems, Man, and Cybernetics, Part C: Applications and Reviews*, vol. 37, no. 6, pp. 1067–1080, 2007.

[7] J. Haverinen and A. Kemppainen, "Global indoor self-localization based on the ambient magnetic field," *Robotics and Autonomous Systems*, vol. 57, no. 10, pp. 1028–1035, 2009.

[8] S. Bozkurt Keser, U. Yayan, A. Yazici, and S. Gunal, "A priori verification and validation study of RFKON database," *International Journal of Computer Science: Theory and Application*, vol. 5, pp. 20–27, 2016.

[9] C. L. Mak, M. A. Hon, W. M. Lau, and W. M. Cheung, "Refined Wi-Fi fingerprinting with tag-less proximity-based positioning technique," in *Proceedings of the 2014 International Conference on Indoor Positioning and Indoor Navigation (IPIN)*, pp. 758–761, Busan, Korea, 2014.

[10] D. Focken and R. Stiefelhagen, "Towards vision-based 3-D people tracking in a smart room," in *Proceedings of the 4th IEEE International Conference on Multimodal Interfaces*, Washington, DC, USA, October 2002.

[11] L. Li and X. Lin, "Apply pedestrian dead reckoning to indoor Wi-Fi positioning based on fingerprinting," in *Proceedings of the IEEE International Conference on Communication Technology*, pp. 206–210, Guilin, China, November 2013.

[12] S. He and S. H. G. Chan, "Wi-Fi fingerprint-based indoor positioning: recent advances and comparisons," *IEEE Communications Surveys & Tutorials*, vol. 18, no. 1, pp. 466–490, 2016.

[13] J. Chung, M. Donahoe, C. Schmandt, I.-J. Kim, P. Razavai, and M. Wiseman, "Indoor location sensing using geo-magnetism," in *Proceedings of the 9th International Conference on Mobile Systems, Applications, and Services*, pp. 141–154, Bethesda, MD, USA, 2011.

[14] M. Zhang, W. Shen, and J. Zhu, "WIFI and magnetic fingerprint positioning algorithm based on KDA-KNN," in *Proceedings of the 2016 Chinese Control and Decision Conference (CCDC)*, pp. 5409–5415, Yinchuan, China, May 2016.

[15] S. Bozkurt Keser, A. Yazici, and S. Gunal, "A hybrid fingerprint based indoor positioning with extreme learning machine," in *Proceedings of the 25th Signal Processing and Communications Applications Conference (SIU)*, Antalya, Turkey, May 2017.

[16] P. Bahl and N. P. Venkata, "RADAR: an in-building RF-based user location and tracking system," in *Proceedings of the Nineteenth Annual Joint Conference of IEEE Computer and Communications Societies (INFOCOM 2000)*, vol. 2, Tel Aviv, Israel, March 2000.

[17] M. Youssef and A. Agrawala, "The Horus WLAN location determination system," in *Proceedings of the 3rd International Conference on Mobile Systems, Applications, and Services*, pp. 205–218, Seattle, WA, USA, June 2005.

[18] C. Laoudias, R. Piche, and C. G. Panayiotou, "Device self-calibration in location systems using signal strength histograms," *Journal of Location Based Services*, vol. 7, no. 3, pp. 165–181, 2013.

[19] J. Torres-Sospedra, R. Montoliu, A. Martínez-Usó et al., "UJIIndoorLoc: a new multi-building and multi-floor database for WLAN fingerprint-based indoor localization problems," in *Proceedings of the Fifth International Conference on Indoor Positioning and Indoor Navigation*, pp. 261–270, Busan, October 2014.

[20] J. Torres-Sospedra, D. Rambla, R. Montoliu, O. Belmonte, and J. Huerta, "Ujiindoorloc-mag: a new database for magnetic field-based localization problems," in *Proceedings of the 2015 International Conference on Indoor Positioning and Indoor Navigation (IPIN)*, pp. 1–10, Alcala de Henares, Spain, October 2015.

[21] B. O. Mohamed and M. B. Hasan, "Decision tree approach to estimate user location in WLAN based on location fingerprinting," in *Proceedings of the National Radio Science Conference*, pp. 1–10, Cairo, Egypt, March 2007.

[22] J. D. Paola and R. A. Schowengerdt, "A detailed comparison of back propagation neural network and maximum-likelihood classifiers for urban land use classification," *IEEE Transactions on Geoscience and Remote Sensing*, vol. 33, no. 4, pp. 981–996, 1995.

[23] S. Bozkurt, G. Elibol, S. Gunal, and U. Yayan, "A comparative study on machine learning algorithms for indoor positioning," in *Proceedings of the 2015 International Symposium on Innovations in Intelligent SysTems and Applications (INISTA)*, pp. 1–8, Madrid, Spain, September 2015.

[24] N. Pritt, "Indoor positioning with maximum likelihood classification of Wi-Fi signals," in *Proceedings of the 2013 IEEE Sensors*, pp. 1–4, Baltimore, MD, USA, November 2013.

[25] F. Tom, "ROC graphs: notes and practical considerations for researchers," *Machine Learning*, vol. 31, pp. 1–38, 2004.

16

A Cloud-Based Mobile System to Improve Project Management Skills in Software Engineering Capstone Courses

Andres Neyem ⓘD, Juan Diaz-Mosquera ⓘD, and Jose I. Benedetto ⓘD

Computer Science Department, Pontificia Universidad Católica de Chile, Santiago, Chile

Correspondence should be addressed to Andres Neyem; aneyem@uc.cl

Academic Editor: María D. Lozano

Capstone project-based courses offer a favorable environment for the development of student skills through an approach incorporating theoretical and practical components. However, it is often difficult to successfully coordinate between students, stakeholders, and the academic team. The absence of suitable tools for addressing this issue, along with time constraints, often prevents students from attaining the expected course outcomes. This raises the question "How can we improve project management skills in computing majors through the use of technology-enhanced learning environments?" This paper presents a Cloud-based mobile system for supporting project management under a framework of best practices in software engineering capstone courses. The Kanban approach was used as a core of the proposed system. Kanban boards are very popular in the software industry today. It has been empirically shown that they provide increased motivation and project activity control due to their inherent simplicity. This helps the students and academic team be aware of the project context as it aids in preventing ambiguities, flaws, or uncertainties in the development of software artifacts.

1. Introduction

Project management is a valuable skill that university students should develop in order to achieve greater success in the industry. To develop this skill, students require enabling environments that establish the conditions necessary to put knowledge into practice. Project-based courses have proven to be the right space for working on this skill; in these courses, students work as a team to face the challenge of developing projects with real-world limitations and needs.

Through project-based learning courses, computer science and software engineering students have the opportunity to exercise the professional skills they will need after obtaining their degree. Projects with real-life constraints based on the needs and specifications of actual clients are particularly relevant as they are the most effective in bridging the gap between industry and academia. Having the abilities to deal with the management of projects is fundamental for success in this context. Through these abilities, it is possible to maintain a greater level of organization of the activities that are being carried out. Beyond that, being able to correctly manage projects allows for the development of better action and contingency plans, since eventual issues that hinder progress may be foreseen in advance.

For universities, teaching project-based courses is a big challenge that involves heavy responsibilities in providing environments where students can develop their professional skills [1]. The IIC2154 course at the Computer Science Department of the Engineering School at Pontificia Universidad Católica de Chile is a capstone course designed around this vision. The intent of the course is to provide a capstone experience that integrates the knowledge gained from all the previous courses in the curriculum by working on a project with realistic challenges. The blended approach between practical and theoretical components in the teaching of software engineering can bring relevant results because it encourages the development of products closer to production quality [2, 3].

In these kinds of courses, as projects move forward, the complexity surrounding them increases substantially. Charts, statistics, metrics, and reports in general become very important to understanding the real state of projects.

Thus, learning tools take on even more importance as they allow academic teams to follow students' learning process. Some investigations are focused on metrics and strategies aligned with the application of project management abilities and software development methodologies such as Scrum, XP, or Kanban [4, 5]. Although these proposals present fresh approaches in the same vein as our work, they lack a unified ecosystem for supporting project management activities. From this perspective, the aid of technological tools that support the activities of students while managing their projects is essential.

There are different ways to provide technological solutions for the implementation of a unified learning environment. Some teams only use shared Excel files for these tasks; however, there are more sophisticated tools available, such as Microsoft Project, that allow for more detailed specification and control of activities. Nevertheless, approaches like those mentioned above lack the flexibility that a Cloud-based mobile system can provide. These systems offer a mobile point of access (mobile devices) to the functions of management and teamwork in projects through a cloud architecture that maintains traditional interfaces such as desktop web browsers.

In this paper, we present a novel approach that tries to fill the gap between software tools and the development of project management skills by providing a technology-enhanced learning environment. This includes tools and analytics for seamless integration with a proven best practice capstone course framework that allows professors and students to derive as much benefit as possible. This approach presents a Cloud-based mobile system that supports the work of students offering tasks, requirements and test cycle management features, tracking, social charts, and other tools for improving project management skills in capstone project teams. Moreover, this platform produces data-rich activity, letting software practitioners and researchers understand how software development teams work using data science for empirical software engineering [6].

The remainder of this article is structured as follows: Section 2 presents the state of the art through a literature review on practices and tools used as support for project management. Section 3 presents the technology-enhanced learning environment for capstone project teams. Section 4 presents an evaluation of the Cloud-based mobile system based on empirical evidence gathered from the capstone course over a year and a half. Finally, Section 5 presents conclusions and future work related to this research.

2. State of the Art

This section presents the results of a literature review on studies in two areas: studies in software engineering education that define the key elements for the development of effective capstone courses and a series of articles that investigate the use of various technologies for supporting project management activities. The results of this literature review provide the foundations for the design and implementation of this proposal.

2.1. Developing Capstone Experiences in Software Engineering. Capstone courses give students the opportunity to demonstrate integrated knowledge and growth in their careers. Real-life projects have become a staple of software engineering capstone courses in several universities due to their pedagogical value in bridging the academic and industry worlds [7]. Moore and Potts wrote one of the first reports of a software engineering capstone course [8]. They propose emulating an industrial organization through the use of a Real World Lab. In this lab, projects are provided by industrial sponsors who act as clients, delivering consulting, reviewing, direction, and resources. The use of Mini-Task, a development process based on the waterfall life cycle, was also suggested to students. However, this cycle is open to changes after the first iteration in order to include methods and techniques they found useful. In the same vein, Sebern [9] also proposes a Real World Lab where students enroll after taking other software engineering courses, therefore ensuring some experience in developing software projects. It relies on the collaboration of real clients who provide students with existing software, aiming for the application of reverse engineering instead of forward engineering, as projects developed from scratch can either be of limited size or have a high risk of not reaching a functional product.

The Software Factory is another capstone course proposal based on the idea of hands-on experience developed across two semesters [10]. In this course, students develop projects either from scratch or undertake maintenance projects. Different roles are assigned to students according to their seniority. The main focus of the course is to meet industry requirements by educating computer science engineers in mastering modern technologies and processes, guaranteeing a solid and persistent knowledge. On the other hand, Robillard et al. [11] propose a capstone course called Studio in software engineering, where all teams have to develop the same project using UPEDU (the Unified Process for Education). To enroll in this course, students have to first take a software engineering course. They reported that most teams' effort was devoted to a few disciplines and that the learning curve of the process only lets development begin in the second half of the course.

Moving onto learning approaches based on agile software practices, Mahnic [12] uses an agile approach in his capstone course. Even though students were enthusiastic about the agile approach, the development would have been more organized and more easily planned if they had not sometimes missed a more detailed up-front design that would have offered them the complete picture. This course encourages the use of agile practices; however, since it always uses real clients, the clients themselves make sure that student teams work toward building the product they want without losing focus. Vanhanen et al. [2] describe another capstone course that uses an ad hoc agile-based process with industrial clients. The interesting feature of this study is the presence of a mentor, who assists and provides orientation during the entire semester. Undergraduate students participate as developers while master's students act as managers. Students considered the course stressful and laborious, but also extremely rewarding.

Weissberger et al. [13] propose a capstone course where students work on software maintenance projects, although in this case, the client is always the university. Before taking this course, students must already be familiar with several software engineering concepts they are expected to take full advantage of (e.g., stand-up meetings, pair programming, and burndown charts). Stettina et al. [14] describe the design, planning, and continuous improvement of another capstone course on software engineering. They conducted a survey to validate their findings. In this survey, they ask students about their satisfaction related to the project, the teamwork and communication within the project, the use of stand-up meetings, and the use of meeting minutes. Finally, Neyem et al. [15] propose a best practices framework for conducting computer science capstone courses, which is the one implemented in our case study. It involves a project-based approach for enhancing the learning experience, as well as an integration of various agile practices and tools. The goal of this framework is to familiarize students with best practices for developing high-quality software on par with commercial standards found in today's top software industries.

2.2. Project Management Tools. Unlike in the industry, project management tools for the development of academic activities are not normally used in educational environments. In some cases, students prefer to take advantage of informal ways of managing and collaborating on projects; an example of this is Google Docs [16, 17]. However, it is known that proper project management tools are essential for the successful planning and managing of projects [18].

When students are working on projects, it is important that all members remain aware of the tasks that others are working on. Kanban tools are an example of tools focused to this end. They allow for the better visualization of tasks through a Kanban Board. These tools have proven to be a suitable option for the coordination of work groups [19]. Generally, these kinds of tools have a focus on general-purpose projects; two examples in this category are Asana and Trello [20]. There are more tools in the market for supporting project management based on the Kanban methodology; the paper in [19] presents an extensive analysis of those. Two of the most known tools are Kanbanery and JIRA [21]. While the first is offered for free to charitable organizations and open source projects, the second requires a hefty subscription fee after 10 users.

Kanban tools are used to manage the flow of materials or information in a process [22]. These tools drive teams to visualize the workflow, limit Work In Progress (WIP) at each workflow stage, measure the cycle time (i.e., average time to complete one task), make process policies explicit, and improve collaboration [23–25]. Kanban tools can be defined as shared workspaces because they support cooperative tasks and provide a virtual space where information can be shared and exchanged [26]. It is very important to recognize the relevance of the virtual space in these tools. The virtual space enables users to clarify team members' awareness of the current production issues and forthcoming tasks [24].

Ahmad et al. [23] present an empirical study that investigates factors that users of Kanban tools consider to be important. It was found that the top two benefits of Kanban tools are the improved visibility of work and improved development flow. These benefits help enhance communication and collaboration within teams and related stakeholders. Under Kanban, a project is broken down into smaller pieces, and team members progressively finish these WIPs one by one, resulting in a constant flow of completed work items to customers. However, it was discovered that teams found it challenging to see the big picture through these smaller pieces. This suggests a need to provide awareness to team members of how each WIP relates to the others and the state of the project as a whole.

Oza et al. [27] performed a study that analyzes how the Kanban process aids software teams' communication and collaboration. This study collected data on a software development team that developed a mobile payment software product in six iterations over seven weeks. Data were collected through questionnaires at the end of each iteration. The team was part of the Software Factory, which is an experimental laboratory that provides an environment for research and education in software engineering established by the Department of Computer Science at the University of Helsinki. The team included eight engineers and a coach from the Software Factory. The study found that the use of Kanban had a high impact on both communication and collaboration in the initial iterations.

Ahmad et al. in [28] conduct a systematic review in order to analyze the current trend of Kanban usage in software development and to identify the benefits and challenges of the tool. They report that there is little existing scientific literature that addresses the use of Kanban tools for software development in spite of increasing interest among developers. Similar to the previous study on the implementation of Kanban at a software development team, the systematic review found that team communication and coordination with stakeholders is improved by using Kanban. Furthermore, the visualization of tasks with Kanban enables developers to understand the overall direction of work and helps them manage the workflow. This differs from the results of the empirical study in [27], where some teams found it challenging to see the big picture.

Cicibas et al. in [18] make an interesting comparison between project management tools, including Assembla, BaseCamp, DotProject, GanttProject, Liquid Planner, Artemis View, and Primavera, among others. This work defines a series of criteria for evaluating each tool, including task scheduling, resource management, collaboration, time tracking, project effort estimation, risk assessment, reporting, document management, communication tools, and so on. It was found that Artemis View is the most complete tool according to the criteria defined. Artemis View is an integrated enterprise project and resource management application that runs on Windows and web-based platforms.

Last but not least, according to Schwaber [29], most people responsible for managing projects have been taught a deterministic approach to project management that uses detailed plans, Gantt charts, and work schedules. The agile

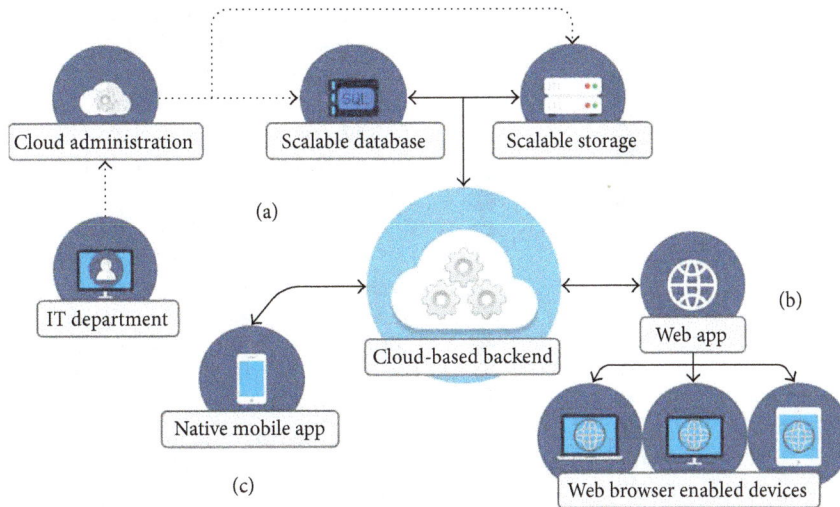

FIGURE 1: General architecture of the proposed Cloud-based mobile system.

methodology Scrum is the exact opposite. Scrum outlines how to guide a project along its optimal course, which unfolds as the project proceeds. Previously mentioned tools such as JIRA and others such as Pivotal Tracker follow this trend. The platform proposed in this work is based on this approach combined with some of the criteria outlined in [30]; this will be shown in more detail in the next section.

In summary, the review of the selected literature shows that Kanban is a promising tool to adopt for software development and project management. Adoption of Kanban has been shown to improve both communication and collaboration within teams and between related stakeholders. Visualization with Kanban is repeatedly mentioned as an important feature to facilitate project management.

3. The Proposed Cloud-Based Mobile System

3.1. The Architecture of the Cloud-Based Mobile System. The architecture of this system incorporates both a server-hosted component and a mobile client for interacting with the system while in the field. We define such mobile and Cloud Computing (MCC) solutions as Cloud-based mobile systems [31]. We opted for this computational model as mobile devices are becoming a popular way to access educational content, and traditional m-learning has certain limitations [32] that can be overcome with an MCC architecture. Additionally, there is evidence in the literature that these applications succeed in promoting teamwork and collaboration [33] in this context. From here onwards, the term mobile devices will refer to both tablets and smartphones.

Figure 1 shows the general architecture of the proposed system. Its functionalities are separated into three main components: (a) a Cloud-based backend service for computational and data storage purposes, (b) a Web application to enhance collaboration and analysis between students and academic teams, and (c) a mobile application that provides the core Kanban functionalities present in the web application.

The mobile component communicates with the web application via a RESTful API to synchronize data; however, it has its own local database to enable offline operation. Its main purpose is to collect personalized information about a group member's work routine and contributions and to notify them of any relevant status change of any task that was assigned to them.

The proposed platform underwent development for two years by the academic team of the capstone course at the Pontificia Universidad Católica de Chile. It seeks to cover all the basic needs of modern agile software projects and to seamlessly integrate them with an academic tool intended for tracking and evaluating individual progress. We draw attention to the latter, as the primary focus of our solution is its pedagogical value. The platform's features include support for formal methodologies, requirements management, team management, software testing, customer collaboration, and documentation [34]. Moreover, in line with our educational necessities, the platform also supports learning analytics that help to monitor the advance of the projects (Section 3.3) and options to generate standardized documentation for arbitrary software projects and for the course and a public API that allows third parties to query this information for more in-depth analysis (e.g., process modeling, data mining, etc.).

Without the proposed platform, it would be very difficult for the academic team to keep track of the progress made by each individual student in each different project. Furthermore, the data recorded in this platform allow offering more complete feedback to the students about the aspects in which they are succeeding or failing. In contrast, previous editions of the capstone course that lacked support of specialized information technologies made it almost impossible for the academic team to offer personalized guidance, and instead their input was mostly limited to global observations on the status of the projects. Additionally, with no tool to centralize all the information relevant to each project, the practical application of formal agile methodologies was rather limited and project management mostly relied on guesswork. On top

TABLE 1: Project management criteria: comparison of web and mobile application.

	Mobile	Web
(1) Task scheduling	●	●
(2) Resource management	●	●
(3) Collaboration	●	●
(4) Time tracking	●	●
(5) Estimation	●	●
(6) Risk assessment	◐	◐
(7) Change management	◐	◐
(8) Project analysis/report	◉	●
(9) Document management	●	●
(10) Communication tools	●	●
(11) Process development method	●	●
(12) Portfolio management	●	●

● Fully supported; ◐ partially supported; ◉ not supported.

of that, the focus on learning analytics of this platform offers a deeper value for both students and teachers, since it enables the platform to be not only a generic project management tool but also a platform enriched with knowledge.

So far, we have placed our focus on the area of software development; however, Conforto et al. show positive experiences in applying agile project management enablers in industries not related to software [35]. With this in consideration, the Cloud-based mobile system is modularly designed to allow for noncomputing industries to disable functionalities irrelevant to their respective fields yet still benefit from the many advantages of agile practices.

In order to determine the project management characteristics of our proposed system, we created a comparative table (Table 1) following the comparison criteria proposed by Ahmad and Laplante [30]. Our main focus is to show what features can be used in the proposed system in the web and mobile application. It is important to highlight that the mobile version does not support reporting, as lack of UI space makes it difficult to display this kind of information in a comprehensive manner. This feature can be better supported in a web browser.

Finally, in both the mobile and the web applications, the core functionalities are centered around "task representations" (from here onwards simply tasks) based on the Kanban board philosophy. It has been empirically shown that they provide increased motivation and project activity control due to their inherent simplicity [24]. While Kanban tasks are no more than granular jobs, our proposed system takes this idea further by allowing participants to enrich them by linking them to other modules in the project. As such, the Kanban board serves as the hub for the entire application. This tool can be found at the following address, under the section "educational tools": http://www.capstonecoursetoolkit.cloud.

3.2. Mobile Application.
The mobile application acts as a companion app for the web platform. Its primary purpose is to notify students whenever the status of a task assigned to them changes. It also features the most commonly used functions of the web platform in a user interface optimized for small screens. This allows students to check the global status of the project, modify a task, or share a comment on

a relevant matter with the group, even while on the move (e.g., meeting with a client offsite).

In order to reach the widest audience possible, the Android platform was chosen for the development of the mobile component, as it had over 85% of the global mobile OS market share in 2016. Multiplatform and mobile web tools were discarded, as we were interested in tapping into the latest Android-exclusive functionalities.

The mobile application features the core functionalities present in the web application (Figure 2), while the least used functions are relegated to the web platform. In the application, users can manage several organizations and projects (Figure 3(a)). The highlight of the application is the Kanban board (Figure 3(b)). It enables users to see the same global projects overview as in the web counterpart. Here, we include functionalities for creating (Figure 3(c)), editing, and deleting tasks, assigning tasks to different users, submitting task progress reports to keep track of the overall time developers have dedicated to them, file sharing, flag raising, and the ability to share comments between teammates.

All user input is immediately recorded in an offline database, enabling lag-free offline operation. Server synchronization is handled by an Android sync adapter. This allows us to bundle and defer sync operations to a later date. In cases where network connectivity is not available, the OS will trigger the synchronization subroutine only once it returns. It will keep doing so until communication with the server is successful. At the same time, the server will send sync requests whenever there is an update to a relevant task via push notifications. The client can then sync the data back and inform the user on what has changed specifically. Events of interest can be customized to ensure the user is not interrupted by any minor change.

Figure 2 shows an overview of the different layers of the mobile application. The entire application uses state-of-the-art technologies to ensure best practices with regards to code structure, performance, and battery life. The UI was developed using Android's data binding library and architecture components. The data layer was made by leveraging ROOM, Google's latest Android ORM. Lastly, synchronization management is delegated to the operating system through Android's sync adapter pattern to allow for network operation bundling with other applications as well.

There are two synchronization cases that warrant further detail: client-server conflicts and large file handling. Since our platform is decentralized, there exists the possibility for conflicts to arise when different users modify the same task from different clients. In this scenario, a default client-first conflict-solving policy eliminates the need for manual conflict solving. While this policy introduces some border cases in which some data may be lost, in practice, these instances are very rare and therefore deemed an acceptable compromise in favor of usability. On the contrast, our application allows users to share arbitrary documents with one another through a file attachment mechanism. No file limit is imposed, so when it comes to media, for example, there is the possibility of dealing with files in the order of several hundred megabytes. This requires handling with care not only because network interruptions may be frequent, but also because high network traffic through a mobile network

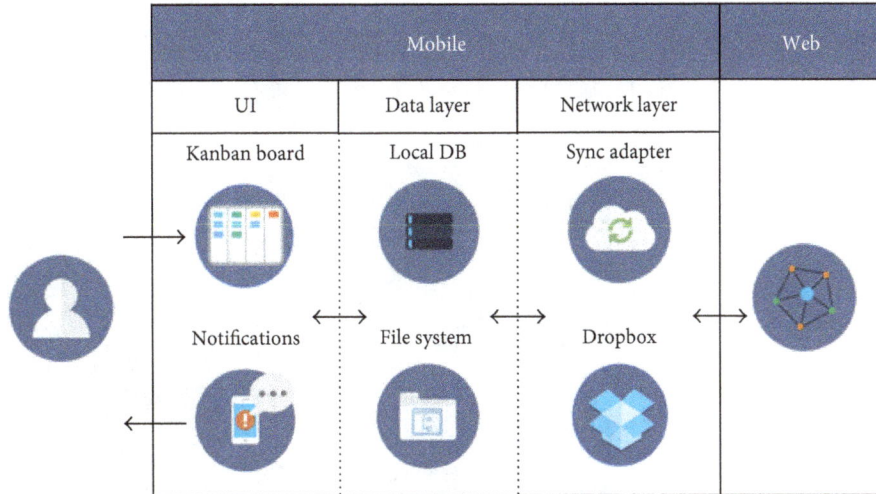

FIGURE 2: Mobile application overview.

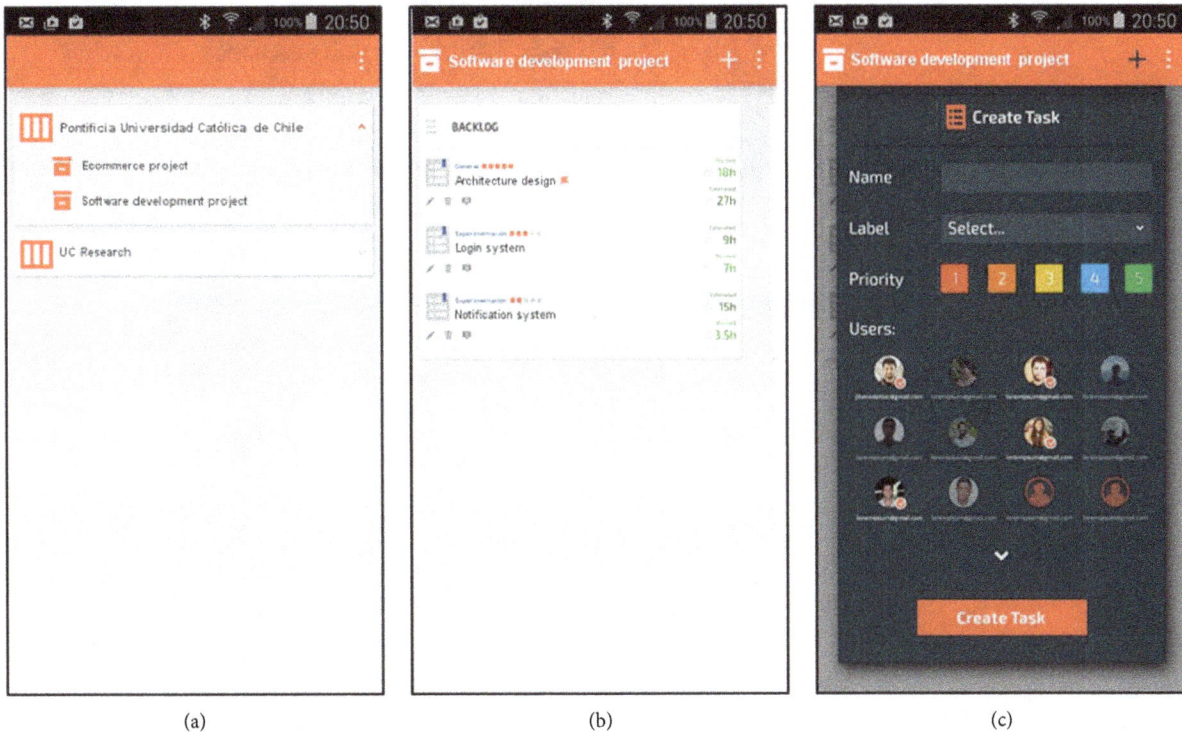

(a)   (b)   (c)

FIGURE 3: Mobile client Kanban flow: (a) organizations and projects listing, (b) Kanban board module, and (c) create task screen.

may result in significant financial costs to the end user. For this reason, our platform works in conjunction with the Dropbox file-sharing application, which already implements reasonable file synchronization policies and a resumable upload protocol. Users wishing to share files through our mobile client are therefore first asked to download the Dropbox application to their devices.

The last module of interest in the mobile client is the time reporting component. Users can access this component from the task options menu (Figure 4(a)) and configure their mobile device through our application to keep track of time when they start working (Figures 4(b) and 4(c)). The user can then safely either close the application or turn off the screen. Once he or she has finished working, he or she can return to the application and choose to stop time tracking so that the client can log the session. If the user does not want to time his or her work session, it is also possible to manually log an arbitrary amount of time at any moment. Users are expected to follow proper work ethics to not abuse the system.

Finally, to enhance productivity, alternating work intervals with short period breaks is also suggested. Studies show frequent short breaks that contribute to improving both productivity and wellbeing [36], especially if accompanied by physical activities. Our mobile client can therefore

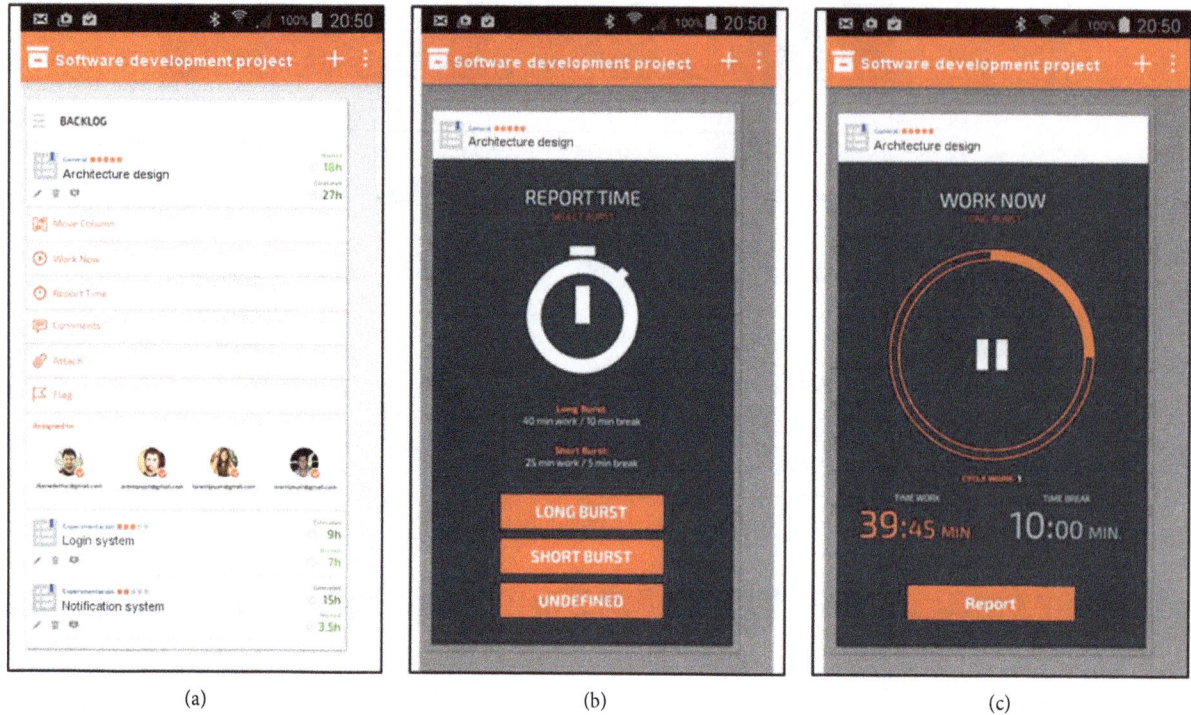

(a)   (b)   (c)

FIGURE 4: Mobile client time reporting flow: (a) task options menu, (b) time reporting module, and (c) burst option selected.

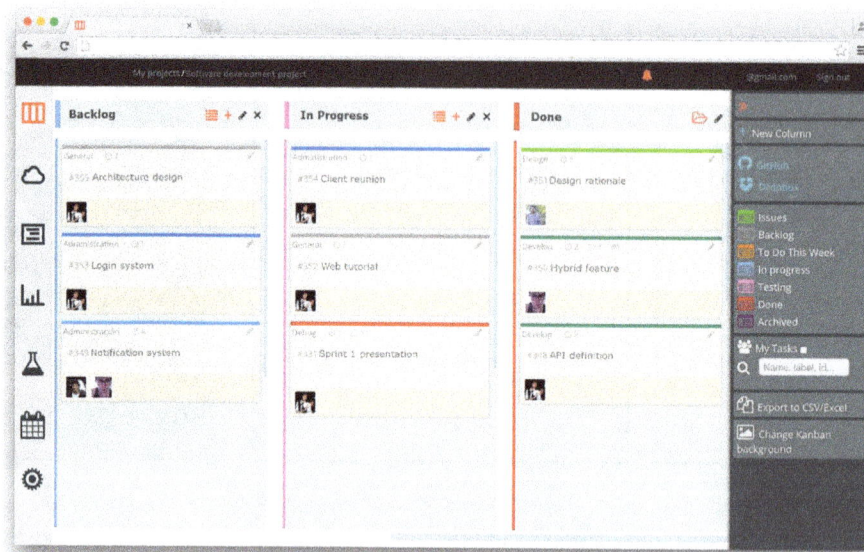

FIGURE 5: Kanban board in the web application.

be configured to notify the user at regular intervals when to take a short break and when to resume working.

3.3. Web Application.

The web application is built using HTML5, CSS, and JavaScript, with Ruby on Rails as a server-side framework. The deployment process is standardized by Capistrano, which, among other benefits, enables fast replication of the application on different servers.

Figure 5 shows the Kanban board: a virtual shared space that offers the entire team a global view of the project, along with all the tasks currently being worked on. Team members can assign tasks to one or more teammates, update their state, share their input via comments, and categorize them through labels. As previously stated, the Kanban board is one of the most effective ways to visualize the workflow and enable student teams to take control over what is to be done and what is currently being worked on.

In the web application, students have the ability to keep track of a project's requirements. These can be associated with tasks registered in the Kanban board. User stories or use cases can be created to reflect the requirements desired by

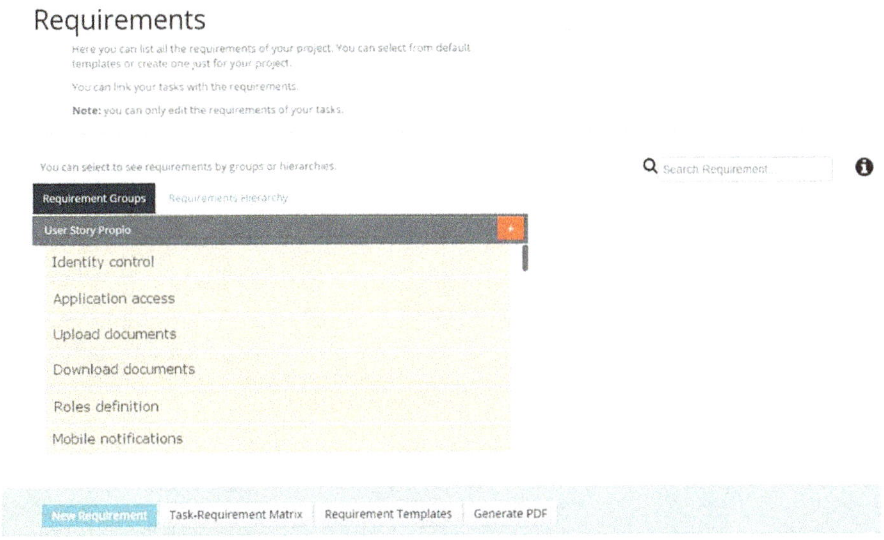

FIGURE 6: Requirements module.

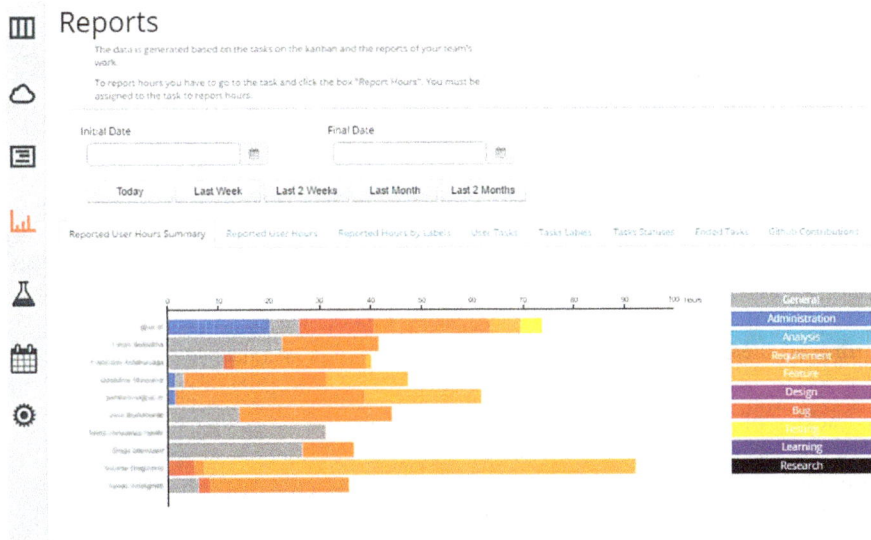

FIGURE 7: Summary of reported user hours.

stakeholders (Figure 6); additionally, tasks can be linked with test cycles with the purpose of ensuring the high quality of the software artifacts delivered.

Figure 7 shows the teamwork-reporting module of the web application. Managers and members of the academic team can use this tool to visualize the contributions of each team member through various charts. Users can see their reported hours by categories such as feature development, testing, analysis, and so on. Moreover, the web application enables visualization of the tasks cataloged as finished and reported hours through labels, among other reports. These options allow users to apply several filters over the information. If users need more detail, they can click on the bars to display related information.

Figure 8(a) shows a breakdown of a teammate's contributions. This is a burndown chart that displays the current rate of work done against an ideal rate that is calculated according to the daily rate of team productivity defined by users.

Additionally, this report allows users to check the number of tasks in the backlog or tasks with a "finished" status that they have reported each day. Figure 8(b) is a well-known chart in the world of Scrum, called a cumulative chart. As the name implies, this chart shows the sum of work accumulated along different stages of a project in progress. This chart allows for visualizing the workload according to different categories, letting users decide where the project needs the most attention.

There is also a feature where students can visualize the main topics in a project in a treemap (Figure 9). It uses text mining techniques to select the most relevant words and shows these according to the level of relevance (tiles of different sizes). This feature can help students and the academic team to determine where they should pay more attention and apply better project management strategies.

In software project management, providing traceability between a task and its associated code becomes a key factor

(a)

(b)

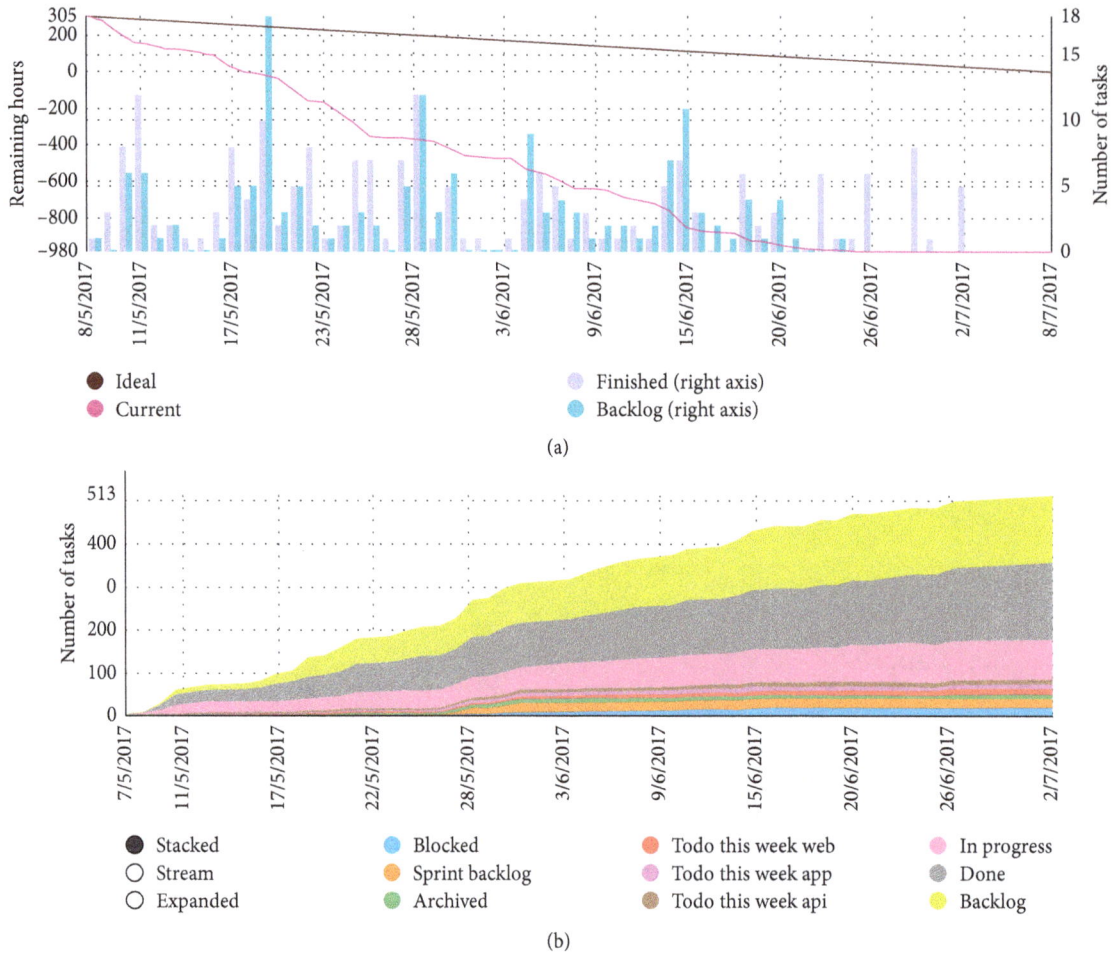

FIGURE 8: Agile reporting: (a) breakdown of a teammate's contributions and (b) cumulative chart of a teammate's contributions.

for maintaining wider tracking of project activities. For this reason, the web application in this work is integrated with the code repository platform GitHub. Through this special feature, users can easily revise code changes, in the same way as on GitHub (Figure 10).

The last feature of interest is the Process Mining- (PM-) based Student Teamwork Assessment. It implements PM techniques [37] to provide descriptive process models (Figure 11(a)) using the information from the proposed system. It helps to understand the workflow inside each project team. The line thickness denotes the level of interaction between activities, and the color of each box represents the frequency of each activity (darker colors mean higher frequency). This view allows the students and the academic team to identify situations out of the ordinary in order to take proper action and also to identify strong points in management behavior.

Likewise, PM allows identifying interactions between team members, as the graph in Figure 11(b) represents. The nodes in this graph show a flow between an administrator figure and members of their team; in other words, it corroborates that the administrator works as the central axis of the project.

3.4. Cloud Service as a Backend. There are multiple ways to deploy backend services in cloud computing architectures.

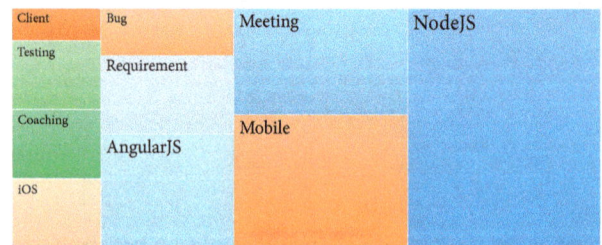

FIGURE 9: Treemap with the main topics in a project.

One is to implement the whole backend logic using the software distribution model IaaS (Infrastructure as a Service). In this scenario, users have total control of their applications, data, runtime environments, and operating systems. In IaaS, users do not need to spend time on the complexity of building and maintaining the physical infrastructure typically associated with developing and launching an app.

Backend as a Service (BaaS) is another option, consisting of a cloud computing approach for providing services that should be able to be easily accessed. It aims for an integrated way of using a shared pool of computing resources or services that can be rapidly distributed with minimal management effort. It reduces the internal complexity that

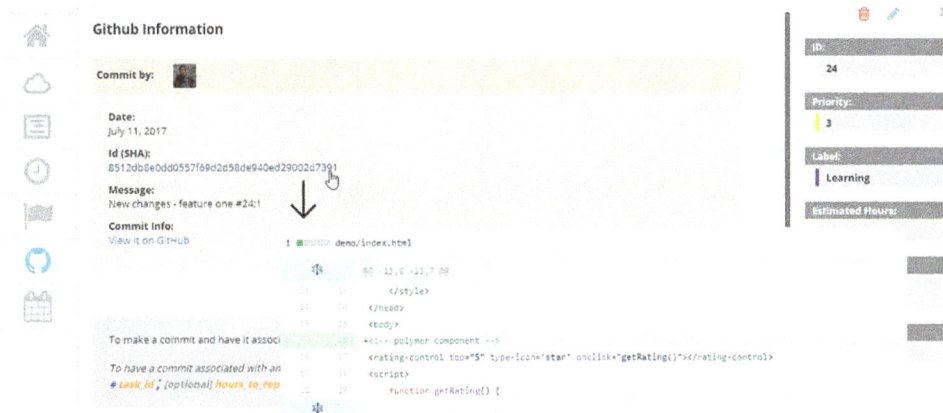

FIGURE 10: GitHub integration.

developers of these services have to face, offering common preset features such as user management, push notifications, storage, social networking integrations, triggers, and so forth. In essence, the advantages of using this approach for mobile and web apps are as follows:

(i) Availability on demand: the services can be used and accessed at any time.

(ii) Agile and fast provisioning: it is not necessary to set up complex environments; furthermore, scalability for apps is easily achievable.

(iii) Multiple access support: the services can be accessed through the Internet from different kinds of devices.

(iv) Security: offers centralized solutions for authentication and authorization.

(v) Better processing capability: complex tasks can be processed in the backend, reducing computational time compared to devices with limited resources.

The backend for our proposed Cloud-based mobile system is built with a two-fold approach. It uses Firebase (as an instance of BaaS) and the Ruby on Rails (RoR) framework (as an instance of IaaS). With Firebase, we share and distribute data to several users and several devices and platforms. To achieve this, we adopted the Firebase Cloud Messaging technology, which was used to send nonsensitive information from the cloud service to mobile devices for providing push notifications.

The IaaS component of our system contains a RESTful API deployed on Amazon EC2 that provides all the services for project management activities. Only authenticated users are allowed to interact with our web services. For storage purposes, our cloud system uses high availability Amazon services such as RDS (Relational Database Service) and S3 (Simple Storage Service). In this way, our whole ecosystem relies on Amazon technologies, which allows for easy horizontal and vertical scalability.

4. Evaluation

The following section serves to demonstrate the positive results of using our Cloud-based mobile system to support the development of project management skills. The evaluation was conducted by carrying out 15 projects over 3 semesters (2016-1, 2016-2, and 2017-1) at Pontificia Universidad Católica de Chile (PUC). Our main focus is to study the participants' perceptions of the project management skills that they acquired using the Cloud-based mobile system proposed in this work. Additionally, we study the relevance of the tool in accomplishing this objective. To complement this study, we also include an analysis of the clients' opinions on the work carried out by our students and the final deliverables they received.

The proposed tool is used throughout the course, but it especially helps students to show their progress in weekly meetings with the academic team. This is a process that is fundamentally based on the information recorded by the students on the platform. The data contained within include the record of hours used in the development of the activities and the registration of requirements and test cycles for the developed software artifacts. All of the above may be further complemented by attaching arbitrary documents that are shared through the application. With this information, the academic team can offer better support to the students (e.g., early detection of risks and establishment of contingency plans). It is important to highlight that our intention is to show the platform as a whole; therefore, we evaluate the experience of the students in regard to the complete approach offered in the course and not only in regard to the web and mobile tool used in an isolated context. In this section, we describe the background, the participants, and the instruments used. Then, we discuss the findings that emerged from the quantitative and qualitative analyzes from the students. Finally, we show the opinions of real clients regarding aspects such as product quality, communication, and their perceptions on the project management abilities developed by students.

4.1. Experiments: Context and Participants. The evaluation of the platform was carried out with a total of 146 students with different levels of expertise in the area of software development. These students belong to the capstone course in computing science major at the Computer Science Department. In this evaluation, we incorporated the opinions of 15 clients on the work done by the students. During the

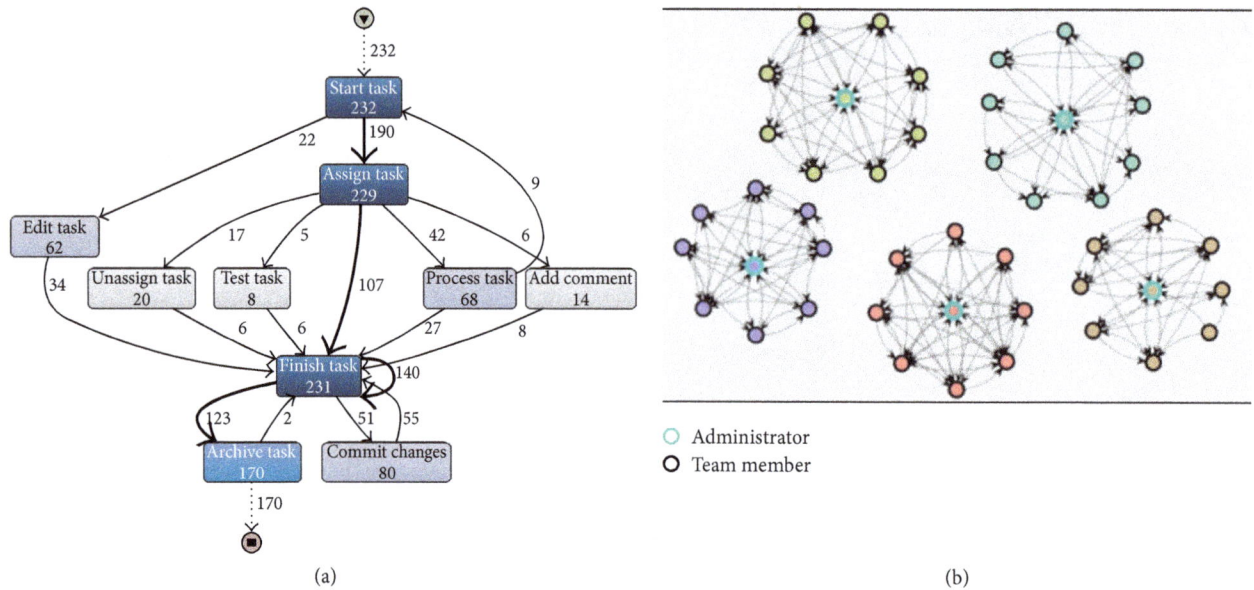

(a)

(b)

FIGURE 11: Process mining approach: (a) PM model from capstone course project and (b) identifying team interactions using PM.

course, students follow a general outline of the best practices framework [15], which consists of five phases: previous work, introduction and three separate software development sprints. The previous work phase should focus on project selection; the introduction phase should emphasize teaching course methodology and applied software engineering principles; and finally, the software development sprints delve into the actual practical development (requirements, design, coding, testing, and deployment).

Over the course of three semesters, the capstone course was able to encompass a large group of projects. These include startups, established businesses, foundations, and university projects. Among those that stand out the most, we have one project consisting of a technological platform to carry out online evaluations and interact with a multimedia encyclopedia (images, videos, and 3D) on anatomy; another one consisting of a system for generating conference-focused websites that allow users to create and manage the flow of events of this type in an intuitive way; and a third one consisting of the development of embedded systems to be used as beacons for a more user-participatory marketing and advertising strategy.

The aforementioned projects, along with 12 others, involve the development of web, mobile, and embedded systems. These projects have an average of 10 members (an administrator included) who share different tasks throughout the duration of the course. These activities demand a high level of management and teamwork as they are frequently interconnected. Moreover, these activities represent different challenges that students have to face. Among the list of challenges are the integration of external systems, the development of APIs, the construction of e-commerce systems, the implementation of payment systems, and the development of notification modules, among other activities.

4.2. Instruments and Data Collection. A survey was conducted at the end of each semester in order to collect quantitative and qualitative data and analyze them according to the focus of this work. This research used a mixed-method approach with a concurrent design, meaning quantitative and qualitative questions were asked in parallel or in the same survey [38]. In this survey, we included three open-ended questions intended to assess the development of project management skills amongst the students. Additionally, clients were asked one more questions intended to evaluate the overall experience of the course. We sent this survey to 146 students (52 in 2016-1, 49 in 2016-2, and 45 in 2017-1) and 15 clients during the three semesters of the capstone course. The level of participation varied in each semester.

The data gathered from the open-ended questions allow us to perform a qualitative analysis with the participants' opinions. Details on the contents of the survey are available in Table 2. Question number 3 evaluates the achievement level that participants perceive for that item and comprises four possible scores. The evaluation scale considers 1 as "unsatisfactory," 2 as "developing," 3 as "satisfactory," and 4 as "exemplary."

4.3. Results. According to the participants' opinions, in overall, the mixed approach between the proposed system and the methodology implemented in the capstone course gave positive results. The qualitative data were analyzed using the grounded theory method in empirical software engineering [39]. The following is a specific analysis for each research-related question asked in the survey.

4.3.1. Question 1: Approach Evaluation. By analyzing the first open-ended question, we discovered that the proposed system played an important role in carrying out the work of task planning. Some participants stated that task planning was carried out weekly and that the proposed system allowed

TABLE 2: Questions in the survey.

Question	Category	Participant
(1) What do you think was the approach that allowed you to exercise the development of the ability "organizational skill" during the course?	Open-ended	Students
(2) What do you think about the proposed tool as a means of support for the development of your project?	Open-ended	Students
(3) At what level were you able to develop project management skills (such as planning, scheduling, tracking, and teamwork) present in a software project?	Multiple choice	Students
(4) What are your opinions about the work done by students in the capstone course (strategies implemented, tools used, students' performance, delivered product, etc.?	Open-ended	Clients

a clear division of tasks through the Kanban approach (Quotes 1 and 2).

Quote 1: *Using a system like the one we used in the course has been fundamental for better task planning.*

Quote 2: *The system's Kanban approach allows a clear division of activities.*

Being able to define tasks and prioritize them and establish their time limits has allowed for a better management of the functionalities that the students have to deliver (Quotes 3 and 4). In addition, the implementation of user stories promoted by the system has also allowed for better organization (Quote 5).

Quote 3: *The number of tasks that are programmed in the system for each week has allowed us to exercise our organizational skills, since we need to prioritize the development of each one.*

Quote 4: *The use of the system and its integrated Google Calendar feature has allowed us to organize the deadlines of the tasks that must be delivered.*

Quote 5: *Using user stories and being able to define several requirements per week has helped us a lot.*

4.3.2. *Question 2: Tool Evaluation.* We addressed this question from three perspectives: the utility, the user experience, and the learnability of the tool. The students' responses and their respective analysis are approached in that order.

From a utility perspective, the responses led to the discovery that regardless of the required time for registering information in the proposed system, students considered it useful as it enabled more control over their projects (Quotes 6, 7, and 8).

Quote 6: *I am aware that the use of tools such as the Kanban board proposed in this course made our project more organized.*

Quote 7: *The proposed system mainly offers order. Another aspect is that the group members can visualize the activities that everyone is performing. Really, if we used it with even more dedication we could obtain greater benefits.*

Quote 8: *With the tasks organized by phases in the tool we have a better perspective of what we have to do.*

Some students highlight the fact that they can easily navigate through the different functionalities of the tool (Quote 9). They also highlight the characteristics related to social awareness such as roles, activities, privileges, and group history that can be easily identified and used in the platform (Quotes 10 and 11).

Quote 9: *The navigation in the tool is simple; having organizations and projects in a hierarchical way makes the management of these very practical. On the other hand, the central functionality, "the Kanban board," is very flexible and allows adapting the projects to the needs of each work team.*

Quote 10: *To know what my teammates are doing, I would read the titles in the tasks on the Kanban board. If I wanted more details, I would open the tasks and look at the description.*

Quote 11: *In my case, being the team leader, I have to constantly check what my teammates are doing. I do that in different ways, for example by chat, face-to-face meetings, shared documents or with the activities registered on the Kanban board.*

In the context of tasks, it is important to know the answers to the following questions: who, what, where, when, and how. Some features of the proposed system allowed students to answer these questions (Quotes 12 and 13).

Quote 12: *The notifications at the start of the system let us see who created or modified the tasks.*

Quote 13: *The visualization of requirements and test cycles is useful to better understand what is going on in projects.*

From a user experience perspective, some students see the Kanban board feature as a way of incentivizing their own work. Watching the work of the other students can generate positive motivation (Quote 14). Moreover, the design of the platform makes them feel comfortable due to the distribution of the content and the style used for the mobile and web application (Quotes 15 and 16).

Quote 14: *I think that the Kanban (board) has a large benefit. On the one hand, I can see how much work the rest of the project members are doing. This can incentivize you to work more.*

Quote 15: *The tool handles a good balance both in the combination of colors and the distribution of its characteristics. It's nice to work on the tool; it even allows some degree of interesting customization.*

Quote 16: *In my opinion, the style of the web and mobile tools is quite modern and intuitive. The tool really allows you to have a real immersion in a project with agile methodologies.*

Some participants highlighted more aspects that they found attractive when using the proposed system (Quotes 17 and 18). Some even recognized that the system could be helpful in a business environment (Quote 19). This is an important finding, because it lets us understand the impact that the system might have on the working experience of the participants.

Quote 17: *The system helps you set short goals and visualize them in the Kanban board.*

Quote 18: *With the system, I can see how well my teammates are doing. Besides watching what they do, I can help them or ask for help if they are doing something related to what I have to do.*

Quote 19: *Actually, I would use the proposed tool for project management in my future working life.*

Regarding the learnability of the tool, it was noted that the first contact of the students with the tool was not difficult. The ease of access, the proper location of the menus, and design similar to tools widely known in the market let the students quickly learn how to use the tool (Quotes 20 and 21).

Quote 20: *The first time I tried the tool it reminded me of Trello. For me it was very easy to understand how to use the tool.*

Quote 21: *I have already had the opportunity to use tools of this style; I can say that the learning curve for this tool is not too steep.*

4.3.3. Question 3: Performance Level.

As we previously mentioned, this question used a four-point scale. *Unsatisfactory* means that the participant felt he or she was not able to demonstrate understanding of the knowledge and skills referred to in this question. *Developing* indicates that the participant felt able to demonstrate understanding of the knowledge and skills referred to in this question, however, only in the case of environments similar to the one seen in the course. *Satisfactory* refers to the ability that the participant has to describe, apply, and solve problems using the knowledge and the skills referred to by this question; moreover, the participant can face and overcome situations in settings different from that of the course. Finally, *Exemplary* means that the participant felt able to show mastery of the knowledge and skills involved in the question, being able to synthesize, organize, plan, manage, evaluate, and/or teach others, in diverse situations covered by the question.

According to the results, the participants considered that they had gained the skills required to manage projects properly. All responses indicated that participants had acquired a satisfactory or exemplary level (Figure 12) of understanding. It is interesting to note that in all cases, the approach used in the capstone course influenced the development of project management skills, and we are certain that the Cloud-based mobile system was a relevant support in that regard.

4.3.4. Question 4: Client Opinions

(1) Product Quality. Throughout the capstone course, different types of projects are carried out, which represent a great technical and coordination challenge both for the

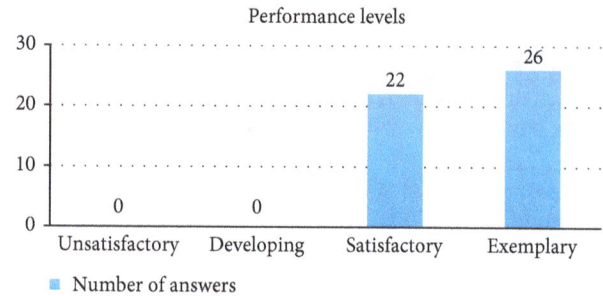

FIGURE 12: Results of question 3.

students and the academic teams. Despite the complexity of the projects, the students have been able to present products with similar quality to that of professional software development companies.

Generally, clients declare that the projects developed by the students meet the objectives expressed throughout the course. In accordance with the requirements, in the large majority of cases, 100% of the requested functionalities are achieved (Quotes 22 and 23). The approach that allows obtaining these results is the clear and timely administration of all the activities that take place in the projects through the management tool that this proposal presents.

Quote 22: *I was surprised with the product delivered by the students. I can assure that they fully complied with everything we had talked about from the beginning.*

Quote 23: *The platform that the students made demonstrates how professional they can be. I have in my hands a product that is ready to be put into production.*

Some clients stated that they received more than they expected. When they came to the capstone course, they had a basic idea of what they are looking for; however, at the end of the course, they were left with the satisfaction of having obtained a product that could put into practice what they had imagined (Quotes 24 and 25). The combination of a framework that coordinates the clients, students, and academic team well and a management tool that allows for a detailed follow-up of the tasks, their managers, and the state of these, constitutes for us the success of our approach.

Quote 24: *I am not an expert in technology, which is why my ideas could be a bit difficult to understand; however, the capstone students managed to understand my requirements and make software with more than I expected.*

Quote 25: *I can say that the students in charge of my project knew how to make and guide me towards a product that exceeded my expectations.*

(2) Communication with Clients. Communication with clients was achieved in different ways, through formal meetings and in-progress presentations, while periodically using the course's project management tool. Clients had the ability to access the Kanban board where all the pending, in-process or completed activities within the project, were recorded (Quotes 26 and 27).

Quote 26: *When I wanted to know what the students were doing, it was enough for me to go to the Kanban board and see the activities registered by them.*

Quote 27: *Through the tool I found an easy way to leave comments about the functionalities that were being developed for my project, in this way I clarified the doubts of the students in a timely manner.*

(3) Perceived Project Management. Clients expressed their opinion on how they viewed the management of the projects carried out by the students. During the three project development cycles, they could see how the activities aimed at delivering functional software artifacts were being managed. Having the requirements, the test cycles and the activities clearly divided by phases, categories, and priorities was a clear indicator of good management (Quotes 28 and 29).

Quote 28: *The way in which I perceived that students were having good project management was noticing that they had clearly defined roles and tasks for each member, and they were quite supported by the platform that manages the course.*

Quote 29: *I thought it was excellent that there was a clearly defined timeline where it was established what was going to be done and under what priority in each cycle of the project.*

In each of the meetings with the clients, the students relied on indicators generated by the tool, which reported the hours of work employed in the management and development of the project. The clients declared that it was easy to see the level of commitment of the students when visualizing the charts that the tool offers (Quotes 30 and 31).

Quote 30: *The students were very transparent in showing and explaining the time spent in the development of the functionalities proposed by us by using the tool.*

Quote 31: *With the indicators shown by the students we could clearly notice the time and commitment with which our project was developed. I consider in general terms that the students knew how to work together and manage the project very well.*

5. Discussion and Conclusions

This paper presents a novel Cloud-based mobile system for assisting students in a capstone course in software engineering. The combination of the tool and a framework of best practices in the capstone course generated positive results for the development of different skills in students. This work places special emphasis on skills relevant to project management, which we consider to be one of the most important aspects in the correct execution and control of tasks in work groups.

The proposed tool involves a series of features that revolve around a Kanban approach, which offers better visualization and monitoring of the activities defined in a project during its different stages. The MCC component of this system is fundamental in more broadly supporting the development of project management skills and therefore teamwork. In this context, the mobile part of the tool is a subset of features that complements the entire proposed system.

The analytics or visualizations presented in this work are an important source of information that allows both the students and the academic team to identify different aspects of work and performance within the projects. It is important to highlight that this work is largely based on the use of learning analytics, which offer a way to use the academic information offered by the capstone course to understand and subsequently optimize learning and the environments where it takes place. Our proposed system is a suitable environment for converting information into knowledge using different techniques in the area of Data Science for Empirical Software Engineering; specifically, we present in this work, an approach using process mining and text mining techniques.

The evaluation instrument used in this study allowed us to recognize that the proposed system had an impact on the students' work experience. However, we must recognize that this would not have been possible without the framework established in the course that motivated a greater organization of the activities they carried out. This framework encourages students to rely on the information registered in the proposed system to be able to better control and track the projects, especially in weekly meetings, where progress in these projects is shared with the academic team.

The investigation bridges two separate worlds: academia and industry. With the proposed platform, greater possibilities are offered to students to recognize and live a software development experience that is more similar to the real world. Web and mobile platforms are a means to facilitate the process of organizing student tasks and carrying out better course coordination by teachers. At the end of each semester, surveys are conducted that allow us to obtain more evidence about the application of our approach. In the future, we hope to integrate additional features with the proposed system, among which we have considered the development of a recommendation system that will offer students assistance in the activities or tasks they want to accomplish. This system will consist of the consolidation of knowledge or experiences of other people who performed similar activities in the past. It will use this information to offer suggestions based on how the affected people were able to overcome such problems before.

Another approach we seek to pursue in the future is machine learning. With information captured from our system and consolidated in a training database, we want to discover special factors for objectively determining, assessing, and predicting student teamwork outcomes by applying machine learning methods to the training database [40]. Moreover, we want to promote the participation of students with the proposed system through gamification strategies that generate better-defined spaces of collaboration.

In summary, the evidence obtained from research data leads us to conclude that systems of this type in educational environments can generate a significant influence on the future work experience of students. This is derived from the analysis of the opinions obtained from the survey. The work conducted in this study presents a technology-enhanced learning environment where a framework of best practices in capstone courses converges with MCC components for the support of student and academic team activities. In essence, this approach constitutes a proposal to improve learning and teaching of critical skills for the proper professional development of students.

Acknowledgments

The authors would like to acknowledge the Computer Science Department, the Education Engineering Unit of the Engineering School, and the VRA FondeDOC for the continued support of this course. Finally, the authors would like to thank all the students from the IIC2154 capstone course who were involved in this educational research project.

References

[1] A. T. Chamillard and K. A. Braun, "The software engineering capstone: structure and tradeoffs," *ACM SIGCSE Bulletin*, vol. 34, no. 1, pp. 227–231, 2002.

[2] J. Vanhanen, T. O. Lehtinen, and C. Lassenius, "Teaching real-world software engineering through a capstone project course with industrial customers," in *Proceedings of the First International Workshop on Software Engineering Education Based on Real-World Experiences*, pp. 29–32, IEEE Press, Zurich, Switzerland, June 2012.

[3] M. Vasilevskaya, D. Broman, and K. Sandahl, "Assessing large-project courses: model, activities, and lessons learned," *ACM Transactions on Computing Education*, vol. 15, no. 4, pp. 1–30, 2015.

[4] L. Alperowitz, D. Dzvonyar, and B. Bruegge, "Metrics in Agile project courses," in *Proceedings of the 38th International Conference on Software Engineering Companion*, pp. 323–326, ACM, Austin, TX, USA, May 2016.

[5] S. Mohan, S. Chenoweth, and S. Bohner, "Towards a better capstone experience," in *Proceedings of the 43rd ACM Technical Symposium on Computer Science Education*, pp. 111–116, ACM, Raleigh, NC, USA, February 2012.

[6] T. Menzies, L. Williams, and T. Zimmermann, *Perspectives on Data Science for Software Engineering*, Morgan Kaufmann, Burlington, MA, USA, 2016.

[7] A. Rusu, A. Rusu, R. Docimo, C. Santiago, and M. Paglione, "Academia-academia-industry collaborations on software engineering projects using local-remote teams," *ACM SIGCSE Bulletin*, vol. 41, no. 1, pp. 301–305, 2009.

[8] M. Moore and C. Potts, "Learning by doing: goals and experiences of two software engineering project courses," in *Proceedings of the 7th SEI CSEE Conference on Software Engineering Education*, pp. 151–164, San Antonio, TX, USA, January 1994.

[9] M. J. Sebern, "The software development laboratory: Incorporating industrial practice in an academic environment," in *Proceedings of the 15th Conference on Software Engineering Education and Training (CSEE&T 2002)*, pp. 118–127, IEEE, Covington, KY, USA, 2002.

[10] J. D. Tvedt, R. Tesoriero, and K. A. Gary, "The software factory: combining undergraduate computer science and software engineering education," in *Proceedings of the 23rd International Conference on Software Engineering (ICSE 2001)*, pp. 633–642, IEEE, Toronto, ON, Canada, May 2001.

[11] P. N. Robillard, P. Kruchten, and P. d'Astous, "Software engineering using the Upedu," Addison-Wesley Longman Publishing Co., Inc., Boston, MA, USA, 2002.

[12] V. Mahnic, "A capstone course on agile software development using Scrum," *IEEE Transactions on Education*, vol. 55, no. 1, pp. 99–106, 2012.

[13] I. Weissberger, A. Qureshi, A. Chowhan, E. Collins, and D. Gallimore, "Incorporating software maintenance in a senior capstone project," *International Journal of Cyber Society and Education*, vol. 8, no. 1, pp. 31–38, 2015.

[14] C. J. Stettina, Z. Zhou, T. Bäck, and B. Katzy, "Academic education of software engineering practices: towards planning and improving capstone courses based upon intensive coaching and team routines," in *Proceedings of the 2013 IEEE 26th Conference on Software Engineering Education and Training (CSEE&T)*, pp. 169–178, IEEE, San Francisco, CA, USA, May 2013.

[15] A. Neyem, J. I. Benedetto, and A. F. Chacon, "Improving software engineering education through an empirical approach: lessons learned from capstone teaching experiences," in *Proceedings of the 45th ACM Technical Symposium on Computer Science Education*, pp. 391–396, ACM, Atlanta, GA, USA, March 2014.

[16] X. Tan and Y. Kim, "Cloud computing for education: a case of using Google Docs in MBA group projects," in *Proceedings of the 2011 International Conference on Business Computing and Global Informatization (BCGIN 2011)*, pp. 641–644, IEEE, Shanghai, China, July 2011.

[17] S. Dekeyser and R. Watson, "Extending Google Docs to collaborate on research papers," Technical Report, University of Southern Queensland, Toowoomba, QLD, Australia, 2006.

[18] H. Cicibas, O. Unal, and K. A. Demir, "A comparison of project management software tools (PMST)," in *Proceedings of the Software Engineering Research and Practice*, pp. 560–565, Las Vegas, NV, USA, July 2010.

[19] E. Corona and F. E. Pani, "A review of lean-Kanban approaches in the software development," *WSEAS Transactions on Information Science and Applications*, vol. 10, no. 1, pp. 1–13, 2013.

[20] H. A. Johnson, "Trello," *Journal of the Medical Library Association*, vol. 105, no. 2, p. 209, 2017.

[21] P. Li, *Jira 7 Essentials*, Packt Publishing Ltd., Birmingham, UK, 2016.

[22] P. Klipp, *Getting Started with Kanban*, Kanbanery, Kraków, Poland, 2011.

[23] M. O. Ahmad, J. Markkula, and M. Oivo, "Insights into the perceived benefits of Kanban in software companies: practitioners' views," in *Proceedings of the International Conference on Agile Software Development*, pp. 156–168, Springer International Publishing, Edinburgh, UK, May 2016.

[24] M. Ikonen, E. Pirinen, F. Fagerholm, P. Kettunen, and P. Abrahamsson, "On the impact of Kanban on software project work: an empirical case study investigation," in *Proceedings of the 2011 16th IEEE International Conference on Engineering of Complex Computer Systems (ICECCS 2011)*, pp. 305–314, IEEE, Las Vegas, NV, USA, April 2011.

[25] E. M. Schön, D. Winter, J. Uhlenbrok, M. J. Escalona, and J. Thomaschewski, "Enterprise experience into the integration of human-centered design and Kanban," in *Proceedings of the ICSOFT-EA*, pp. 133–140, Lisbon, Portugal, July 2016.

[26] A. Neyem, S. Ochoa, and J. Pino, "Designing mobile shared workspaces for loosely coupled workgroups," in *Proceedings of the International Conference on Groupware: Design, Implementation, and Use (CRIWG 2007)*, pp. 173–190, Bariloche, Argentina, September 2007.

[27] N. Oza, F. Fagerholm, and J. Munch, "How does Kanban impact communication and collaboration in software engineering teams?," in *Proceedings of the 2013 6th International Workshop on Cooperative and Human Aspects of*

Software Engineering (CHASE 2013), pp. 125–128, IEEE, San Francisco, CA, USA, May 2013.

[28] M. O. Ahmad, J. Markkula, and M. Oivo, "Kanban in software development: a systematic literature review," in *Proceedings of the 2013 39th EUROMICRO Conference on Software Engineering and Advanced Applications (SEAA 2013)*, pp. 9–16, IEEE, Santander, Spain, September 2013.

[29] K. Schwaber, *Agile Project Management with Scrum*, Microsoft Press, Redmond, WA, USA, 2004.

[30] N. Ahmad and P. A. Laplante, "Software project management tools: making a practical decision using AHP," in *Proceedings of the 30th Annual IEEE/NASA Software Engineering Workshop (SEW'06)*, pp. 76–84, IEEE, Washington, DC, USA, April 2006.

[31] D. Carver, W. K. Chan, C. K. Chang, and H. Yang, "Software engineering and applications for cloud-based mobile systems," *IEEE Transactions on Services Computing*, vol. 9, no. 5, pp. 742–744, 2016.

[32] A. M. Rosado da Cruz, *Modern Software Engineering Methodologies for Mobile and Cloud Environments*, IGI Global, Hershey, PA, USA, 2016.

[33] S. Abolfazli, Z. Sanaei, M. H. Sanaei, M. Shojafar, and A. Gani, "Mobile cloud computing," in *Encyclopedia of Cloud Computing*, S. Murugesan and I. Bojanova, Eds., John Wiley & Sons, Ltd., Chichester, UK, 2016.

[34] N. Uikey and U. Suman, "An empirical study to design an effective agile project management framework," in *Proceedings of the CUBE International Information Technology Conference and Exhibition*, pp. 385–390, ACM, Pune, India, September 2012.

[35] E. C. Conforto, F. Salum, D. C. Amaral, S. L. da Silva, and L. F. M. de Almeida, "Can agile project management be adopted by industries other than software development?," *Project Management Journal*, vol. 45, no. 3, pp. 21–34, 2014.

[36] R. A. Henning, P. Jacques, G. V. Kissel, A. B. Sullivan, and S. M. Alteras-Webb, "Frequent short rest breaks from computer work: effects on productivity and well-being at two field sites," *Ergonomics*, vol. 40, no. 1, pp. 78–91, 1997.

[37] W. M. Van der Aalst, *Process Mining: Data Science in Action*, Springer, Berlin, Germany, 2016.

[38] J. W. Creswell and V. L. P. Clark, *Designing and Conducting Mixed Methods Research*, Sage Publications, Thousand Oaks, CA, USA, 2007.

[39] S. Adolph, W. Hall, and P. Kruchten, "Using grounded theory to study the experience of software development," *Empirical Software Engineering*, vol. 16, no. 4, pp. 487–513, 2011.

[40] D. Petkovic, "Using learning analytics to assess capstone project teams," *Computer*, vol. 49, no. 1, pp. 80–83, 2016.

17

Trustworthy Event-Information Dissemination in Vehicular Ad Hoc Networks

Rakesh Shrestha and Seung Yeob Nam

Department of Information and Communication Engineering, Yeungnam University, 280 Daehak-Ro, Gyeongsan-si,
Gyeongsangbuk-do 712-749, Republic of Korea

Correspondence should be addressed to Seung Yeob Nam; synam@ynu.ac.kr

Academic Editor: Francesco Gringoli

In vehicular networks, trustworthiness of exchanged messages is very important since a fake message might incur catastrophic accidents on the road. In this paper, we propose a new scheme to disseminate trustworthy event information while mitigating message modification attack and fake message generation attack. Our scheme attempts to suppress those attacks by exchanging the trust level information of adjacent vehicles and using a two-step procedure. In the first step, each vehicle attempts to determine the trust level, which is referred to as truth-telling probability, of adjacent vehicles. The truth-telling probability is estimated based on the average of opinions of adjacent vehicles, and we apply a new clustering technique to mitigate the effect of malicious vehicles on this estimation by removing their opinions as outliers. Once the truth-telling probability is determined, the trustworthiness of a given message is determined in the second step by applying a modified threshold random walk (TRW) to the opinions of the majority group obtained in the first step. We compare our scheme with other schemes using simulation for several scenarios. The simulation results show that our proposed scheme has a low false decision probability and can efficiently disseminate trustworthy event information to neighboring vehicles in VANET.

1. Introduction

Vehicular networks are expected to be used for traffic control, accident avoidance, parking management, and so on [1]. Communication security between vehicles needs to be addressed carefully due to the safety requirements of vehicular network applications [2]. There is a lot of ongoing research on security topics, which aims to provide secure communications and verification of data to thwart malicious attackers. One of the major issues in vehicular ad hoc network (VANET) is message trust, which can be used to secure VANET communications. It is essential to periodically evaluate the trustworthiness of event information based on trust metrics. Generally, trust computation in a static network is relatively simple, because the trust level can be calculated based on the behavior of the nodes with sufficient observations [3]. However, message trust computation in VANET is challenging due to the ephemeral nature of the network topology.

The wireless access in vehicular environments (WAVE) protocol is based on the IEEE 802.11p standard and provides the basic radio standard for dedicated short range communication (DSRC) operating in the 5.9 GHz frequency band [4]. Vehicular communications can be achieved in the infrastructure domain for vehicle-to-infrastructure (V2I) communications or in the ad hoc domain for vehicle-to-vehicle (V2V) communications. We mainly focus on V2V communications because road side units (RSUs) [1] may not be available in some parts of the country during the initial stages of deployment of the vehicular communications infrastructure. Vehicles communicate with other vehicles through on-board units (OBUs) forming mobile ad hoc networks that allow communications in a completely distributed manner [5]. We note that some event information (e.g., accident reports) needs to be disseminated quickly and accurately, with minimum delay. Failure in timely and accurate dissemination of such time-critical information might lead to collateral damage to neighboring vehicles.

Some of the issues in vehicular networks include simple routing problems and application-oriented problems like Sybil attacks and false data dissemination [6]. The traditional reputation systems may not work efficiently in vehicular networks [7]. Public key infrastructure (PKI) may not be available everywhere during the initial stages of vehicular network deployment around a country, because some regions may not be covered due to deployment costs or budget issues. Generally, cryptography-based verification of message trustworthiness is computationally expensive. It can protect against a few types of attack from external nodes. However, it will not protect against malicious nodes in the network, which already have the required cryptographic keys, and may not be suitable for V2V ephemeral network communications. Our scheme does not use cryptography and centralized servers, and, thus, it does not have a single point of failure. Most VANET models assume that the system is up and running, where all vehicles have a certain trust score. However, it is not easy to know the trustworthiness of vehicles without having had any interaction with those vehicles. In highly distributed vehicular networks, vehicles can join and leave a network frequently [8, 9]. When a new vehicle joins the network for the first time, there is no information about it. One of the challenges faced by VANET is that the trust model of the VANET should consider the requirement for anonymity of vehicles. The trust model should have minimal overhead in terms of computation complexity, as well as storage. The trust model should be robust to data-centric attacks and be able to detect those attacks [10–12]. VANET security frameworks should be light, scalable, reliable, and secure.

Our proposed scheme investigates the trustworthiness of event information received from adjacent vehicles, which serves as multiple pieces of evidence. We use truth-telling probability as a measure for the trustworthiness of a vehicle. The vehicles communicate through safety messages to report events, such as accident information, safety warnings, information on traffic jams, weather reports, and reports of ice on the road. In our proposed scheme, all vehicles are assumed to have a pseudo identity (PID), which is independent of the node identity. Each vehicle broadcasts an event message to adjacent vehicles from the time it collects information about that event. Every vehicle maintains the trust level of its neighbors in a distributed manner to cope with the propagation of false information. We introduce an enhanced K-means clustering technique to minimize the effect of malicious nodes on trust level calculation. We use a modified threshold random walk algorithm with a single threshold to make a final decision about the occurrence of an event, while supporting real-time decision. We focus on determining the trustworthiness of the event information in the received messages by considering reports from neighboring vehicles differently with a truth-telling probability.

The main contributions of our work can be summarized as follows:

(i) Our proposed scheme can contribute to dissemination of trustworthy information since each vehicle makes a decision on the trustworthiness of information in the received message individually, while dropping packets containing fake information.

(ii) Since all the decisions are made based on the information received from the neighbor vehicles, our proposed scheme can work in an infrastructure-less environment as well.

(iii) Our proposed scheme can make a better decision on the trustworthiness of a given message compared to a simple voting mechanism, since the modified threshold random walk (TRW) can give a higher weight on the opinion of a vehicle, which makes more true statements than false statements.

The remainder of this paper is organized as follows. In Section 2, we discuss the related work. In Section 3, we propose a trustworthy event-information dissemination scheme for VANET. In Section 4, we evaluate the performance of the proposed scheme using simulation. Section 5 concludes the paper with future work.

2. Related Work

Several trust management systems have been proposed for VANET [13–15]. Trust management systems evaluate the trust values of the neighbor nodes to prevent them from interacting with the malicious nodes. The authors in [16] provide a quantitative and systematic review of existing trust management schemes for VANET. They address comprehensive trust model concepts, problems, and solutions related to VANET trust management. There are several works on trust management scheme based on infrastructure framework and cryptography techniques. Trust management schemes can be divided into four categories based on the use of infrastructure and cryptographic measures such as public key infrastructure (PKI) as shown in Table 1. The first category represents the trust management techniques based on infrastructure such as RSU and PKI. In the second category, the nodes rely on infrastructure for trust management without using PKI. In the third category, each node handles the issue of message trustworthiness based on PKI without using any infrastructure. In the fourth category, the nodes are fully decentralized and operate in infrastructure-less environment, and they do not depend on PKI.

In the first category, trust management systems are based on infrastructure such as RSU and PKI and can be effective in identifying malicious nodes with some accuracy [17–23]. However, trust management schemes in this category may not work if infrastructure is not available. The trust management based on PKI is computationally expensive and cannot secure VANET against insider attack, where the malicious nodes have already acquired the cryptographic keys [5, 23]. Some researchers used group signatures (GS) techniques [17–21] to authenticate message sender and guarantee message integrity such as Identity-Based Group Signatures (IBGS) in [18], GSIS in [19], and Identity-based Batch Verification (IBV) in [20]. However, GS schemes are usually based on PKI, and message sender authentication cannot prevent legitimate nodes from sending malicious messages.

TABLE 1: Trust management category based on infrastructure and PKI.

	PKI based	Non-PKI based
Infrastructure based	DTE [5], ESA [17], IBGS [18], GSIS [19], IBV [20], EGSS [21], iTrust [22], PMBP [23]	STRM [13], TRIP [24], RaBTM [25]
Non-infrastructure based	ETM [26], MTM [27], LSOT [28], BTM [29]	CTID [14], ITM [15], RMCV [30], ERM [31], VARS [32], TMEP [33], CRMS [34]

Trust management schemes in the second category require infrastructure including roadside unit or central authority without using PKI [13, 24, 25]. Schemes such as Trust and Reputation Infrastructure-based Proposal (TRIP) [24] and road side unit (RSU) and Beacon-Based Trust Management (RaBTM) [25] may not work if infrastructure is not available. In the third category, the issue of message trustworthiness was investigated based on PKI in an infrastructure-less environment [26–29]. In [27], the authors proposed a multidimensional approach for trust management in decentralized VANET environments using four different types of roles for different types of vehicles. The sender needs to authenticate itself to the receiver node using PKI for verifying each other's role implemented in a distributed manner. However, VANET faces several issues while deploying the PKI scheme due to key distribution in a real world.

In order to overcome the limitation of existing approaches, some researchers investigated trust management without using PKI in an infrastructure-less environment, which corresponds to the fourth category [14, 15, 30–34]. Trust management schemes in this category such as Vehicle Ad Hoc Reputation System (VARS) [32] are more suitable for distributed VANET architecture. Our proposed scheme also belongs to this category, since the decision on the trustworthiness of event information received from neighbor nodes is made in an infrastructure-less environment without using PKI.

The existing trust management systems are established on specific application domain implementing different trust-based models to enhance intervehicular communication. The trust-based models can be classified into three main categories. They are entity based, data-centric based, and hybrid trust models [35]. Entity based trust model deals with the trustworthiness of each node considering the opinions of the peer nodes [26–28, 36]. In [24], the authors proposed a fuzzy approach for the verification of the trustworthiness of the nodes by using feedback from their neighbors. However, the trustworthiness of a message may not always agree with the trustworthiness of the node itself. Thus, this model cannot resolve the issue of message trustworthiness properly.

On the other hand, in data-centric trust model, the trustworthiness of the reported events from the neighbor vehicles is evaluated rather than the trust of the entities or the node itself [5, 30, 31, 35]. In [5], the authors used a Bayesian inference decision module to evaluate the received event reports. But, the inference module uses the prior probability, which is not easy to obtain due to dynamic topology of VANET. In [28], the authors proposed a trust model called a Lightweight Self-Organized Trust (LSOT), which contains trust certificate-based and recommendation-based trust evaluations. However, it did not distinguish between the trust value of a node and that of the reported message. The trustworthiness of the nodes does not guarantee the trustworthiness of the message as the trustworthy nodes can send fake or faulty messages, if attackers compromise them. In [30], the authors proposed Real-time Message Content Validation (RMCV) scheme in an infrastructure-less mode. This scheme assigns a trust score to a received message based on three metrics, that is, message content similarity, content conflict, and message routing path similarity. The message trustworthiness is based on the maximum value of final trust scores collected from the neighbor nodes. However, this scheme does not consider high mobility of the vehicles and its time complexity is high.

Hence, a hybrid trust model is introduced that combines the entity based and data-centric trust models to evaluate the trustworthiness of a message [32–34]. The authors in [29] proposed a hybrid trust management mechanism called Beacon-based Trust Management (BTM) system, which constructs entity trust from beacon messages and computes data trust from crosschecking the plausibility of event messages and beacon messages. However, their trust model is based on PKI and digital signature, which incurs overhead while signing and authenticating each beacon message before broadcasting.

Thus, we attempt to overcome the limitation of the existing schemes by improving the hybrid trust model for message trustworthiness. As a first step, we use initialization-step enhanced K-means clustering algorithm (IEKA) for clustering of the vehicles into normal and malicious vehicle groups to determine the trustworthiness of each neighbor node. As a second step, we use a modified threshold random walk (TRW) algorithm to decide the trustworthiness of a given message. Thus, our scheme is based on a hybrid trust model. Although RMCV is based on data-centric trust model, it belongs to the fourth category, that is, trust management scheme that requires neither PKI nor infrastructure, according to the classification in Table 1. Thus, we compare our proposed scheme with the RMCV scheme. The detailed comparison and performance evaluation is discussed in Section 4.

3. System Model

Each vehicle collects sufficient information to assess the validity and correctness of a message. Notations explains the parameters and variables used in this paper.

When an event occurs on the road, a vehicle that is near that event sends the safety event message, M_E, to neighboring

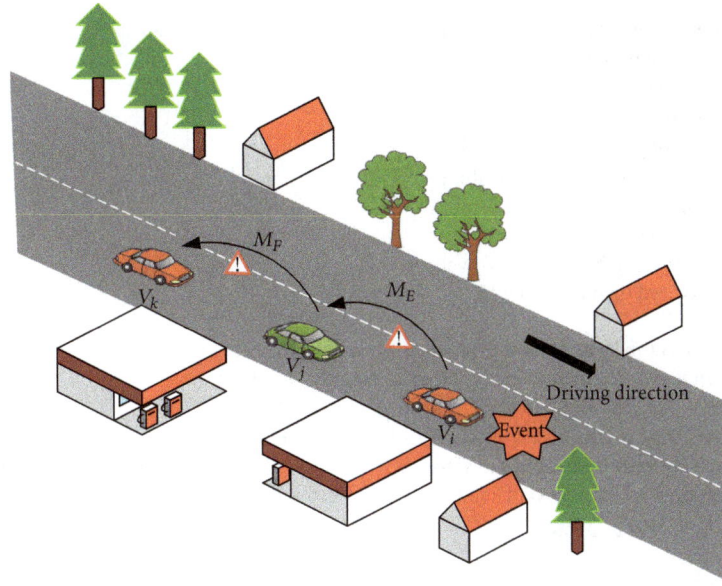

FIGURE 1: Trustworthy message dissemination scheme.

vehicles. Let us suppose that vehicle V_j wants to know the true information about the event reported by vehicle V_i in Figure 1.

The vehicle V_j manages an information pair $(\mathbf{p_i}, \boldsymbol{\theta_i})$ for each neighbor vehicle V_i, where $\mathbf{p_i}$ is the pseudo identity of the ith neighbor vehicle and $\boldsymbol{\theta_i}$ is the trust level, that is, truth-telling probability, of vehicle V_i. We assume that the transportation authority preloads the pseudo ID of the vehicles during vehicle registration, and it should be renewed periodically. To maintain the privacy in VANET, the pseudonym should change over time to achieve unlinkability that protects the vehicle from location tracking. Only privileged authorities are allowed to trace or resolve a pseudonym of the vehicle to a real identity under specific condition [37]. The truth-telling probability $(\boldsymbol{\theta_i})$ is the ratio of the number of true event reports propagated by vehicle V_i to the total number of event reports sent by vehicle V_i over a specific time period.

3.1. Proposed Trustworthy Information Dissemination Scheme. An outline of the proposed scheme to determine the trustworthiness of event information in the received message is shown in Algorithm 1. The vehicle parameters such as pseudo ID (PID) and default trust level are initialized at the beginning. All the vehicles periodically broadcast beacon messages, along with status information such as speed and location to neighboring vehicles. If there is no event triggered, then the vehicles will gather information from the neighboring vehicles. If a vehicle encounters any event by itself, then it broadcasts a safety message along with the trust levels of neighboring vehicles that it knows. Each vehicle accumulates the trust levels of the neighboring vehicles based on the collected safety messages. V_j creates a trust matrix based on the trust level opinions from other vehicles. Thus, the trust matrix manages the trust levels of each neighboring vehicle. Sometimes, vehicles misbehave by

sending false information due to selfish motives like getting easier and faster access to the road, or due to faults. To prevent such false information that can corrupt the trust level of legitimate vehicles, we use a clustering algorithm. Our proposed clustering algorithm attempts to separate the trust level opinions of normal vehicles from the trust level opinions of malicious vehicles. The vehicle will calculate the aggregated trust level of adjacent vehicles belonging to the majority group of normal vehicles from the trust matrix. It will update the trust matrix using the average of trust levels. Then, a modified TRW is applied to know whether the event has actually occurred or not. The modified TRW can provide better decision on the trustworthiness of an event information by giving higher weights on the true event messages. After the trustworthiness of the event information has been verified, the event message is disseminated to other neighboring vehicles along with the updated trust levels. If the event information contained in the message turns out to be untrustworthy, then the message is dropped.

When new vehicles join the VANET, they are not likely to have enough information to infer the trust levels of neighboring vehicles at the beginning. We need a trust level bootstrapping procedure to assign a default trust level for this situation [38]. The trust level, that is, the truth-telling probability, ranges from 0 to 1. If vehicle A does not have any information on vehicle B, then the truth-telling probability of vehicle B is set to 0.5 at vehicle A. We assume that each vehicle sets the truth-telling probability for itself to 1 by default.

We mainly deal with two types of messages: beacon messages and safety messages. The vehicles use beacons to periodically broadcast and advertise status information to neighboring vehicles at intervals of 100 ms. The sender reports its speed, position, and so on to neighboring vehicles with beacon messages via one-hop communications [39]. On the other hand, safety messages support vehicles on the road

---

//The process is executed by a receiver vehicle upon receiving safety message
// p_{ij}: Pseudo ID of ith neighbor vehicle of V_j
// θ_j: truth-telling probability of V_j
// θ_{ij}: estimator of θ_i by V_j
// Θ_j: trust level opinion generated by V_j
// $\hat{\theta}_i$: estimator for truth-telling probability of V_i
Input: $Y = \{\Theta_i\}$ $(i = 1, 2, \ldots, n)$
Output: $\{\hat{\theta}_i\}$, updated trust matrix
(1)   Information gathering from neighbor vehicles
(2)   If event is triggered then goto step (5)
(3)   Else goto step (2).
(4)   If event source is the V_j itself then goto step (13).
(5)   Else V_j accumulates the trust levels opinions of neighbor vehicles
           $\Theta_j = ((p_{1j}, \theta_{1j}), (p_{2j}, \theta_{2j}), \ldots, (p_{jj}, \theta_{jj}), \ldots, (p_{nj}, \theta_{nj}))$
(6) V_j generates a trust matrix based on the trust level opinions.
(7)   Use modified clustering algorithm to separate trust level opinions of normal from malicious vehicles.
(8)   Calculate aggregated trust level of adjacent vehicles belonging to majority group from trust matrix
           $\hat{\theta}_i = (1/n) \sum_{j=1}^n \theta_{ij},$
(9)   Update the trust matrix
(10)   Use modified TRW to know if the event has actually occurred or not.
(11)   If we decide that the event message is trustworthy, then goto step (13).
(12)   Else drop the message.
(13)   Broadcast safety message and trust level to neighboring vehicles.

ALGORITHM 1: Determining trustworthiness of event information in the received message.

by delivering time-critical information so that proper action can be taken to prevent accidents and to save people from life-threatening situations. Safety messages include different types of events, E_x, such as road accidents, traffic jams, slippery roads, road constructions, poor visibility due to fog, and emergency vehicle warnings. Vehicles broadcast a safety message to neighboring vehicles when they encounter events on the road [1]. The message payload includes information about the vehicle's position, message sending time, direction, speed, and road events [19]. Each vehicle gathers information about the neighboring vehicles within its communication range.

One advantage of our proposed message dissemination scheme is to avoid a central trusted third party for trust accumulation in a distributed vehicular networking environment. We consider VANET without infrastructure such as RSUs. Vehicles communicate with each other in V2V mode using DSRC [40]. This allows fast data transmission for critical safety applications within a short range of 250 m. A basic safety application contains vehicle safety-related information, such as speed, location, and other parameters, and this information is broadcast to neighboring vehicles [41–43]. Let us consider two vehicles: V_i and V_j. The truth-telling probability of V_i depends on whether vehicle V_i is truthful when relaying event information. According to Velloso et al. [44], the more positive experiences vehicle V_j has with vehicle V_i, the higher the trust vehicle V_j will have towards vehicle V_i.

Let us suppose that vehicle V_i has a pseudo ID p_i and broadcasts a safety warning message M_E, which is defined in (1), when event E_x, where x represents an event type, is detected. If the vehicle itself detects the event, then it

broadcasts the safety message along with the trust levels of neighboring vehicles. If a vehicle receives a safety message from other vehicles, it will accumulate the safety message along with trust levels from neighboring vehicles. When vehicle V_j collects event information from vehicle V_i, it finds the type and location of event from the message. Let event message M_E be given by

$$M_E = (p_i, t, L_E, l_i),\qquad(1)$$

where p_i is the pseudo ID of vehicle V_i, t is the message generation time, L_E is the location of event E_x, and l_i is the location of V_i at time t.

In addition to this, each vehicle periodically broadcasts a beacon message defined as $M_B = (p_i, t_i, l_i, s_i)$, where p_i is the pseudo ID of V_i, t_i is the beacon generation time, l_i is the location of V_i, and s_i is the speed of V_i.

Let θ_i be the trust level, that is, truth-telling probability, of vehicle V_i. Truth-telling probability θ_i is defined as the ratio of the number of true events reported by vehicle V_i divided by the total number of events reported by vehicle V_i over a specific period of time. Let m denote the total number of true events reported by V_i and let n denote the total number of events reported by the vehicle up to the current time. Then, the truth-telling probability is

$$\theta_i = \frac{m}{n}.\qquad(2)$$

A value for θ_i approaching 1 indicates reliable behavior of the corresponding vehicle, whereas a value close to zero indicates a high tendency towards providing false information [45].

3.2. Calculation of Trust Level of Neighbor Vehicles.

When an event occurs, the nearby vehicles broadcast safety messages with additional data, such as pseudo IDs and truth-telling probabilities of other vehicles. Based on the safety messages from the neighboring vehicles, trust matrix $[\theta_{ij}]$ can be obtained, where θ_{ij} is estimation of θ_i by vehicle V_j. The trust matrix manages the truth-telling probability of each neighboring vehicle from the viewpoint of other vehicles. We assume that each vehicle sets its own truth-telling probability to 1. If the trust matrix is constructed, the aggregated trust level, that is, truth-telling probability of vehicle V_i, is calculated from the trust matrix by

$$\hat{\theta}_i = \frac{1}{n}\sum_{j=1}^{n}\theta_{ij}, \tag{3}$$

where $\hat{\theta}_i$ is the estimator for the truth-telling probability of V_i.

3.2.1. Estimation of Truth-Telling Probability Based on the Correctness of Message Information.

If we can decide whether specific event information received from a vehicle is correct, this information can be used to estimate the truth-telling probability of the reporting vehicle more accurately. The reliable information about a specific event might be obtained from direct observation of an event spot, or announcement from a public and reliable group.

We explain how the truth-telling probability can be estimated more accurately if we collect more evidence to decide the correctness of messages generated by a given vehicle. We can estimate the truth-telling probability θ_i, defined in (2), based on the correctness of recent N messages from V_i. We introduce a random variable X_n to estimate the number of true reports among the recent N reports from V_i on arrival of the nth report from V_i. Then, the truth-telling probability of V_i can be estimated by X_n/N. We attempt to estimate X_n from X_{n-1} using the following relation:

$$X_n$$
$$= \begin{cases} 1 + \left(1 - \dfrac{1}{N}\right)X_{n-1}, & \text{when the report } n \text{ is correct,} \\ \left(1 - \dfrac{1}{N}\right)X_{n-1}, & \text{otherwise.} \end{cases} \tag{4}$$

Then, we can show that X_n/N approaches the truth-telling probability θ_i of V_i under the assumption that the correctness of one message is independent of the correctness of other messages. By taking expectation on (4), we can obtain $E[X_n]$ as

$$E[X_n] = \Pr[\text{message } n \text{ is correct}]$$
$$\cdot E[X_n \mid \text{message } n \text{ is correct}]$$
$$+ \Pr[\text{message } n \text{ is incorrect}]$$
$$\cdot E[X_n \mid \text{message } n \text{ is incorrect}]$$

$$= \theta_i E\left[1 + \left(1 - \frac{1}{N}\right)X_{n-1}\right] + (1 - \theta_i)$$
$$\cdot E\left[\left(1 - \frac{1}{N}\right)X_{n-1}\right] = \theta_i + \left(1 - \frac{1}{N}\right)E[X_{n-1}].$$
$$\tag{5}$$

By solving the recursive relation in (5), we can obtain

$$E[X_n] = \left(1 - \frac{1}{N}\right)^n \{E[X_0] - \theta_i N\} + \theta_i N. \tag{6}$$

Thus, regardless of the initial condition on X_0, we have $\lim_{n\to\infty}E[X_n] = \theta_i N$, and $\lim_{n\to\infty}E[X_n/N] = \theta_i$ from (6). In other words, we can say that X_n/N approaches the truth-telling probability θ_i asymptotically, and we use the estimator of X_n/N and the relation in (4) to update the truth-telling probability of some vehicle whenever we have some evidence to determine the correctness of a message from that vehicle.

3.3. Clustering Algorithm.

If there is no evidence to determine the truth of a given message, then the truth-telling probability of vehicle V_i will be calculated using (3). However, malicious vehicles can modify the trust levels of neighboring vehicles to mislead vehicles in a vehicular network. Thus, we need a clustering algorithm that can separate the trust levels of normal vehicles from the trust levels of malicious vehicles. It can reduce the effect of malicious vehicles on the trust levels of normal vehicles. In this subsection, we propose a new clustering algorithm to tackle this issue.

The main goal of our modified clustering algorithm is outlier detection. Our modified clustering algorithm classifies the trust level (truth-telling probability) opinions of the vehicles into two groups, one with the trust level opinions of normal vehicles and the other with the trust level opinions of malicious vehicles. We will select the majority group and neglect the outliers corresponding to the minority group.

Let us assume that an event has occurred on the road and the vehicles near the event location send event messages along with trust level opinions to neighbor vehicles. The vehicle V_j gathers reports about a specific event from neighbor vehicles and manages the trust level opinions of other vehicles as follows. Each vehicle maintains a sorted vehicle list (SVL), which manages pseudo IDs of all the adjacent vehicles in an ascending order, and the vehicle index will be assigned based on the sequence in the sorted list as shown in Table 2. Whenever a vehicle V_j needs to disseminate its own trust level opinion to its neighbors, it sends its trust level opinion Θ_j defined as

$$\Theta_j = \left((p_{1j}, \theta_{1j}), (p_{2j}, \theta_{2j}), \ldots, (p_{jj}, \theta_{jj}), \ldots, (p_{nj}, \theta_{nj})\right), \tag{7}$$

where p_{kj} is the pseudo ID of the kth neighbor vehicle of the vehicle j and θ_{jj} is likely to be set to 1 because every node will trust itself.

If V_i receives trust level opinion Θ_j, then V_i updates its own SVL by adding the vehicles that are in Θ_j, but are not in the SVL. After updating SVL, V_i derives $\tilde{\Theta}_{j'}$ from the received Θ_j as

$$\tilde{\Theta}_{j'} = \left(\tilde{\theta}_{1j'}, \tilde{\theta}_{2j'}, \ldots, \tilde{\theta}_{j'j'}, \ldots, \tilde{\theta}_{n'j'}\right), \tag{8}$$

TABLE 2: Example of sorted vehicle list (SVL).

Vehicle index	Pseudo ID
1	1135
2	2056
3	2079
4	2146
5	3012
...	...

where j' is the new index of the vehicle j according to its sequence in the updated SVL and n' is the total number of vehicles in the updated SVL. When p_{kj} in the received Θ_j agrees with the k'th pseudo ID in the updated SVL, $\tilde{\theta}_{k'j'}$ in $\tilde{\Theta}_{j'}$ is updated as

$$\tilde{\theta}_{k'j'} = \theta_{kj}. \tag{9}$$

In this case, n' is always larger than or equal to n since $\tilde{\Theta}_{j'}$ accommodates all the vehicles in Θ_j. If $n' > n$, then it means that there is some pseudo ID that is in the SVL, but not in Θ_j. If an index l corresponds to such a pseudo ID, $\tilde{\theta}_{lj'}$ will be set to 0.5 since the vehicle j' does not know the vehicle l. If $\tilde{\Theta}_{j'}$ is derived, then the trust matrix table is updated by adding the transpose of $\tilde{\Theta}_{j'}$ as the j'th column.

If each vehicle includes the pseudo IDs and the truth-telling probabilities of all the vehicles that it knows in the trust level opinion message defined in (7), then the traffic overhead due to this message can be excessively large. However, we can reduce the message overhead by omitting trivial information. For example, if θ_{kj} in Θ_j is 0.5, this means that the vehicle j does not know the vehicle k since 0.5 is the default value used to initialize the truth-telling probability of a new vehicle. In this case, the vehicle j need not advertise this probability because this default value can be easily filled up by the neighbor vehicles according to the trust matrix updating rule mentioned above with (7), (8), and (9). The vehicles on the road are likely to be ignorant of each other in terms of the trust matrix table, since they need not exchange the trust level opinions if there is no event. Thus, we expect that the policy of omitting trivial information can significantly reduce traffic overhead due to trust level opinion messages.

If V_j collects trust level opinions Θ_i ($i \neq j$) from other vehicles along with event information, then V_j can construct trust matrix Γ, defined as

$$\Gamma = \begin{bmatrix} \theta_{11} & \theta_{12} & \cdots & \theta_{1n} \\ \theta_{21} & \theta_{22} & \cdots & \theta_{2n} \\ \vdots & \vdots & \ddots & \vdots \\ \theta_{n1} & \theta_{n2} & \cdots & \theta_{nn} \end{bmatrix}. \tag{10}$$

We use a simple example to show our proposed clustering algorithm illustrated by Table 3 and Figure 2. Table 3 shows an example of trust matrix defined in (10), when $n = 3$. Three columns in Table 3 correspond to points A, B, and

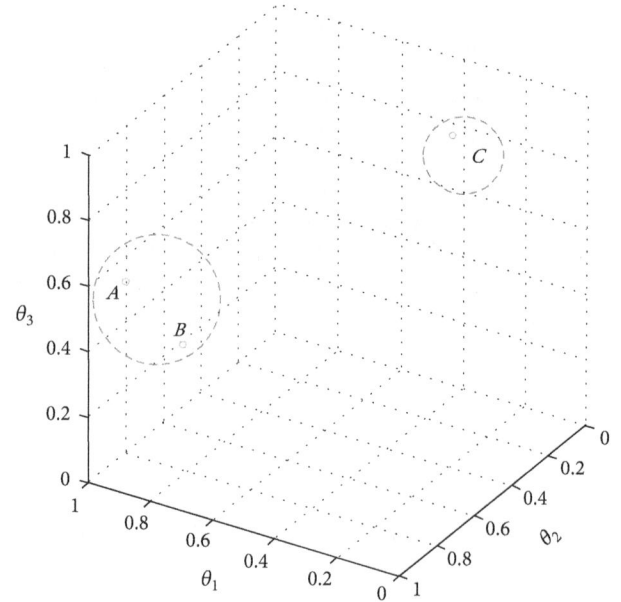

FIGURE 2: Clustering example for the proposed clustering algorithm.

TABLE 3: Trust matrix table of vehicle.

V_j	V_i		
	1	2	3
1	1	0.7	0.2
2	0.8	1	0.4
3	0.5	0.5	1
	A	B	C

C in Figure 2, respectively. The three probability values in the first column of Table 3 correspond to θ_{11}, θ_{21}, and θ_{31}, respectively, and these values are estimation of θ_1, θ_2, and θ_3 by V_i. This tuple of probability values is represented as point A in the three-dimensional space of Figure 2, where each axis represents θ_1, θ_2, and θ_3, respectively. If there is no attacker, that is, all vehicles tell the truth, then all the points will be close to each other. The trust level opinions (Θ_i's) with similar characteristics are likely to form the same cluster. If there is an attacker that tells a lie, then the corresponding point will deviate from the majority group, and this point can be distinguished as an outlier. Even if the attacker tries to change or give higher trust levels by using collusion attack, it can be detected as an outlier. Let us suppose that point C represents an attacker that tells a lie by changing the trust level, as shown in Table 3. The clustering algorithm will separate the trust level opinions into two groups, one group with normal-vehicle trust levels (i.e., A and B) and the other group with malicious vehicle trust levels (i.e., C). Figure 2 describes the outcome of one possible clustering algorithm. The final aggregated trust level is calculated based on the trust level opinions corresponding to the majority group using (3). The final aggregated trust level based on the majority group will be used to update the trust level of the vehicle itself. The resulting trust level is appended to the message during message propagation.

Input: $\mathbf{Y} = \{\boldsymbol{\Theta}_i\}$ $(i = 1, 2, \ldots, n)$
Output: $\mathbf{C} = \{\mathbf{c}_j\}$ $(j = 1, 2)$
Initialize: Calculate unique centroid as initial cluster center,
$$\boldsymbol{\mu} = \frac{\sum_{i=1}^{n} \boldsymbol{\Theta}_i}{|\mathbf{Y}|}$$

(1)   for each $\mathbf{c}_j \in \mathbf{C}$ do
(2) $\mathbf{c}_j \leftarrow \boldsymbol{\Theta}_i \in \mathbf{Y}$
(3) $\mathbf{c_1} = \underset{\boldsymbol{\Theta}_i \in \mathbf{Y}}{\arg\max} \|\boldsymbol{\mu} - \boldsymbol{\Theta}_i\|$
(4) $\mathbf{c_2} = \underset{\boldsymbol{\Theta}_i \in \mathbf{Y}}{\arg\max} \|\mathbf{c_1} - \boldsymbol{\Theta}_i\|$
(5)   end
(6)   While two centroids are not converged, do
(7)   for each $\boldsymbol{\Theta}_i \in \mathbf{Y}$ do
(8)   Assign $\boldsymbol{\Theta}_i$ to nearest centroid,
(9) $\mathbf{c}_j = \underset{j}{\arg\min} \|\mathbf{c}_j - \boldsymbol{\Theta}_i\|^2$
(10)   end
(11)   Update cluster centroid;
(12)   Calculate new centroid $\mathbf{c}_j$ as
(13)   for each $\mathbf{c}_j \in \mathbf{C}$ do
(14) $\mathbf{c}_j = (1/|\mathbf{c}_j|) \sum_{\boldsymbol{\Theta}_i \in c_j}^{i} \boldsymbol{\Theta}_i,$
(15)   end
(16)   end

ALGORITHM 2: Proposed modified clustering algorithm.

We propose a modified K-means clustering algorithm. The main problem with a K-means algorithm lies in the initialization step, so we introduce an enhanced K-means clustering technique by modifying the initialization step, which is called initialization-step enhanced K-means clustering algorithm (IEKA). We use the IEKA to cluster the trust level opinions, while reducing the effect of malicious vehicles on trust levels for other vehicles. Our proposed clustering algorithm can be described in more detail as follows.

After generating a trust matrix, IEKA partitions the trust level opinions into $K(\leq n)$ groups $\mathbf{C} = \{\mathbf{c}_1, \mathbf{c}_2, \ldots, \mathbf{c}_k\}$. We designate $\mathbf{Y}$ to be a set of the $\boldsymbol{\Theta}_i$ vectors; that is, $\mathbf{Y} = \{\boldsymbol{\Theta}_1, \boldsymbol{\Theta}_2, \ldots, \boldsymbol{\Theta}_n\}$. We consider only two clusters for our scheme. Initially, we take the mean of all the data points in $\mathbf{Y}$ to find a unique centroid, that is, $\boldsymbol{\mu}$.

$$\boldsymbol{\mu} = \frac{\sum_{i=1}^{n} \boldsymbol{\Theta}_i}{|\mathbf{Y}|}. \tag{11}$$

We calculate the Euclidean distance between $\boldsymbol{\mu}$ and each vector in $\mathbf{Y}$. Choose the point that has the maximum distance from the unique centroid; that is, the selected point is at the farthest distance from the unique centroid. We consider this point as the first centroid, $\mathbf{c}_1$, for the first cluster:

$$\mathbf{c}_1 = \underset{\boldsymbol{\Theta}_i \in \mathbf{Y}}{\arg\max} \|\boldsymbol{\mu} - \boldsymbol{\Theta}_i\|. \tag{12}$$

Similarly, we compute the Euclidean distance between first centroid $\mathbf{c}_1$ and the remaining points in $\mathbf{Y}$ and select the

point with the maximum distance from $\mathbf{c}_1$. Then, this point becomes the second centroid, $\mathbf{c}_2$:

$$\mathbf{c}_2 = \underset{\boldsymbol{\Theta}_i \in \mathbf{Y}}{\arg\max} \|\mathbf{c}_1 - \boldsymbol{\Theta}_i\|. \tag{13}$$

As a next step, we run the conventional K-means clustering algorithm, with $\mathbf{c}_1$ and $\mathbf{c}_2$ being the centroids of two separate groups. Update centroids $\mathbf{c}_1$ and $\mathbf{c}_2$ by calculating the mean value for each group. This gives new centroids $\mathbf{c}_1$ and $\mathbf{c}_2$ and then reassigns each data point to the cluster to which it is closest. We will repeat this process until those two centroids converge. The proposed modified clustering algorithm is given in Algorithm 2.

After clustering using Algorithm 2, which is based on (11), (12), and (13), the aggregated trust level of each neighbor vehicle is calculated based on the trust level opinions belonging to the majority group. We assume that the number of malicious vehicles is less than that of normal vehicles. The aggregated trust level is used for TRW calculation. In the next subsection, we discuss the decision on the event based on hypothesis testing using TRW.

3.4. Event Decision Based on Threshold Random Walk (TRW).

Sequential hypothesis testing is usually used to determine if a specific hypothesis is true or not based on sequential observations [46]. Among the sequential hypothesis testing schemes, threshold random walk has been used to detect scanners with a minimal number of packet observations, while guaranteeing false positives and false negatives [47]. Since we are interested in determining whether a given message is true or not, if true message constitutes one of

TABLE 4: Aggregated trust level table.

Neighbor PID	Trust level (truth-telling prob.)	Event observations (X_i)
p_1	$\hat{\theta}_1$	$X_1 = E_1$
p_2	$\hat{\theta}_2$	$X_2 = \overline{E}_1$
p_3	$\hat{\theta}_3$	$X_3 = \overline{E}_1$
...	...	...
p_n	$\hat{\theta}_n$	$X_n = E_1$

the two hypotheses, then threshold random walk might be applied to this problem. The threshold random walk scheme in [47] uses two thresholds, that is, one upper bound and one lower bound, and the decision is made when a likelihood ratio reaches either threshold. However, in this threshold random walk scheme, we cannot know the number of samples required to reach either threshold in advance. This means that real-time decisions may not be possible if we cannot collect a sufficient number of samples in a short interval. In this paper, we use a modified threshold random walk scheme to determine the validity of a given event, while resolving the issue of real-time decision. We resolve this issue by applying threshold random walk with a single threshold instead of two thresholds. Hereafter, we explain the threshold random walk (TRW) scheme applied to our problem in more detail.

E_1 represents one of the events that can happen on a road. After clustering trust level opinions of neighbor vehicles using IEKA, each vehicle determines the occurrence of event E_1 based on the aggregated trust level table. The aggregated trust level table consists of vehicle PIDs, aggregated trust levels, and event observations, as shown in Table 4.

In Table 4, X_i is the report received from the ith neighbor vehicle about event E_1. Event E_1 represents the occurrence of an event, and $\overline{E}_1$ represents nonoccurrence of event E_1. We need a rule to make a decision about the occurrence of the event. We assume that X_i's are independent of each other among different vehicles. For a given event, suppose random variable Y_i can take only two values (0 and 1); that is,

$$Y_i = \begin{cases} 0; & \text{if } X_i = E_1 \\ 1; & \text{if } X_i = \overline{E}_1. \end{cases} \tag{14}$$

After collecting a sufficient number of reports, we wish to determine whether the event (E_1) has really occurred using sequential analysis [46].

Let us consider two hypotheses: one is null and the other is an alternate hypothesis (i.e., H_0 and H_1), where H_0 is the hypothesis that event E_1 has occurred and H_1 is the hypothesis that event E_1 has not occurred, that is, $\overline{E}_1$. We also assume that conditionals on the hypothesis $Y \mid H_j$ where $j = 0, 1$ are independent. From the definition of the truth-telling probability and (14), we obtain

$$\Pr(Y_i = 0 \mid H_0) = \theta_i,$$

$$\Pr(Y_i = 1 \mid H_0) = 1 - \theta_i,$$

$$\Pr(Y_i = 0 \mid H_1) = 1 - \theta_i,$$

$$\Pr(Y_i = 1 \mid H_1) = \theta_i, \tag{15}$$

where $\Pr(Y = k \mid H_j)$ is the conditional probability that the observation of Y, given hypothesis H_j, is k. Then, $\Pr(Y_i = 0 \mid H_0) = \theta_i$ becomes the truth-telling probability, and $\Pr(Y_i = 1 \mid H_0) = 1 - \theta_i$ becomes the lying probability. In order to make a timely decision, we collect report samples from neighbor vehicles during an interval of fixed duration T. Let N denote the number of report samples collected during this interval. Following the approach of Wald [46], we use collected report samples to calculate the likelihood ratio by

$$\Lambda = \prod_{i=1}^{N} \frac{\Pr(Y_i \mid H_1)}{\Pr(Y_i \mid H_0)}. \tag{16}$$

Although the TRW scheme in [47] makes a decision based on two thresholds, the upper and lower bounds, we use a single threshold to make a decision without the issue of long waiting time. When the threshold is η, the decision rule is as follows:

If $\Lambda \geq \eta$, then accept hypothesis H_1.

If $\Lambda < \eta$, then accept hypothesis H_0.

In this paper, the threshold η will be set to 1, and the truth-telling probability θ_i of an unknown vehicle i will be set to 0.5. When a vehicle receives N report messages, if the Nth report has come from a vehicle with no information on the truth-telling probability, $\Pr(Y_N \mid H_1) = \Pr(Y_N \mid H_0)$ since $\theta_i = 1 - \theta_i$. Thus, the report from the unknown vehicle will not affect the likelihood ratio by (15) and (16). Furthermore, if all the report messages are from the vehicles with no history information, then the likelihood ratio in (16) becomes 1, and, thus, it is fair to put $\eta = 1$, since it is not easy to make a decision in this case.

The advantage of our threshold random walk compared to a simple voting scheme can be described with a simple example as follows. Let us consider a case where an event E_1 is true, and a vehicle receives 5 report messages. Among them, only two report that E_1 is true, and the other three claim that E_1 did not happen. If we make a decision based on a simple voting, then the decision will be $\overline{E}_1$. However, if we apply threshold random walk considering the truth-telling probability of each node, the decision can be different as follows. If the truth-telling probability of the two nodes claiming E_1 is 0.8 and the truth-telling probability of the three nodes claiming $\overline{E}_1$ is 0.6, then the likelihood ratio defined in (16) becomes

$$\Lambda(X) = \frac{0.2}{0.8} \times \frac{0.2}{0.8} \times \frac{0.6}{0.4} \times \frac{0.6}{0.4} \times \frac{0.6}{0.4} = 0.21$$

$$< 1 \, (= \eta). \tag{17}$$

Thus, we will select the hypothesis H_0 according to the decision rule mentioned above, since the likelihood ratio calculated in (17) is less than the threshold η. This means

(a)

(b)

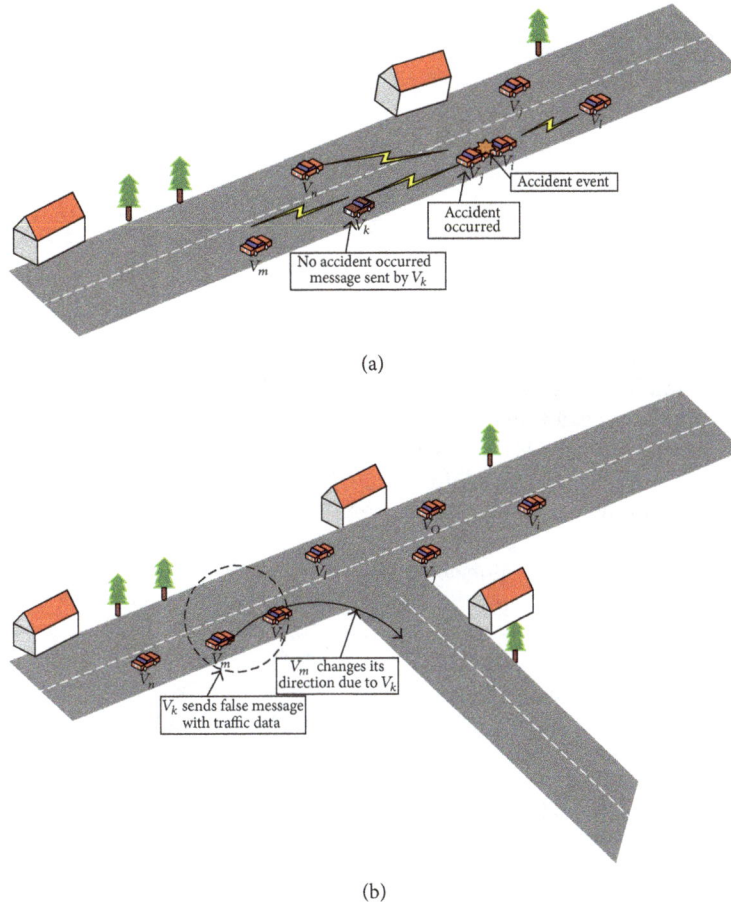

FIGURE 3: Two types of attack patterns considered in this paper: (a) message modification attack and (b) fake message generation attack.

the correct decision of E_1 is made by the proposed threshold random walk. This advantage comes from the fact that the likelihood ratio in (16) gives a higher weight to the opinion of vehicles with a high truth-telling probability.

After the decision on the actual occurrence of the event is made, vehicle j will forward the received message to its neighboring vehicles (with aggregated trust levels) within radio range, which is denoted by

$$M_F = \left(p_j, t, M_E, \Theta_j \right), \tag{18}$$

where p_j is the PID of vehicle j, which forwards the message, t is the time at which M_F was sent, and Θ_j denotes the trust level opinion of vehicle j defined in (7).

3.5. Attack Model. We consider two types of attacks: message modification attack and fake message generation attack in a VANET environment. Figure 3 shows an example of both message modification and fake message generation attack. A malicious vehicle might modify warning messages, either with malicious intent or due to an error in the communications system. In the message modification attack, malicious vehicles can modify message information at any time and falsify the parameters.

In Figure 3(a), when an accident event occurs on the road, the vehicles in an accident or the vehicles which are

close to that accident broadcast the accident event message. After vehicle V_j sends an accident report to other vehicles, a malicious vehicle V_k modifies the message and sends the modified no-accident message as M_F, defined in (18), with the intent to affect decisions taken by other vehicles. Similarly, in a fake message generation attack, malicious vehicles generate a false warning message. For example, in Figure 3(b) [48], a malicious vehicle might send an accident message to neighboring vehicles, even when there is no such event on the road, to clear the route it wants to take. In this case, the malicious vehicle wants to convince other vehicles that an event has occurred. In this scenario, the attacker may have already compromised one or more vehicles and launches attacks by generating a fake message for neighboring vehicles. We assume that the number of malicious vehicles is less than the number of normal vehicles [12]. In simulation, we vary the number of malicious vehicles from 5% to 50% of overall vehicles to evaluate the performance of our proposed scheme in an adversarial environment.

4. Performance Evaluation

4.1. Simulation Setup. The performance of our proposed scheme was evaluated through simulation. We used the Vehicles in Network Simulation (VEINS) framework version

TABLE 5: Simulation parameters.

Parameters	Value
Network simulation package	OMNET++
Vehicular traffic generation tool	SUMO
Wireless protocol	802.11p
Simulation time	300 s
Scenario	Urban/highway
Transmission range	250 m

4a2 [49], which is based on both OMNeT++ version 4.6 [50], a discrete event-driven network simulator, and Simulation of Urban Mobility (SUMO) version 22 for road traffic simulation [51]. VEINS connects OMNeT++ and SUMO through Traffic Control Interface (TraCI). VEINS provides realistic models for IEEE 802.11p networks. It provides OMNeT with a set of application programming interfaces to connect the SUMO platform and to dynamically access information about SUMO simulated objects. SUMO allows the creation of scenarios that include realistic mobility patterns, such as vehicle movement and overtaking, as well as lane changing.

We use the default map of Erlangen, Germany, from the VEINS framework with the map size of 2500 m × 2500 m for our simulation. We evaluated our scheme under different traffic densities to consider diverse situations. When the vehicles reach the edge of the road, the vehicles reroute their path and can meet other vehicles multiple times during simulation. The number of vehicles increases linearly with time from 0 s to 300 s. The average vehicle speed changes from 40 km/h in an urban scenario to 110 km/h for highway scenarios. The key parameters considered in our simulation are summarized in Table 5.

We considered two scenarios (urban and highway) by varying parameters such as speed, vehicle density, and percentage of malicious vehicles, as shown in Table 6. The number of malicious vehicles was varied considering the mobility of vehicles in a realistic simulation environment by adjusting vehicle densities and vehicle speeds. We assume that the normal vehicles and the malicious vehicles are uniformly distributed on the roads for each ratio of malicious vehicles [52].

4.2. Simulation Results. In this section, we analyze the simulation results based on OMNet++. The traffic density increases from free-flow traffic (5 vehicles/km^2) to congested traffic (100 vehicles/km^2) where vehicles can meet multiple times. The simulation scenarios are summarized in Table 6. For performance evaluation, we have considered false decision probability and message overhead. We compared our scheme with other schemes under different scenarios. In order to evaluate our proposed scheme, we considered the message modification attack and the fake message generation attack one by one, while increasing the number of malicious vehicles from 5% to 50% in both scenarios. The positions of normal vehicles and the initial distribution of the attackers were randomly determined. We calculated the average false decision probability by averaging the simulation results for

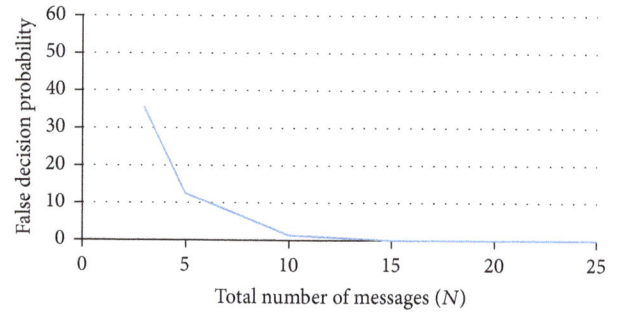

FIGURE 4: False decision probability versus total number of messages (N).

30 simulation runs. A decision is regarded as a false decision when the decision result does not agree with the true status of the event at the time of the decision. In other words, a false decision probability is the ratio of the number of incorrect decisions to the total number of decisions.

In order to update the truth-telling probability of vehicle V_i based on the truth of a given message according to (4), we need to decide the parameter N, that is, the number of recent messages from V_i that will be considered in this estimation. In order to decide N, we run 20 simulations under fake message attack with 30% of malicious vehicles in a highway scenario. We calculate the average false decision probability for various values of N, and Figure 4 shows the result. As the value of N increases, the false decision probability tends to decrease. The false decision probability reaches zero at $N = 15$ and does not change for larger values of N. Thus, N is fixed to 15 hereafter based on this result.

In Figure 4, the false decision probability is high for lower values of N. Let us consider an example to explain worse performance for lower values of N. Let us take an extreme case of $N = 1$. Then, this means that when V_j receives a message from V_i, it decides the truth-telling probability of V_i only based on the last message, since $N = 1$. Thus, if V_j finds that the last message from V_i was false, then V_j will think that the truth-telling probability of V_i is 0, according to the updating rule described in (4). On the other hand, if V_j finds that the last message from V_i was true, then V_j will think that the truth-telling probability of V_i is 1. Thus, the estimated truth-telling probability of each vehicle is either 0 or 1. However, if the truth-telling probability of a given vehicle is different from 0 or 1, then this updating rule (with $N = 1$) will never find the accurate truth-telling probability, since the truth-telling probability is always 0 or 1 according to the updating rule. Hence, the truth-telling probability can be significantly different from the correct truth-telling probability for lower values of N, especially when $N = 1$.

For our modified TRW scheme, we need to determine an optimal value of message collection time T to achieve a good decision accuracy. We run several simulations under fake message generation attack with 30% of malicious vehicle in highway scenario. We calculated the average false decision probability against the message collection time under the simulation parameters given in Table 5. In the sparse network, we received report message as low as five messages

TABLE 6: Simulation scenarios.

Scenario number	Type	Total vehicles	Malicious vehicle ratio	Average speed	Std. of speed
(1)	Urban	100	0~50%	40 km/h	4.76
(2)	Highway	100	0~50%	110 km/h	7.52

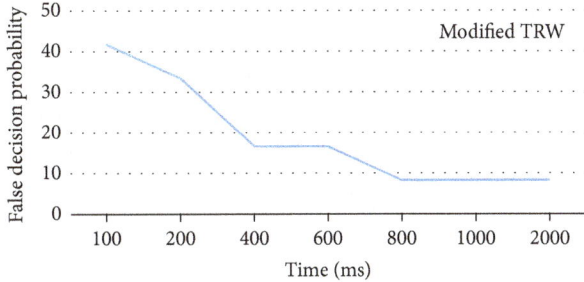

FIGURE 5: False decision probability for various values of message collection time (T).

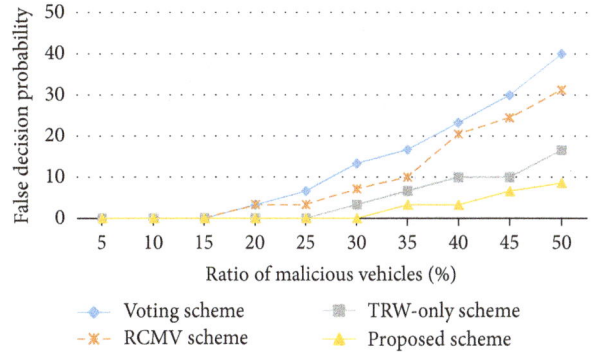

FIGURE 6: Comparison of the proposed scheme with other schemes under a message modification attack in a highway scenario.

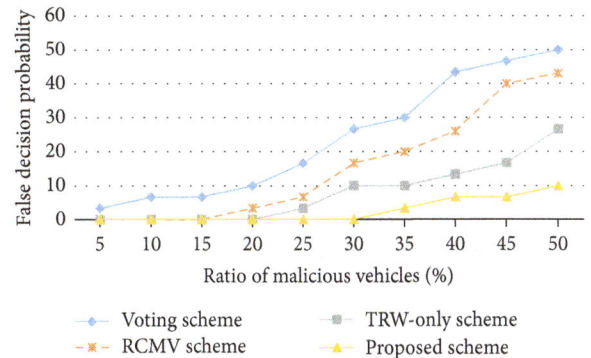

FIGURE 7: Comparison of the proposed scheme with other schemes under a fake message generation attack in a highway scenario.

with report collection time less than 100 ms which results in high false decision probability. The simulation result is shown in Figure 5. As the report collection time interval increases, the false decision probability decreases and, from 800 ms, the false decision probability does not decrease anymore. Based on this, we set the value of T to 1 sec, and this value will be used for T hereafter.

We now compare our proposed scheme with other schemes: RMCV scheme, a simple voting scheme, and TRW-only scheme. RMCV is an information oriented trust model and the outcome of the scheme is a trustworthiness value associated with each received message. In RMCV scheme, we consider the message trustworthiness based on the content similarity. The message trustworthiness is likely to increase as the message contents are similar among different vehicles. In the TRW-only scheme, we used the modified threshold random walk to make a decision about the event in the warning message without applying our proposed clustering algorithm. Several voting methods have been proposed to estimate the trustworthiness of each report message [53–55]. In the simple voting mechanism, each vehicle collects a fixed number of warning messages from the neighboring vehicles regarding an event and makes a decision by following the opinion of the majority group [55]. For the voting scheme, we collected 15 messages to make a decision, as this was the optimal number according to our simulation.

We compare our proposed scheme with other schemes in terms of false decision probability for various ratios of malicious vehicles under the message modification attack in a highway scenario as shown in Figure 6. We can see that our proposed scheme yields a lower false decision probability compared to the other mechanisms, even when the number of malicious vehicles increases. The simple voting mechanism performs worst among the four schemes. The performance of the RMCV scheme is close to TRW-only scheme when the malicious vehicle ratio is low. However, it degrades significantly compared to our proposed scheme as the malicious vehicle ratio increases. Our proposed scheme

has a false decision probability of 0% when the ratio of malicious vehicles is 30% in highway scenario.

We compare our proposed scheme with other schemes in terms of false decision probability under a fake message attack in a highway scenario as shown in Figure 7. We consider a case where the attacker generates messages about a fake event. Our proposed scheme yields better performance compared to the RMCV, simple voting, and TRW-only schemes with a low false decision probability of less than 10%. The false decision probability of RMCV and voting scheme exceed 40% when the ratio of malicious vehicles increased to 50%. In Figure 7, the false decision probability increases as the ratio of malicious vehicles increases, with a tendency similar to Figure 6.

We compare our proposed scheme with other schemes in terms of false decision probability under a message modification attack in an urban scenario in Figure 8. Our proposed scheme exhibits better performance compared to other schemes. The false decision probability does not exceed

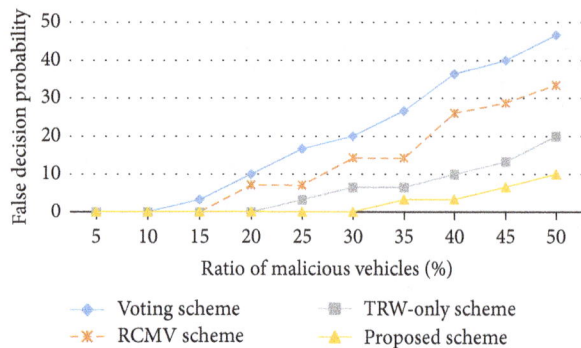

FIGURE 8: Comparison of the proposed scheme with other schemes under a message modification attack in an urban scenario.

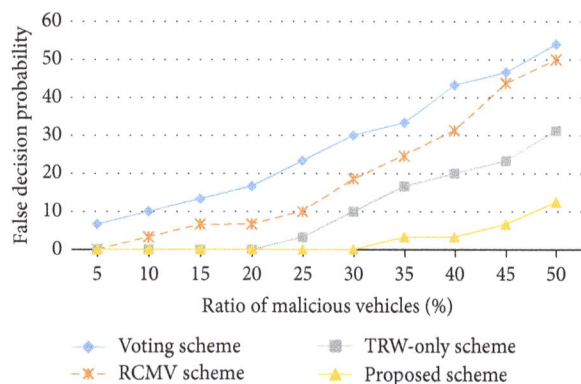

FIGURE 9: Comparison of the proposed scheme with other schemes under a fake message generation attack in an urban scenario.

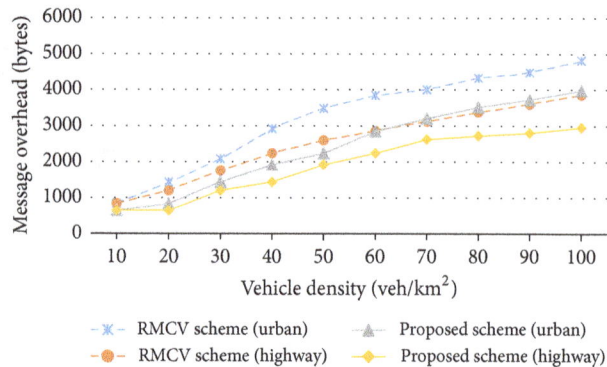

FIGURE 10: Message overhead for various values of vehicle densities under no malicious vehicle.

message overhead is the cost incurred due to the extra message that is exchanged with neighboring vehicles. In terms of message overhead, as the vehicle density increases, the message overhead also increases in both scenarios, as shown in Figure 10. In the beginning, when the vehicle density is low, our proposed scheme has low message overhead as the scheme does not advertise the default trust level of new neighbor vehicles; however the pseudo IDs in the trust level opinion pair cause some message overhead. We can see that the message overhead is higher for the RMCV as compared to our scheme in both scenarios because in their scheme the vehicle nodes send query messages to the neighboring vehicles and then receive response messages regarding the accident event. On the contrary, there is no query message, but only one-way report messages are sent in our scheme. The message overhead in the urban scenario for both schemes is slightly higher than the highway scenario, because the speed of the vehicles in the urban scenario is less than that of highway scenario, and, thus, each vehicle accumulates more messages, compared to the highway scenario.

We also compare our scheme with RMCV in terms of message overhead in the presence of malicious vehicles under fake message attack. The average vehicle density increases from 1 vehicle per km^2 to 100 vehicles per km^2 throughout the simulation time in urban and highway scenarios. We run several simulations by increasing the ratio of the malicious vehicles from 5% to 50% in both scenarios. Our scheme collects warning messages from neighboring vehicles to detect the trustworthiness of event information contained in the received messages. In both schemes, the message overhead increases as the ratio of the malicious vehicles increases because the vehicles accumulate more messages due to the presence of the malicious vehicles. The message overhead for different ratios of malicious vehicles is shown in Figure 11. Our scheme has a lower message overhead compared to the RMCV scheme in both scenarios.

5. Conclusion and Future Work

In this paper, we proposed a trustworthy event-information dissemination scheme in VANET. We determine and disseminate only the trustworthy event messages to neighbor

10% for our proposed scheme. However, it reaches 20% for the TRW-only scheme. The RMCV and the simple voting schemes exhibit much higher false decision probabilities compared to our proposed scheme.

We compare our proposed scheme with other schemes in terms of false decision probability under a fake message attack in an urban scenario in Figure 9. The false decision probability of the proposed scheme increases when the density of the malicious vehicles generating the false message increases, resulting in a false decision probability slightly greater than 10%. In an urban scenario, the high density of vehicles and low speeds help the propagation of false event messages generated by attackers. Thus, the false decision probability in this case is slightly higher than that for the highway scenario. In this scenario, the RMCV and the simple voting schemes exhibit higher false decision probabilities compared to the proposed scheme, with a tendency similar to Figure 8.

We now compare our scheme with the RMCV scheme in terms of message overhead. We considered a situation where there is an actual accident without malicious vehicles. In Figure 10, we present the simulation results of the message overhead with respect to the varying density of vehicles per square kilometer in both urban and highway scenarios. The

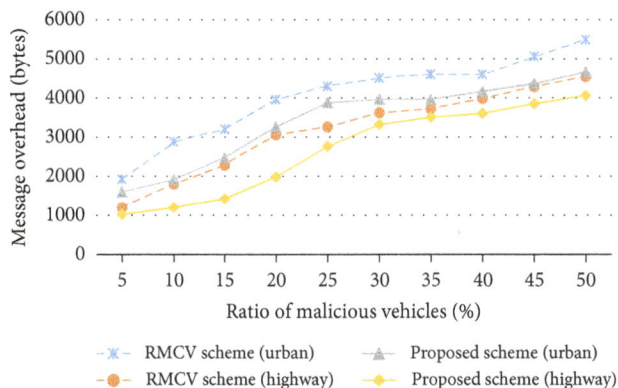

FIGURE 11: Message overhead for various ratios of malicious vehicles under fake message generation attack.

vehicles. We introduced a modified K-means clustering algorithm to reduce the effect of malicious vehicles on the trust levels (i.e., the truth-telling probabilities) of other vehicles. In other words, the issue of node trustworthiness is resolved through a modified K-means clustering algorithm in our proposed scheme. In the next step, the issue of message trustworthiness is resolved by applying a modified TRW to the report messages received from neighbor vehicles along with the information on node trustworthiness. We compared our proposed scheme with RMCV, simple voting, and TRW-only schemes through simulation. The simulation results show that our proposed scheme has a lower false decision probability compared to other schemes as well as low message overhead compared to the RMCV scheme. The simulation results also show that our proposed scheme can effectively cope with message modification attack and fake message generation attack as long as the number of benign vehicles is larger than the number of malicious vehicles. Our scheme has an additional advantage that the decision on the trustworthiness of a given message is made in an infrastructure-less environment without using PKI.

In this paper, we assumed that the malicious vehicles are uniformly distributed on the roads. However, this assumption may not be valid if colluding malicious vehicles move as a group to increase their influence on the nearby vehicles. Such a complicated issue will be studied in more detail in our future work.

Notations

p_i: Pseudo ID of vehicle V_i
θ_i: Trust level (truth-telling prob.) of vehicle V_i
E_x: Event of type x
M_E: Event message
M_B: Beacon message
M_F: Forwarded message
L_E: Location of event E_x
l_i: Location of vehicle V_i
s_i: Speed of vehicle V_i
θ_{ij}: Estimation of trust level θ_i for vehicle i by vehicle j.

Acknowledgments

This research was supported in part by Basic Science Research Program through the National Research Foundation of Korea (NRF) funded by the Ministry of Education (2013R1A1A2012006 and 2015R1D1A1A01058595), and by the MSIT (Ministry of Science and ICT), Korea, under the ITRC (Information Technology Research Center) support program (IITP-2017-2016-0-00313) supervised by the IITP (Institute for Information & communications Technology Promotion).

References

[1] H. Hartenstein and L. Kenneth, *VANET: Vehicular Applications and Inter-Networking Technologies*, Wiley, New Jersey, NJ, USA, 1st edition, 2009.

[2] P. Papadimitratos, L. Buttyan, T. Holczer et al., "Secure vehicular communication systems: design and architecture," *IEEE Communications Magazine*, vol. 46, no. 11, pp. 100–109, 2008.

[3] K. Govindan and P. Mohapatra, "Trust computations and trust dynamics in mobile adhoc networks: a survey," *IEEE Communications Surveys & Tutorials*, vol. 14, no. 2, pp. 279–298, 2012.

[4] Y. L. Morgan, "Notes on DSRC & WAVE standards suite: its architecture, design, and characteristics," *IEEE Communications Surveys & Tutorials*, vol. 12, no. 4, pp. 504–518, 2010.

[5] M. Raya, P. Papadimitratos, V. D. Gligor, and J.-P. Hubaux, "On data-centric trust establishment in ephemeral ad hoc networks," in *Proceedings of the 27th IEEE Communications Society Conference on Computer Communications (INFOCOM '08)*, pp. 1912–1920, Arizona, Ariz, USA, April 2008.

[6] R. Shrestha, S. Djuraev, and S. Y. Nam, "Sybil attack detection in vehicular network based on received signal strength," in *Proceedings of the 3rd International Conference on Connected Vehicles and Expo, ICCVE '14*, pp. 745-746, November 2014.

[7] J. Doucer, "The Sybil Attack," in *Proceedings of the IPTPS '01 Revised Papers from the First International Workshop on Peer-to-Peer Systems*, 2002.

[8] G. Martuscelli, A. Boukerche, and P. Bellavista, "Discovering traffic congestion along routes of interest using VANETs," in *Proceedings of the 2013 IEEE Global Communications Conference, GLOBECOM '13*, IEEE, Atlanta, GA, USA, December 2013.

[9] Z. Huang, S. Ruj, M. Cavenaghi, and A. Nayak, "Limitations of trust management schemes in VANET and countermeasures," in *Proceedings of the IEEE 22nd International Symposium on Personal, Indoor and Mobile Radio Communications (PIMRC '11)*, pp. 1228–1232, IEEE, Toronto, Canada, September 2011.

[10] M. Raya, P. Papadimitratos, and J.-P. Hubaux, "Securing vehicular communications," *IEEE Wireless Communications Magazine*, vol. 13, no. 5, pp. 8–15, 2006.

[11] Z. Li and C. T. Chigan, "On joint privacy and reputation assurance for vehicular ad hoc networks," *IEEE Transactions on Mobile Computing*, vol. 13, no. 10, pp. 2334–2344, 2014.

[12] F. Ishmanov, S. W. Kim, and S. Y. Nam, "A secure trust establishment scheme for wireless sensor networks," *Sensors*, vol. 14, no. 1, pp. 1877–1897, 2014.

[13] N. Yang, "A similarity based trust and reputation management framework for vanets," *International Journal of Future Generation Communication and Networking*, vol. 6, no. 2, pp. 25–34, 2013.

[14] K. Rostamzadeh, H. Nicanfar, N. Torabi, S. Gopalakrishnan, and V. C. M. Leung, "A context-aware trust-based information dissemination framework for vehicular networks," *IEEE Internet of Things Journal*, vol. 2, no. 2, pp. 121–132, 2015.

[15] R. A. Shaikh and A. S. Alzahrani, "Intrusion-aware trust model for vehicular ad hoc networks," *Security and Communication Networks*, vol. 7, no. 11, pp. 1652–1669, 2014.

[16] S. A. Soleymani, A. H. Abdullah, W. H. Hassan et al., "Trust management in vehicular ad hoc network: a systematic review," *EURASIP Journal on Wireless Communications and Networking*, vol. 2015, no. 1, article 146, 2015.

[17] M. Raya, A. Aziz, and J. P. Hubaux, "Efficient secure aggregation in VANETs," in *Proceedings of the 3rd ACM International Workshop on Vehicular Ad Hoc Networks (VANET '06)*, pp. 67–75, ACM, New York, NY, USA, September 2006.

[18] B. Qin, Q. Wu, J. Domingo-Ferrer, and L. Zhang, "Preserving security and privacy in large-scale VANETs," in *Proceedings of the 13th International Conference Information and communications security (ICICS '11)*, vol. 7043 of *Lecture Notes in Computer Science*, pp. 121–135, Springer Berlin Heidelberg, Berlin, Germany, 2011.

[19] X. Lin, X. Sun, P.-H. Ho, and X. Shen, "GSIS: a secure and privacy-preserving protocol for vehicular communications," *IEEE Transactions on Vehicular Technology*, vol. 56, no. 6 I, pp. 3442–3456, 2007.

[20] C. Zhang, R. Lu, X. Lin, P.-H. Ho, and X. Shen, "An efficient identity-based batch verification scheme for vehicular sensor networks," in *Proceedings of the 27th IEEE Communications Society Conference on Computer Communications (INFOCOM '08)*, pp. 816–824, IEEE INFOCOM, Arizona, Ariz, USA, April 2008.

[21] A. Wasef and X. Shen, "Efficient group signature scheme supporting batch verification for securing vehicular networks," in *Proceedings of the IEEE International Conference on Communications (ICC '10)*, pp. 1–5, Cape Town, South Africa, May 2010.

[22] H. Zhu, S. Du, Z. Gao, M. Dong, and Z. Cao, "A probabilistic misbehavior detection scheme toward efficient trust establishment in delay-tolerant networks," *IEEE Transactions on Parallel and Distributed Systems*, vol. 25, no. 1, pp. 22–32, 2014.

[23] D. Tian, Y. Wang, H. Liu, and X. Zhang, "A trusted multi-hop broadcasting protocol for vehicular ad hoc networks," in *Proceedings of the 2012 1st International Conference on Connected Vehicles and Expo, ICCVE '12*, pp. 18–22, December 2012.

[24] F. Gómez Mármol and G. Martínez Pérez, "TRIP, a trust and reputation infrastructure-based proposal for vehicular ad hoc networks," *Journal of Network and Computer Applications*, vol. 35, no. 3, pp. 934–941, 2012.

[25] Y.-C. Wei and Y.-M. Chen, "An efficient trust management system for balancing the safety and location privacy in VANETs," in *Proceedings of the 11th IEEE International Conference on Trust, Security and Privacy in Computing and Communications (TrustCom '12)*, pp. 393–400, Liverpool, UK, June 2012.

[26] U. F. Minhas, J. Zhang, T. Tran, and R. Cohen, "Towards expanded trust management for agents in vehicular ad-hoc networks," *International Journal of Computational Intelligence: Theory and Practice*, vol. 5, no. 1, pp. 3–15, 2010.

[27] U. F. Minhas, J. Zhang, T. Tran, and R. Cohen, "A multifaceted approach to modeling agent trust for effective communication in the application of mobile Ad Hoc vehicular networks,"

IEEE Transactions on Systems, Man, and Cybernetics, Part C: Applications and Reviews, vol. 41, no. 3, pp. 407–420, 2011.

[28] Z. Liu, J. Ma, J. Zhang, H. Zhu, and Y. Miao, "LSOT: a lightweight self-organized trust model in VANETs," *Mobile Information Systems*, vol. 2016, no. 1, Article ID 7628231, pp. 1–15, 2016.

[29] Y.-M. Chen and Y.-C. Wei, "A beacon-based trust management system for enhancing user centric location privacy in VANETs," *Journal of Communications and Networks*, vol. 15, no. 2, Article ID 6512239, pp. 153–163, 2013.

[30] S. Gurung, D. Lin, A. Squicciarini, and E. Bertino, "Information-oriented trustworthiness evaluation in vehicular ad-hoc networks," *Lecture Notes in Computer Science (including subseries Lecture Notes in Artificial Intelligence and Lecture Notes in Bioinformatics): Preface*, vol. 7873, pp. 94–108, 2013.

[31] Q. Ding, X. Li, M. Jiang, and X. Zhou, "Reputation-based trust model in vehicular ad-hoc networks," in *Proceedings of the International Conference on Wireless Communications And Signal Processing (WCSP)*, 2010.

[32] F. Dötzer, L. Fischer, and P. Magiera, "VARS: a vehicle ad-hoc network reputation system," in *Proceedings of the 6th IEEE International Symposium on a World of Wireless Mobile and Multimedia Networks, WoWMoM '05*, pp. 454–456, June 2005.

[33] C. Chen, J. Zhang, R. Cohen, and P.-H. Ho, "A trust modeling framework for message propagation and evaluation in VANETs," in *Proceedings of the 2nd International Conference on Information Technology Convergence and Services (ITCS '10)*, IEEE, Cebu, Phillippines, August 2010.

[34] A. Patwardhan, A. Joshi, T. Finin, and Y. Yesha, "A data intensive reputation management scheme for vehicular ad hoc networks," in *Proceedings of the 2006 3rd Annual International Conference on Mobile and Ubiquitous Systems: Networking and Services, MobiQuitous*, July 2006.

[35] J. Zhang, "A survey on trust management for VANETs," in *Proceedings of the 25th IEEE International Conference on Advanced Information Networking and Applications (AINA '11)*, pp. 105–112, Biopolis, Singapore, March 2011.

[36] P. Wex, J. Breuer, A. Held, T. Leinmüller, and L. Delgrossi, "Trust issues for vehicular ad hoc networks," in *Proceedings of the IEEE 67th Vehicular Technology Conference-Spring (VTC '08)*, pp. 2800–2804, Singapore, May 2008.

[37] J. Petit, F. Schaub, M. Feiri, and F. Kargl, "Pseudonym schemes in vehicular networks: a survey," *IEEE Communications Surveys & Tutorials*, vol. 17, no. 1, pp. 228–255, 2015.

[38] Z. Malik and A. Bouguettaya, "Reputation bootstrapping for trust establishment among web services," *IEEE Internet Computing*, vol. 13, no. 1, pp. 40–47, 2009.

[39] W. R. J. Joo and D. S. Han, "An enhanced broadcasting scheme for IEEE 802.11p according to lane traffic density," in *Proceedings of the 20th International Conference on Software, Telecommunications and Computer Networks (SoftCOM '12)*, September 2012.

[40] W. Fehr, *Security system design for cooperative vehicleto-vehicle crash avoidance applications using 5.9 GHz Dedicated Short Range Communications (DSRC) wireless communications*, 2012.

[41] H. Xiong, K. Beznosov, Z. Qin, and M. Ripeanu, "Efficient and spontaneous privacy-preserving protocol for secure vehicular communication," in *Proceedings of the 2010 IEEE International Conference on Communications, ICC 2010*, May 2010.

[42] F. Jiménez, J. E. Naranjo, and Ó. Gómez, "Autonomous manoeuvring systems for collision avoidance on single carriageway roads," *Sensors*, vol. 12, no. 12, pp. 16498–16521, 2012.

[43] X. Tang, D. Hong, and W. Chen, "Data Acquisition Based on Stable Matching of Bipartite Graph in Cooperative Vehicle–Infrastructure Systems," *Sensors*, vol. 17, no. 6, p. 1327, 2017.

[44] P. B. Velloso, R. P. Laufer, D. D. O. O. Cunha, O. C. M. B. Duarte, and G. Pujolle, "Trust management in mobile ad hoc networks using a scalable maturity-based model," *IEEE Transactions on Network and Service Management*, vol. 7, no. 3, pp. 172–185, 2010.

[45] F. Bao, I. Chen, M. Chang, and J. Cho, "Hierarchical trust management for wireless sensor networks and its applications to trust-based routing and intrusion detection," *IEEE Transactions on Network and Service Management*, vol. 9, no. 2, pp. 169–183, 2012.

[46] A. Wald, *Sequential Analysis*, John wiley ad Sons, New York, NY, USA, 7th edition, 1965.

[47] J. Jung, V. Paxson, A. W. Berger, and H. Balakrishnan, "Fast portscan detection using sequential hypothesis testing," in *Proceedings of the IEEE Symposium on Security and Privacy*, IEEE, California, Calif, USA, 2004.

[48] G. Peter, S. Zsolt, and A. Szilard, "Highly automated Vehilce Systems," Mechatronics Engineer MSc Curriculum Development, BME MOGI, 2014.

[49] C. Sommer, R. German, and F. Dressler, "Bidirectionally coupled network and road traffic simulation for improved IVC analysis," *IEEE Transactions on Mobile Computing*, vol. 10, no. 1, pp. 3–15, 2011.

[50] A. Varga and R. Hornig, "An overview of the OMNeT++ simulation environment," in *Proceedings of the 1st International ICST Conference on Simulation Tools and Techniques for Communications, Networks and Systems (SIMUTools '08)*, March 2008.

[51] D. Krajzewicz, J. Erdmann, M. Behrisch, and L. Bieker, "Recent Development and Applications of SUMO - Simulation of Urban MObility," in *Proceedings of the Recent Development and Applications of SUMO - Simulation of Urban MObility*, vol. 5, pp. 128–138, December 2012.

[52] W. Ben Jaballah, M. Conti, M. Mosbah, and C. E. Palazzi, "The impact of malicious nodes positioning on vehicular alert messaging system," *Ad Hoc Networks*, vol. 52, pp. 3–16, 2016.

[53] A. Tajeddine, A. Kayssi, and A. Chehab, "A privacy-preserving trust model for VANETs," in *Proceedings of the 10th IEEE International Conference on Computer and Information Technology, CIT-2010, 7th IEEE International Conference on Embedded Software and Systems, ICESS-2010, 10th IEEE Int. Conf. Scalable Computing and Communications, ScalCom '10*, pp. 832–837, July 2010.

[54] J. Petit and Z. Mammeri, "Dynamic consensus for secured vehicular ad hoc networks," in *Proceedings of the 2011 IEEE 7th International Conference on Wireless and Mobile Computing, Networking and Communications, WiMob '11*, pp. 1–8, October 2011.

[55] B. Ostermaier, F. Dötzer, and M. Strassberger, "Enhancing the security of local danger warnings in VANETs - A simulative analysis of voting schemes," in *Proceedings of the 2nd International Conference on Availability, Reliability and Security, ARES '07*, pp. 422–431, April 2007.

18

Modeling and Optimization for Collaborative Business Process Towards IoT Applications

Yongyang Cheng [ID],[1] Shuai Zhao [ID],[1] Bo Cheng [ID],[1] Shoulu Hou,[2] Yulong Shi,[3] and Junliang Chen[1]

[1]*State Key Laboratory of Networking and Switching Technology, Beijing University of Posts and Telecommunications, Beijing 100876, China*
[2]*Data61, Commonwealth Scientific and Industrial Research Organization, Marsfield, NSW 2122, Australia*
[3]*Middleware Systems Research Group, University of Toronto, Toronto, Canada M5S 2E4*

Correspondence should be addressed to Yongyang Cheng; zhuifeng@bupt.edu.cn

Academic Editor: Francesco Gringoli

The rapid development of Internet of Things (IoT) attracts growing attention from both industry and academia. IoT seamlessly connects the real world and cyberspace via various business process applications hosted on the IoT devices, especially on smart sensors. Due to the discrete distribution and complex sensing environment, multiple coordination patterns exist in the heterogeneous sensor networks, making modeling and analysis particularly difficult. In addition, massive sensing events need to be routed, forwarded and processed in the distributed execution environment. Therefore, the corresponding sensing event scheduling algorithm is highly desired. In this paper, we propose a novel modeling methodology and optimization algorithm for collaborative business process towards IoT applications. We initially extend the traditional Petri nets with sensing event factor. Then, the formal modeling specification is investigated and the existing coordination patterns, including event unicasting pattern, event broadcasting pattern, and service collaboration pattern, are defined. Next, we propose an optimization algorithm based on Dynamic Priority First Response (DPFR) to solve the problem of sensing event scheduling. Finally, the approach presented in this paper has been validated to be valid and implemented through an actual development system.

1. Introduction

IoT brings together physical sensors which have never been connected before. These sensors are the fundamental building blocks for the creation of smart applications and communicate via device-to-device connections [1, 2]. As we know, the IoT-aware business process applications are event-driven by nature [3]. Therefore, the business processes hosted on the wide-spread geographical distribution sensors should respond to different sensing events gathered in the dynamic IoT environment as fast as they could. The real-time event handling not only makes the cooperation among cross-sensor business processes more efficient, but also makes it possible to invoke corrective processes before emergencies snowball into disasters (e.g., traffic congestion, fire hazard, power emergency, etc) [4]. It is worth mentioning that a typical cross-sensor business process application, especially in large-scale application scenarios, is usually dispersed in different monitoring areas. Due to the discrete distribution of sensors and the complexity of execution environment, they need to collaborate to accomplish specific tasks [5–8]. Therefore, different kinds of coordination patterns exist in the heterogeneous smart sensor networks, which make modeling more difficult. In general, the common coordination patterns in the IoT environment include event unicasting pattern, event broadcasting pattern, and service collaboration pattern.

In order to comprehensively respond to external sensing events, massive data gathered by the smart sensors are routed, forwarded, and processed among IoT-aware business process applications in the heterogeneous sensor networks. However, the available resource, especially the computing resource, is often limited in a particular scene [9–12].

For example, in the modern forest-protection, numerous smart sensors are discretely deployed in multiple monitoring areas. These sensors communicate via sensing events and share the same remote computing resource. These sensing events not only have a huge number, but also have complex data formats, including text, image, and video. Furthermore, different kinds of sensing events are assigned different priorities. But, all of them tend to get response from the IoT system as fast as possible. Thus, it is particularly important to reasonably schedule all kinds of sensing events based on certain principles with limited computing resource.

As a tool to model discrete event systems, Petri nets [13–18] have been widely used to model the business process applications, which have shown great power in dealing with concurrences and conflicts. Therefore, in this paper, the modeling methodology for collaborative business process towards IoT applications is investigated on the basis of Petri nets. Although Petri nets could provide theoretical supports for the modeling and analysis, challenges still remain to be solved. First of all, the business processes towards IoT applications are event-driven by nature. But Petri nets only concern about the control structures, including sequence structure, AND-Split/Join structure, XOR-Split/Join structure, and Loop structure, without taking into account the sensing event factor involved in the process interactions. Thus, we need to extend Petri nets with this factor. Furthermore, due to the discrete distribution and complex sensing environment, multiple coordination patterns exist in the heterogeneous sensor networks. However, to our best knowledge, most of the current work pays little attention to the systematic description and formal definition for coordination patterns in the business process modeling phase. Finally, massive sensing events need to be routed, forwarded, and processed in the resource-constrained IoT environment. All of them tend to be scheduled as fast as possible. Most of the sensing event scheduling algorithms [9, 19, 20, 21] only focus on the technical implementation details, lacking a qualitative and quantitative analysis that combines the specific IoT application scenario.

The main contributions of our work contain the following:

(i) We extend the traditional Petri nets with sensing event factor, making it support the direct modeling of collaborative business process towards IoT applications.

(ii) The formal modeling specification of IoT sensor tasks is investigated, and its coordination patterns, including event unicasting pattern, event broadcasting pattern, and service collaboration pattern, are defined.

(iii) We propose a novel optimization algorithm based on Dynamic Priority First Response (DPFR) to solve the problem of sensing event scheduling.

(iv) We qualitatively and quantitatively evaluate our proposed algorithm. In addition, the work presented in this paper has been implemented through an actual system.

The rest of this paper is organized as follows: We discuss the related work in Section 2. Section 3 describes a scenario of collaborative business process towards IoT application in forest-protection. In Section 4, the formal modeling specification of sensor tasks is investigated and its coordination patterns are defined. We propose a novel algorithm to optimize sensing event scheduling and consider both qualitative and quantitative aspects for the performance evaluation in Section 5. Based on our proposed approach, we design a real development system and give the implementation details in Section 6. Concluding remarks are made in Section 7.

2. Related Work

In this section, we mainly compare our proposed approach with other existing approaches. We will clearly point out the similarities and differences with their works.

2.1. Modeling of IoT-Aware Business Processes. Tan et al. [13] proposed an approach for modeling an e-commerce system with a third-party payment platform from the view point of a business process. This approach integrated both data and control flows based on Petri nets. Rationality and transaction consistency were defined and validated to guarantee the transaction properties of an e-commerce business process. van der Aalst [22] considered that business process applications were deployed over a lot of organizations. He addressed two valuable questions in his work. The first was how to decide the minimal requirements for the in-organization business process and the second was how to judge an in-organization business process modeled with Petri nets was consistent based on an interaction structure specified through an event sequence. In [23], Zeng et al. proposed a method extended with message and resource factors to model the cross-department collaborative business process. Based on a cross-department medical diagnosis scenario, different coordination patterns were described and integrated modeling approach was illustrated.

2.2. Optimization Algorithms. Munir et al. [24] proposed an opportunistic, reliable, and realistic QoS mechanism for bulk data transfers in grids, which maximized acceptance rate and network resource utilization by using an event-based priority queue. The metrics that had been studied in the evaluation of elastic scheduling heuristics were the acceptance percentage and the mean flow time of requests. Li et al. [25] proposed a scheduling algorithm of events with uncertain timestamps, which could support effective scheduling of reading events and writing events in CPS. The experiments verified that the scheduling algorithm could guarantee providing correct feedback of the event sequences to CPS. In [20], Aziz et al. proposed a local search heuristic which handled an event selection namely Event Selection based on Soft Constraint Violation (ESSCV) applied in a modified PSO algorithm to solve class scheduling problems. Erbas et al. [21] studied static priority scheduling of recurring real-time tasks. They focused on the non-preemptive uniprocessor case and obtained schedule-theoretic results for this case.

2.3. Summary of Related Work. The above-mentioned approaches are generally still in their infancies. The work [13] and [22] only focus the discrete business processes without considering coordination patterns among different process tasks. Our approach takes full account of the complexity of IoT sensing environment. We give the formal modeling specification of sensor tasks and define different coordination patterns. The work [23] also extends Petri nets with message and resource factors. But the premise of this extension is not to consider resource conflicts. However, the available computing resource is often limited in a particular scene. Therefore, we propose the corresponding algorithm to optimize sensing event scheduling. Although the studies [20, 21, 24–26] have presented some sensing event scheduling algorithms, most of their approaches are based on software simulation, which lacks theory basic that combines the specific IoT application scenario. Different to them, we initially verify the feasibility and effectiveness of our proposed optimization algorithm at the theoretical level. Then, we consider both qualitative and quantitative aspects for the performance evaluation of the algorithm.

3. Running Example

A large area of forest is widely distributed in North China. The work of forest-protection used to be done by the way of human mountaineering in the past. However, this way not only costs a lot of resources, but also could not deal with emergencies in real-time. Using our proposed approach, numerous smart sensors are discretely deployed in multiple monitoring areas. These sensors are in charge of collecting their surrounding information and sending the corresponding sensing events to gateways regularly. Owing to inconsistency of the raw data, event manager will initially do some basic filtering or aggregation operations. Then, encapsulated events are routed, forwarded, and computed among business process applications hosted on the distributed servers. Finally, valuable information would be submitted to officers for further processing and analyzing. In general, a typical scenario of modern forest-protection usually involves the following sensors: smoke sensor, infrared thermal imaging sensor, tracking camera sensor, light sensor, and sound sensor. Furthermore, the common physical resources include the fire-fighters, fire-engines, and fire-airplanes. Officers would allocate these resources according to the need. Figure 1 illustrates the whole business process application of this case.

In general, this scenario includes the following steps:

(1) The smoke sensor collects its surrounding information and regularly sends these raw data to the smart gateway for further filtering, classifying, and aggregating.

(2) When the fire emergency occurs in a monitoring area, the smoke sensor will publish a smoke event.

(3) Event manager sets the event-priority according to the smoke concentration information.

(4) Distributed servers route this processed event and publish the corresponding command event to the infrared thermal imaging sensor, making it detect whether there are moving targets (e.g., explorers, campers, etc.).

(5) If the moving targets are detected, the infrared thermal imaging sensor will publish a detected event in time.

(6) Distributed servers subscribe this event and publish a command event to the tracking camera sensor.

(7) The tracking camera sensor keeps tracking of the targets, transmits the actual information, and requires entity resources to the remote dispatch center.

(8) The officer allocates a certain amount of fire-fighters, fire-engines, or fire-airplanes according to the need. Meanwhile, an alarm event is automatically published to the light sensor and sound sensor.

(9) Once the light and sound sensor have subscribed the alarm event, they will send light and sound warning to guide the moving targets to escape the hazardous area.

The interaction interface between the discrete business processes is event-based loosely coupled. When emergencies occur, the appropriate business processes could be invoked to handle the corresponding events timely. According to this specific description, we could summarize the following characteristics of these business processes.

(1) The whole IoT-aware application involves smoke sensor, infrared thermal imaging sensor, tracking camera sensor, light sensor, and sound sensor. Each sensor has its respective business process and tasks.

(2) The execution of this IoT-aware cross-sensor business process application is driven by different kinds of priority-based sensing events. These sensors need to collaborate with each other to complete the rescue.

(3) Generally speaking, a smart sensor task is usually composed of sensor name, task item, subscribed event, and published event. Taking this collaborative IoT-aware business process application in the forest-protection scenario as an instance, sensor tasks and their corresponding elements are illustrated in Table 1.

4. Modeling for the Business Processes

To better analyze and optimize the collaborative business process towards IoT applications, we should first have a complete modeling method. In this section, we initially introduce the formal modeling specification for single-sensing sensor node. Then, different kinds of coordination patterns are investigated and formally defined. Finally, we model the above-mentioned forest-protection business process application based on our proposed approach.

4.1. Formal Modeling Specification for Single Sensor Node. In the traditional Petri nets-based business process modeling, a task is usually composed of task name and task attribute. Considering that the business processes usually involve all

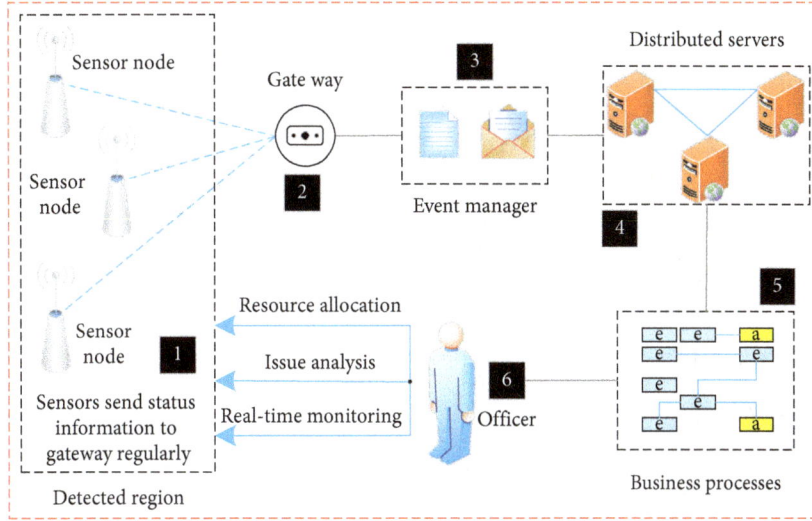

FIGURE 1: A case of collaborative cross-sensor business process application.

TABLE 1: Task of cross-sensor business process.

Sensor name	Task item	Subscribed event	Published event
Sm	Smoke detection	Φ	Smoke event
I	Moving-target detection	Smoke event	Detected event
T	Moving-target tracking	Detected event	Alarm event
L	Alarm target	Alarm event	Φ
So	Alarm target	Alarm event	Φ

Sm = smoke sensor, I = infrared thermal imaging sensor, T = tracking camera sensor, L = light sensor, So = sound sensor.

kinds of sensing events in the dynamic IoT environment, we extend the traditional Petri nets with this factor.

4.1.1. Petri Nets.
A Petri net is a model that could be represented by a graph.

Definition 1: $N = (P, T, F)$ is a net if and only

(1) $P = \{p_1, p_2, \ldots, p_m\}$ is a finite set of places

(2) $T = \{t_1, t_2, \ldots, t_m\}$ is a finite set of transitions

(3) $F \subseteq (P \times T) \cup (T \times P)$ is a finite set of directed arcs

(4) $P \cap T = \phi$ and $P \cup T \neq \phi$

Definition 2: Let $x \in P \cup T$ denote any element of a net N. Given $x \in P \cup T$, $\cdot x = \{y \mid (y, x) \in F\}$ is called the input set or preset of x, and $x \cdot = \{y \mid (x, y) \in F\}$ is the output set or postset of x. $N = (P, T, F)$ is called a pure net if it satisfies $\forall t \in T : \cdot t \cap t \cdot = \phi$.

Definition 3: A 4-tuple $\sum = (P, T, F, M)$ is a Petri net if and only if

(1) $N = (P, T, F)$ is a pure net

(2) $M : P \longrightarrow N$ is a marking function and M_0 is the initial marking

(3) $\sum$ has the following transition firing rules:

(a) $t \in T$ is enabled in M, denoted as $M [t>$ if $\forall p \in \cdot t : M(p) \geq 1$

(b) If $M [t>$, then t may fire in M, and its firing generates a new marking $M \cdot$, denoted as $M [t > M \cdot$, where

$$M \cdot (p) = \begin{cases} M(p) + 1, & \text{if } p \in t \cdot - \cdot t, \\ M(p) - 1, & p \in \cdot t - t \cdot, \\ M(p), & \text{otherwise.} \end{cases} \quad (1)$$

4.1.2. Extended Petri Nets.
In the traditional Petri nets-based modeling, the transition T is used to represent the tasks. Considering that the IoT-aware business processes usually involve sensing events, we extend the traditional Petri nets with this factor.

Definition 4: A 5-tuple $\sum_{ER} = (P, T, F, M, R)$ is an extended Petri net if the following conditions hold.

(1) There is one input source $i \in P$ and one output source $o \in P$, such that $\cdot i = \phi$ and $o \cdot = \phi$

(2) $\forall x \in P \cup T$ is on a path from i to o

(3) $\forall p \in P$ satisfies Equation (1)

(4) $P = P_C \cup P_E, P_C \cap P_E = \phi$; P_C represents the common places, P_E represents sensing event places

(5) $F = F_C \cup F_E, F_C = (P_C \times T) \cup (T \times P_C), F_E = (P_E \times T) \cup (T \times P_E)$; F_C represents the common control structure, F_E represents the published or subscribed sensing events

(6) $R(t)$ is the resource waiting time. $\forall t \in T$, with the arrival time of the first resource-token in the pre-places of transition t as the start time

Compared with the traditional Petri nets, the main differences defined in Definition 4 are as follows:

(i) We separate sensing events P_E from the traditional places P_C. Based on this, we could more clearly understand their roles in the whole processes.

(ii) We define a variable $t \in R(t)$ representing the waiting time for resources. An embedded timer starts counting when the token fires from the source place i.

(iii) All resources are in the ready state when they are allocated. In addition, sensing events are generated during the execution phase of business processes.

4.1.3. Modeling for Single Sensor Node.

As mentioned above, a sensor task is usually composed of sensor name, task item, subscribed event, and published event. Thus, the specification of a single sensor node is formally defined as follows.

Definition 5: A Sensor task is a four-tuple *SensorTask* = <*Name, Item, SubEvent, PubEvent*>. Then, each item of the sensor task is interpreted as follows.

(1) Name is the name of a smart sensor task, representing the content of the task

(2) Item represents the specific task of a smart sensor to be performed, which is usually an atomic operation

(3) SubEvent is the sensing event that is needed to drive the smart sensor node to work

(4) PubEvent is the sensing event that is published when the task of a smart sensor node has been finished

Using our extended Petri nets model, a sensor task is described by a transition which has one input place and one output place, representing the start and end of the sensor task, respectively. Due to the importance of sensing events, their corresponding places are added in this model. Thus, a formal model for single sensor node is illustrated in Figure 2. P_{Start} is the start place and P_{End} is the end place. $P_{SubEvent}$ is the sensing event place which represents the subscribed sensing event when a sensor task starts. Similarly, $P_{PubEvent}$ is the sensing event place which represents the published sensing event when a task of the smart sensor has been finished. It is worth to note that each sensing event has its own priority, which is defined as a constant in the sensing event place and could be modified by the event manager.

To distinguish the common place and sensing event place, they are drawn as normal circle and double circle with dashed border. Moreover, directed arc drawn with full border represents the control flow and while directed arc drawn with dashed border represents the data flow. Taking the scenario we mentioned above as an instance, the tracking camera sensor is denoted as t_t = <*tracking camera sensor, moving − target tracking, moving − target detected, alarm*>. Its corresponding model is illustrated in Figure 3.

4.2. Coordination Patterns among the IoT-Aware Collaborative Business Processes.

Due to the discrete distribution of sensors and the complexity of IoT sensing environment, they need to collaborate to accomplish specific tasks. Therefore, different kinds of coordination patterns exist among business processes, including event unicasting pattern, event broadcasting pattern, and service collaboration pattern. In this section, we define these coordination patterns based on extended Petri nets.

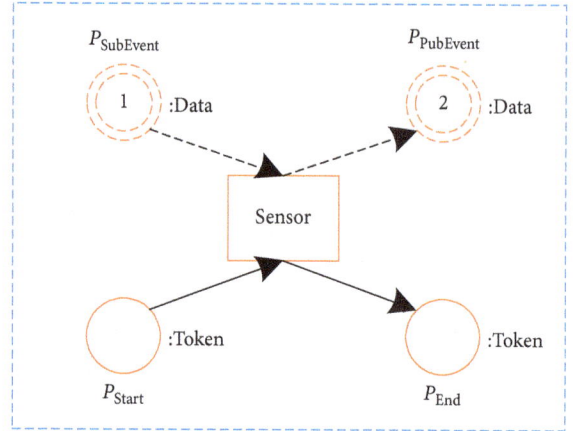

FIGURE 2: Extended Petri nets-based model of a sensor task.

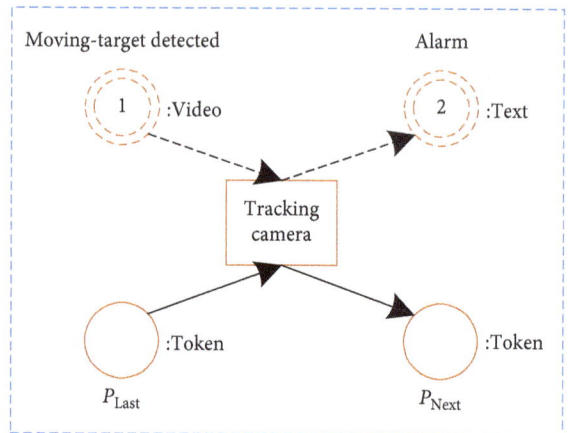

FIGURE 3: Extended Petri nets-based model of the tracking camera sensor.

4.2.1. Event Unicasting Pattern.

If two tasks affiliate to the different smart sensors, one sensor publishes an event, and the other one just needs to subscribe this event during execution phase. Then, we say that event unicasting pattern exists between them. Definition 6 describes this coordination pattern.

Definition 6: Let S_1 and S_2 be two different smart sensor nodes. Then <S_1, I_1, $SubEvent_1$, $PubEvent_1$> and <S_2, I_2, $SubEvent_2$, $PubEvent_2$> are two different four-tuples. Event unicasting pattern exists between these two smart sensors if the following conditions hold.

(1) $S_1 \cap S_2 = \phi$

(2) $I_1 \cap I_2 = \phi$

(3) $PubEvent_1 \cap SubEvent_2 \neq \phi$
    or $PubEvent_2 \cap SubEvent_1 \neq \phi$

We take above-mentioned scenario of collaborative business processes in forest-protection as an example. Task infrared thermal imaging sensor is denoted as t_i = <*infrared thermal imaging sensor, moving − target detection, smoke event, detected event*>, and the task smoke sensor is formalized as t_{sm} = <*smoke sensor, smoke detection, ϕ, smoke event*>. Obviously, these two business processes belong to different kinds of sensors; therefore, conditions (1) and (2) in Definition 6 are satisfied. In addition, the smoke event published by smoke senor is subscribed by infrared thermal imaging sensor; therefore, condition (3) in Definition 6 is also satisfied. In this way, according to Definition 6, event unicasting pattern exists between them. Figure 4 illustrates this coordination pattern.

4.2.2. Event Broadcasting Pattern. If tasks affiliate to the different sensors, one sensor task publishes a sensing event, which needs to be subscribed by two or more sensor tasks. Then, we claim that event broadcasting pattern exists between them. Definition 7 describes this coordination pattern.

Definition 7: Let S_1, S_2, ..., and S_n be different smart sensor nodes. Then $<S_1, I_1, SubEvent_1, PubEvent_1>$ and $<S_n, I_n, SubEvent_n, PubEvent_n>$ are different four-tuples. Event broadcasting pattern exists between these smart sensors if the following conditions hold.

(1) $S_1 \cap S_2 \cap, ..., \cap S_n = \phi$

(2) $I_1 \cap I_2 \cap, ..., \cap I_n = \phi$

(3) $PubEvent_1 \cap PubEvent_2 \cap, ..., \cap PubEvent_n \neq \phi$

We take above-mentioned scenario of collaborative business processes in forest-protection as an example. Task tracking camera sensor is denoted as $t_t = <tracking\ camera\ sensor, moving-target\ tracking, detected\ event, alarm\ event>$, task light sensor is formalized $t_l = <light\ sensor, alarm\ target, alarm\ event, \phi>$, and task sound sensor is formalized $t_{so} = <sound\ sensor, alarm\ target, alarm\ event, \phi>$. Obviously, these three sensors belong to different monitoring areas; therefore, conditions (1) and (2) in Definition 7 are satisfied. In addition, the light sensor and sound sensor both subscribe the alarm event published by tracking camera sensor; therefore, condition (3) in Definition 7 is also satisfied. In this way, according to Definition 7, event broadcasting pattern exists among them. Figure 5 illustrates this coordination pattern.

4.2.3. Service Collaborative Pattern. If one specific task requires multiple sensor tasks to collaborate with each other, then we claim that service collaboration pattern exists between them. Definition 8 describes this coordination pattern.

Definition 8: Let S_1 and S_2 be two different smart sensor nodes. Then $<S_1, I_1, SubEvent_1, PubEvent_1, ReqResource_1>$ and $<S_2, I_2, SubEvent_2, PubEvent_2, ReqResource_2>$ are two different four-tuples. Service collaboration pattern exists between these two smart sensors if the following conditions hold.

(1) $S_1 \neq S_2$

(2) $I_1 = I_2$

(3) $SubEvent_1 = SubEvent_2$

(4) $PubEvent_1 = PubEvent_2$

We take above-mentioned scenario of collaborative business processes in forest-protection as an example. Task light sensor is formalized as $t_l = <light\ sensor, alarm\ target,$

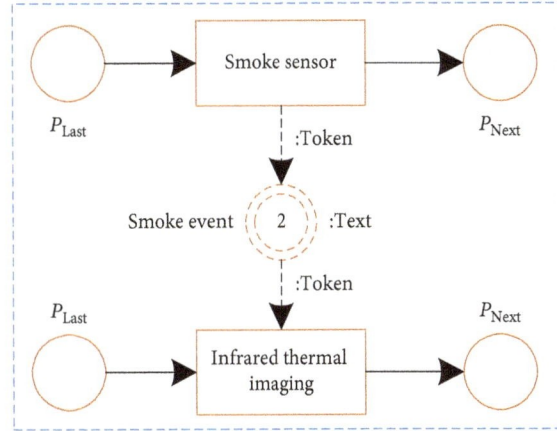

FIGURE 4: Event unicasting pattern in collaborative business processes.

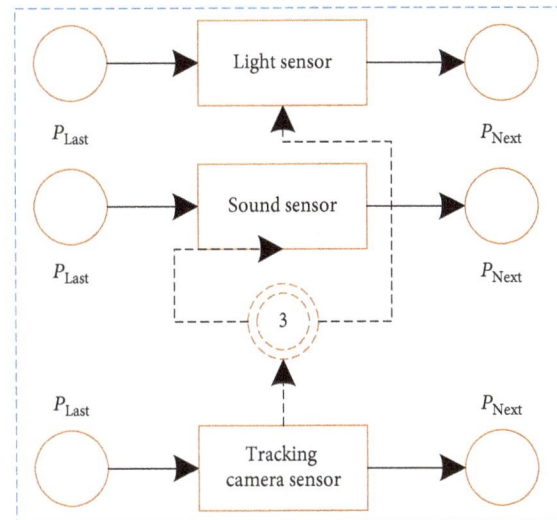

FIGURE 5: Event broadcasting pattern in collaborative business processes.

$alarm\ event, \phi>$, and the task sound sensor is formalized as $t_{so} = <sound\ sensor, alarm\ target, alarm\ event, \phi>$. Obviously, the task item, subscribed event, and published event are same, but the task of alarm target is executed by light sensor and sound sensor together. Therefore, according to Definition 8, service collaboration pattern exists between them. The collaboration service is drawn as a rectangle with dashed borders. Figure 6 illustrates this coordination pattern.

4.3. Modeling the Whole Business Processes. In general, the modeling for collaborative business processes towards IoT applications in the forest-protection scenario includes the following steps:

(1) According to Table 1, the whole business processes include five kinds of sensors: smoke sensor, infrared thermal imaging sensor, tracking camera sensor, light sensor, and sound sensor. Modeling all these sensor nodes using the formal specification.

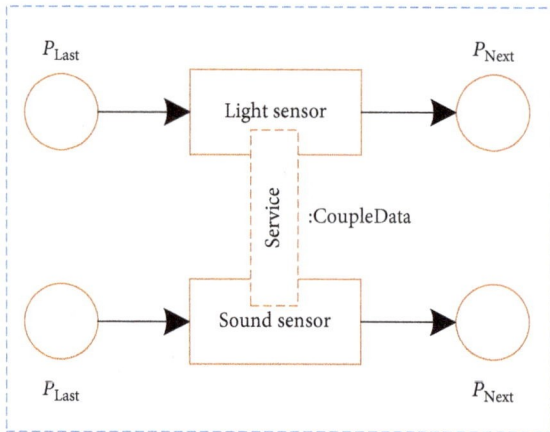

FIGURE 6: Service collaboration pattern in collaborative business processes.

(2) Modeling the common control structures, including start/end structure, And-Split/Join structure, and Xor-Split/Join structure. Their corresponding Business Process Model and Notation (BPMN) objects and Petri nets-based models are given in Figure 7.

(3) Modeling the existing coordination patterns, including event unicasting pattern, event broadcasting pattern, and service collaboration pattern.

So far, the complete modeling approach of this collaborative business process towards IoT applications in the forest-protection scenario has been obtained and depicted in Figure 8. Based on this model, we could further analyze and optimize this collaborative business process towards IoT application.

5. Optimization Algorithms

In the previous section, we have introduced the formal modeling method for collaborative business process towards IoT applications in the forest-protection scenario using extended Petri nets. Based on this extended Petri nets model, its corresponding sequence diagram could be easily obtained, which is illustrated in Figure 9. As we see, the response sequence for tracking camera senor, light sensor, and sound sensor is uncertain. Furthermore, besides the fire-related sensors, other sensors (e.g., temperature sensor, humidity sensor, carbon dioxide sensor, etc) are also deployed in the forest and send sensing events to the event manager. All these sensing events tend to get response from the IoT system as fast as possible.

However, the available computing resources, usually memory and CPU, are often limited in this particular scene. It is particularly important to reasonably schedule computing resources in a resource-constrained IoT execution environment. The performance (e.g., response time, throughout, event loss ratio, etc) of the entire system is directly determined by the sensing event scheduling algorithm of choice. Thus, we propose a DPFR-based optimization algorithm to solve the problem of sensing event scheduling. Furthermore, we consider both qualitative and quantitative aspects for the evaluation of our proposed optimization algorithm.

5.1. Sensing Events Scheduling

5.1.1. Sensing Event Format.
Multiple types of smart sensors are widely deployed in the forest, which are in charge of collecting the environmental information regularly. The gathered raw data not only has a huge number, but also has complex data formats. We take this forest-protection application as an instance, the smoke sensor collects data in text format, the infrared thermal imaging sensor collects data in image format, and the tracking camera sensor collects data in video format. In addition, different sensing events require different response time. For example, the smoke and fire events tend to get a quicker response than the temperature or humidity events. Thus, sensing events need to be assigned corresponding priorities according to certain rules. The sensing events with higher priorities could be first responded and allocated the computing resources.

To better schedule sensing events, we encapsulate them before they are submitted to the event manager. The formal format of encapsulated sensing event is illustrated in Figure 10. Then, each item is interpreted as follows:

(i) *Topic*: it is a hierarchical string that could be filtered based on a finite number of expressions. The client could get all the interesting messages from the Broker based on the subscribed event topic.

(ii) *DataType*: the storage form of information gathered by smart sensors. In this forest-protection scenario, it mainly includes text, image, and video.

(iii) *Content*: the details of sensing events. It is usually compressed before transmission. The event size varies from a few kilobytes to several megabytes.

(iv) *Priority*: it is an integer that is used to indicate the urgency of the sensing event. The higher the priority, the faster the event gets response. Its value can be modified by the event manager according to certain rules.

(v) $R(t)$: we have introduced it in Section 6. It is a variable that indicates the waiting time for computing resource. An embedded timer starts timing when the business process application begins.

(vi) *Timeout*: a constant that represents the threshold of waiting time for every sensing event. Once this constant is exceeded, the sensing event will be abandoned and its corresponding storage resources will be released immediately. The value of timeout could be set by users.

5.1.2. Event Manager.
The first purpose of event manager is to act as the communication substrate supporting event-based interaction among the collaborative business process towards IoT applications. This kind of interaction is commonly referred to as publish-subscribe messaging

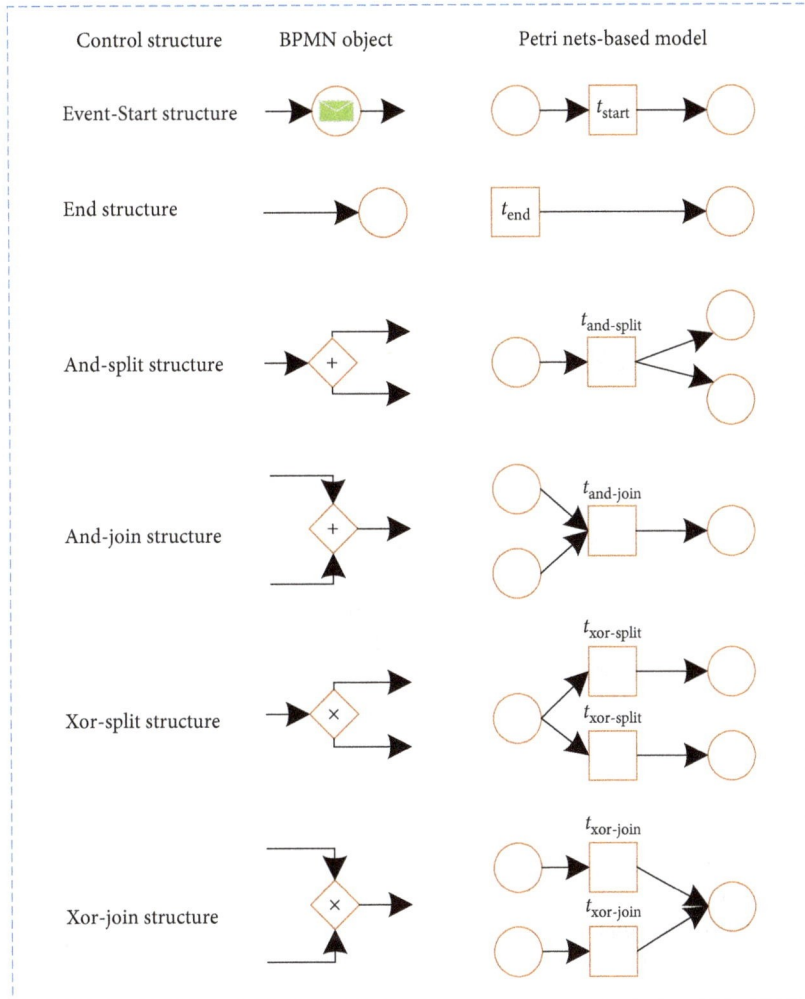

FIGURE 7: Modeling control structure using Petri nets model.

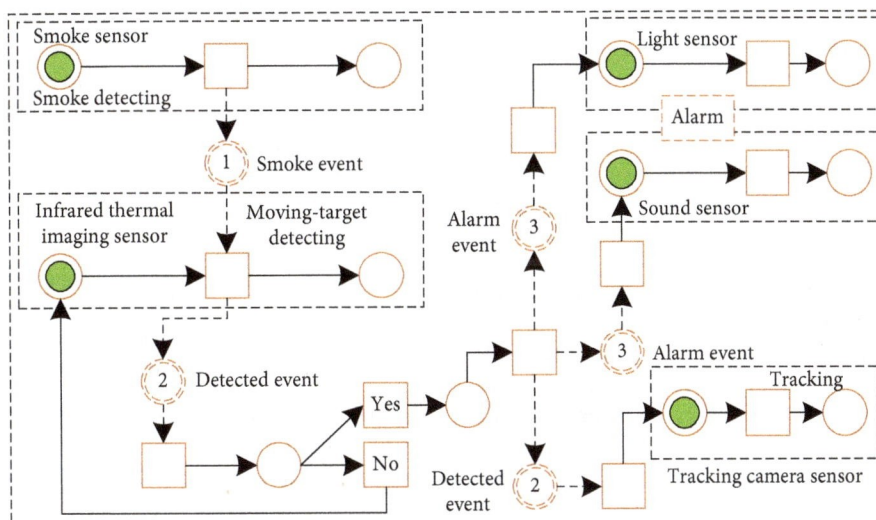

FIGURE 8: The model of collaborative business process towards IoT applications in forest-protection using extended Petri nets.

networks [27]. The publish-subscribe messaging networks are frameworks where publishers publish structured messages (events) to an event service agent (event manager) and subscribers express interest in different events to the event service agent through Lightweight Directory Access Protocol (LDAP) topic tree which is responsible for managing the

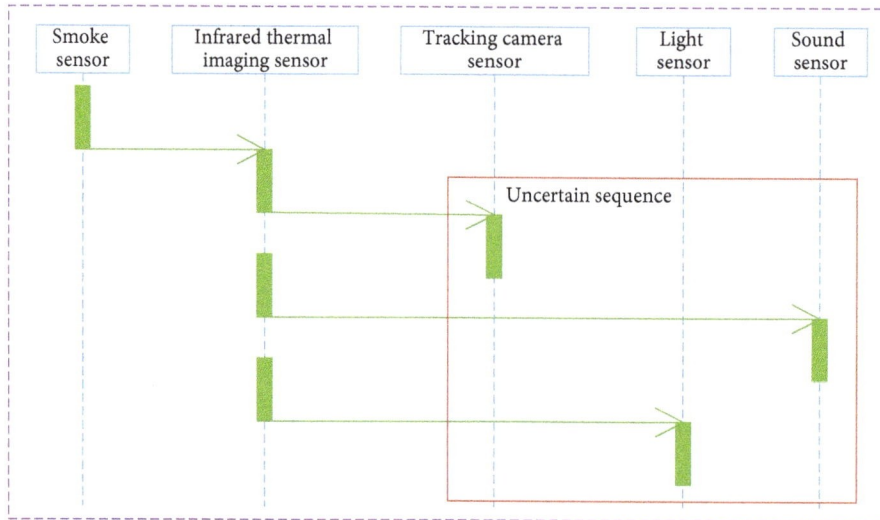

FIGURE 9: The sequence diagram of collaborative business process towards IoT applications in forest-protection.

FIGURE 10: The formal format of encapsulated sensing event.

event topics. e.g., the business process in tracking camera sensor could subscribe a detected event from the event manager, which is published by the business process in infrared thermal imaging sensor. A main property of the publish-subscribe messaging networks is that they are space-decoupled, meaning that the publisher does not need to know the identity of the subscribers. This allows highly flexible and dynamic architecture.

Besides being the basis of communication, event manager is also the core of event scheduling. Due to the huge number of sensing events and limited computing resources, usually CPU and memory, there is no guarantee that each sensing event could be responded immediately. Thus, we design a scheduling mechanism based on memory and secondary storage. When a new event arrives, it first enters the buffer queue of the secondary storage. Then, event manager schedules sensing events between memory and secondary storage based on certain scheduling algorithms. Only the sensing events that are in the ready queue of the memory could get the computing resource and be responded immediately. However, once the waiting time exceeds the threshold, the sensing event will be abandoned and its corresponding storage resources will be released immediately. Figure 11 illustrates the role of event manager in the event scheduling.

5.2. The Goal and Evaluation Standard for Sensing Event Scheduling Algorithm.
The limited computing resource is the most important resource in the modern IoT system. Sensing event scheduling is mainly to allocate computing resource to each sensing event based on some scheduling algorithms. The throughput and response time of the IoT system are directly affected by the selected scheduling algorithm.

5.2.1. The Goal for Sensing Event Scheduling.
The goal for sensing event scheduling is the final result that needs to be achieved. In general, the common goals are represented as follows:

(i) *Fairness*: to ensure that each sensing event could obtain a reasonable computing resource. We could not make some of the sensing events unavailable for a long time. We should maximize fairness on the basis of priority.

(ii) *Low Response Time*: to ensure that the sensing events could be responded timely. For the high-priority sensing event, it should get response within the specified time. It is an important guarantee for solving emergencies.

(iii) *High Throughput*: to ensure that the IoT system could respond to as many sensing events as possible within a unit time. It is an important measure of efficiency.

5.2.2. The Evaluation Standard for Sensing Event Scheduling.
There are a variety of sensing event scheduling algorithms that could be chosen in the phase of design. In order to allow users to select the appropriate algorithm based on demand, we should clearly give some evaluation standards.

(i) *Turnaround Time*: it is the total time taken between the submission of a sensing event for response and the return of the complete output to the user. It usually consists of three parts: the waiting time in the buffer queue of the secondary storage, the waiting time in the ready queue of the memory, and the execution time in the computing resources. To more comprehensively evaluate the sensing event scheduling algorithms, we propose the concept of Average Turnaround Time (ATT).

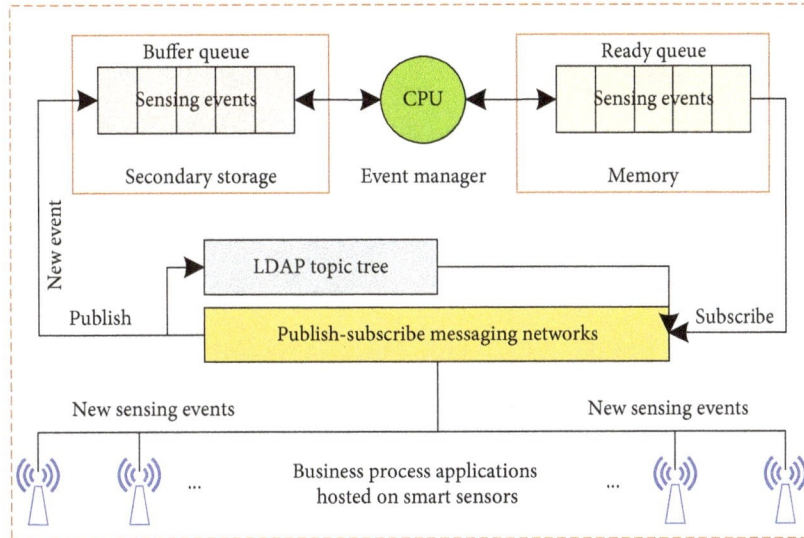

FIGURE 11: The role of event manager in the sensing event scheduling.

$$ATT = \frac{1}{n}\left[\sum_{i=1}^{n} ATT_i\right]. \qquad (2)$$

(ii) Furthermore, we define the Weighted Turnaround Time (WTT) equals the ratio of ATT to Execution Time (ET). Thus, the Average Weighted Turnaround Time (AWTT) is defined as follows:

$$AWTT = \frac{1}{n}\left[\sum_{i=1}^{n} \frac{ATT_i}{ET_i}\right]. \qquad (3)$$

(iii) *Response Time*: response time is the total amount of time it takes to respond to a request for execution. Ignoring transmission time for a moment, the response time is the sum of the execution time and waiting time.

(iv) *Throughput*: throughput is the total numbers of sensing events that have been responded by the IoT system within a unit time. When dealing with large sensing events, such as image and video format events, throughput might be only one per minute. On the contrary, throughput might be more than a thousand per second when the IoT system deals with text format events.

(v) *Loss Tolerance*: as mentioned above, we introduce the concept of timeout to ensure that the storage resources could not be occupied by the unavailable sensing events for a long time. Then, we define Event Loss Rate (ELR) to evaluate the fairness of the sensing event scheduling algorithm. The ELR vaule equals to the ratio of abandoned sensing events to total sensing events.

5.3. The Sensing Event Scheduling Algorithm

5.3.1. First Come First Response (FCFR). The approaches mentioned in the work [28–30] could be summarized as a First Come First Response (FCFR) algorithm that schedules sensing events based on the sequence. The sensing event entering the buffer queue in the secondary storage earlier will be first responded.

The data structure in the secondary storage is queue by nature. This data structure has the natural characteristics of "First In First Out." Thus, FCFR algorithm could be realized easily. However, this algorithm is conducive to the long sensing events, but not conducive to the short ones. The short sensing events behind the long sensing events have to wait a long time for scheduling, resulting in poor performances in turnaround time and throughput. Furthermore, FCFR algorithm does not consider the priority of the sensing event, which is an important factor in the IoT environment.

Then, we consider an example to quantitatively analyze the performance of FCFR algorithm. We suppose that there are five different kinds of sensing events in the buffer queue of the secondary storage. These sensing events have different arrival time, execution time, and priorities. The detailed information is illustrated in Table 2.

We could obtain the corresponding sequence diagram of these sensing events based on FCFR algorithm, which is illustrated in Figure 12. The final turnaround time and weighted turnaround time are depicted in Table 3.

5.3.2. Static Priority First Response (SPFR). The approaches mentioned in the work [21, 31, 32] could be summarized as a Static Priority First Response (SPFR) algorithm that schedules sensing events based on the static priority. The sensing event with higher priority will be first responded. Each sensing event is assigned an integer when it enters the buffer queue in the secondary storage. This integer could not be changed during the execution period.

As mentioned above, there are a variety of sensing events that tend to get response from the IoT system as fast as possible. We initially assign static priority according to the

TABLE 2: The detailed information of Sensing Events.

Sensing event name	Arrival time	Execution time	Priority
E_1	0	3	3
E_2	1	6	5
E_3	2	1	1
E_4	3	4	4
E_5	4	2	2

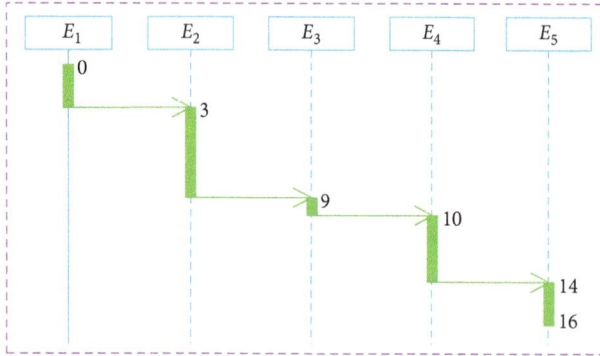

FIGURE 12: The scheduling sequence diagram based on FCFR algorithm.

TABLE 3: The Evaluation results based on FCFR algorithm.

Sensing event name	Turnaround time	Weighted turnaround time
E_1	3	1
E_2	8	1.33
E_3	8	8
E_4	11	2.75
E_5	12	6
Average	8.4	3.82

urgency of the sensing event. We take this collaborative IoT-aware business process application in the forest-protection scene as an instance, the fire-related sensing events should have higher priorities than the temperature or humidity sensing events. SPFR algorithm could ensure that the sensing event with higher priority could be first responded. As we know, priority-based sensing event scheduling algorithm is often nonpreemptive, meaning that even if the low-priority sensing event enters the buffer queue earlier than the high-priority sensing event, it has to wait a long time for the execution of the latter one. However, SPFR algorithm is conducive to the sensing event that could be assigned a high priority at the beginning of the IoT system, but not conducive to the sensing event that is assigned a low priority. This is not only contrary to the principle of fairness but also results in poor performances in turnaround time and throughput.

Then, we also consider the example in FCFR algorithm to quantitatively analyze the performance of SPFR algorithm. The corresponding sequence diagram of these sensing events based on SPFR algorithm is illustrated in Figure 13. The final turnaround time and weighted turnaround time for each sensing event are depicted in Table 4.

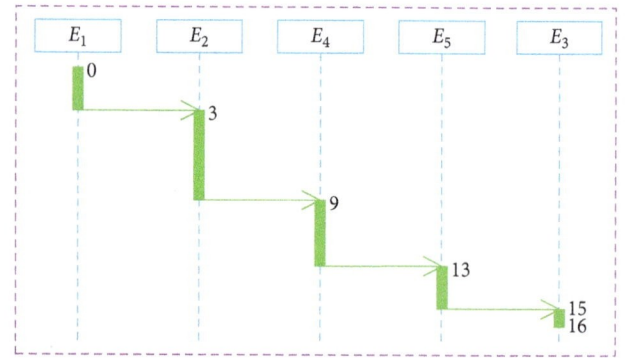

FIGURE 13: The scheduling sequence diagram based on SPFR algorithm.

TABLE 4: The Evaluation results based on SPFR algorithm.

Sensing event name	Turnaround time	Weighted turnaround time
E_1	3	1
E_2	8	1.33
E_3	14	14
E_4	10	2.5
E_5	11	5.5
Average	9.2	4.87

5.3.3. Dynamic Priority First Response (DPFR). In this paper, we propose a Dynamic Priority First Response (DPFR) algorithm that schedules sensing events based on the dynamic priority. The sensing event with higher priority will be first responded. Different to SPFR algorithm, the priority could be modified by the event manager according to certain rules during the execution period of IoT system.

In order to maximize fairness on the basis of priority, we propose this dynamic priority-based sensing event scheduling algorithm. The core of DPFR algorithm is the principle to be followed for priority modification. We should ensure that every sensing event could obtain a reasonable computing resource from the IoT system. Thus, the concept of Response Ratio (RR) is proposed and defined as follows:

$$RR = \frac{\text{response time}}{\text{execution time}}. \tag{4}$$

As we know, the response time equals the sum of waiting time and execution time. Thus, Equation (4) could be presented in the following format:

$$RR = 1 + \frac{\text{waiting time}}{\text{execution time}}. \tag{5}$$

Based on DPFR algorithm, the event manager regularly calculates and modifies the values of all the sensing event priorities. The sensing event whose RR is largest will be first responded. We could draw the following conclusions through analyzing Equation (5).

(i) If the waiting time is the same, the shorter of the execution time, the larger of the RR. This ensures that IoT system could respond to as many sensing events as possible within a unit time.

(ii) If the execution time is the same, the longer of the waiting time, the larger of the RR. This ensures that the low-priority sensing event, which has waited a long time in the buffer queue of the secondary storage, could also get the response from the IoT system.

Then, we consider the example discussed earlier to quantitatively analyze the performance of DPFR algorithm. We assume that none of the sensing events have timed out. With Algorithm 1, we could get the priority sequence for each scheduling unit. At the time 3.0, the sensing event E_1 has been responded and the sensing event E_5 has not entered the buffer queue. According to DPFR algorithm, the RR of sensing event E_2 is $1 + (3 - 1)/6 = 1.33$, the RR of sensing event E_3 is $1 + (3 - 2)/1 = 2$, and the RR of sensing event E_4 is 1. Thus, the sensing event E_3 should be responded. We could calculate the RRs of the remaining sensing events at the time 4.0 using the same method. The RR of sensing event E_2 is $1 + (4 - 1)/6 = 1.5$, the RR of sensing event E_4 is $1 + (4 - 3)/4 = 1.25$, and the RR of sensing event E_5 is 1. Thus, the sensing event E_2 should be responded. At the time 10.0, the RR of sensing event E_4 is $1 + (10 - 3)/4 = 2.75$, and the RR of sensing event E_5 is $1 + (10 - 4)/2 = 4$. Thus, the sensing event E_5 should be responded. The corresponding sequence diagram of these sensing events based on DPFR algorithm is illustrated in Figure 14. The final turnaround time and weighted turnaround time for each sensing event are depicted in Table 5.

5.4. Further Simulation Experiments. By comparing the results of turnaround time and weight turnaround time among these three algorithms, we could find that DPFR algorithm has the better performance. To further evaluate our proposed optimization algorithm, we conduct simulation experiments on the basis of larger data sets.

The simulation experiments were conducted on a PC, which had 6 G of RAM, 3.07 GHz of CPUs, and 500 G of disk space. We allocated 1 G of RAM and 2 G of disk space for the following experiments. To simulate the sensors to collect data in the IoT environment, we sent ten data packets to the buffer queue in the secondary storage every 1 millisecond. In addition, we randomly specified the data size and initial priority of the data packet. The data size ranged from 1k to 100k, and the initial priority ranged from 1 to 5. We repeatedly conducted the experiments in the condition that the values of timeout were set to 5 milliseconds and 10 milliseconds, respectively. The sampling time was 1000 milliseconds, and the final results were illustrated as follows:

Throughput is the total numbers of sensing events that have been responded by the IoT system within a unit time. The value of throughout is directly affected by the average response time. Due to the different scheduling principles, the response time of sensing events in DPFR algorithm was the shortest. Thus, it could respond to more sensing events than FCFR and SPFR algorithms within a unit time. Figure 15 illustrated the results of throughput based on three different sensing event scheduling algorithms. In FCFR algorithm, the sensing events at the end of the buffer queue had to wait a long time for scheduling the sensing events before them.

Similar to FCFR algorithm, the sensing events with low-initial priority had to wait a long time for scheduling the high-priority sensing events in SPFR algorithm. However, both algorithms considered only a single factor and were not conducive to the certain parts of the sensing events in the buffer queue. Thus, the waiting time and execution time of these sensing events were significantly increased. Different to them, in our proposed DPFR algorithm, sensing events were scheduled based on their RR values. According to the definition of RR, we could know that the priority was determined by the waiting time and execution time together. If the waiting time was the same, the shorter of the execution time, the larger of the RR value. If the execution time was the same, the longer of the waiting time, the larger of the RR value. This algorithm took into account both factors of the sensing events, which ensured that each sensing event could be responded as soon as possible. Thus, the average response time of sensing events in DPFR algorithm was shortest and the value of throughput in DPFR was largest. Figure 16 showed the average response time based on three sensing event scheduling algorithms.

In order to improve the system utilization, we should try our best to keep the CPU busy and reduce the waiting time. Figure 17 illustrated CPU utilization of the system with different sensing event scheduling algorithms. The results showed that DPFR algorithm performed better than FCFR and SPFR algorithms at lower traffic intensity. As CPU was considered to be a limited resource, DPFR algorithm could better adapt to the resource-constricted IoT environment.

As mentioned above, timeout was set to ensure the buffer queue in the secondary storage to be fully utilized. As the sampling time increased, the waiting time of sensing events, especially the later arrival and lower initial priority sensing events, was significantly increased, resulting in the increase of abandoned sensing events. Figures 18 and 19 illustrated the amount of abandoned sensing events when the timeout was set to 10 milliseconds and 5 milliseconds. We could see that the abandoned sensing events in DPFR algorithm were less than in FCFR and SPFR algorithms. Moreover, the curve increased more gently for DPFR algorithm than the other algorithms with increasing sampling time. Furthermore, the shorter of the timeout threshold, the more obvious of the difference among these three different algorithms.

Fairness was used to evaluate whether the IoT system could reasonably allocate limited resources. We should try our best to ensure that all kinds of sensing events could get response form the IoT system. In order to further compare the performances of three sensing event scheduling algorithms, we defined that the sensing events, whose data size were larger than 50k, as long events and the rest as short events. In addition, we defined that the sensing events, whose initial priority were larger than 2, as high-priority events and the rest as low-priority events. Figure 20 illustrated the number of IoT system responses to various sensing events based on different scheduling algorithms. We could see that the number of responded long events was much larger than short events in the FCFR algorithm.

Input: The sensing event buffer queue $Q = (E_1, E_2, \ldots, E_n)$ in the secondary storage
Output: The highest-priority sensing event E_{max}
  The event manager sets the timeout value $t_{timeout}$
  for Q is not empty **do**
    Set the temporary variable E_{max}
    Get the waiting time $t_{waiting}$ of each event
  if $t_{waiting} < t_{timeout}$ **then**
    Get the execution time $t_{execution}$ of each event
    Calculate the response ratio E_{RR}
    if $E_{max} \leq E_{RR}$ **then**
      $E_{max} = E_{current}$
      current++
    else current++
  else Abandon current event, release storage space
  return The highest-priority sensing event E_{max}

ALGORITHM 1: DPFR Sensing Event Scheduling Algorithm.

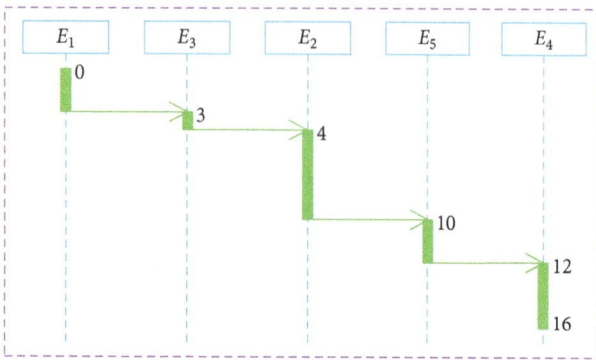

FIGURE 14: The scheduling sequence diagram based on DPFR algorithm.

TABLE 5: The Evaluation results based on DPFR algorithm.

Sensing event name	Turnaround time	Weighted turnaround time
E_1	3	1
E_2	9	1.5
E_3	2	2
E_4	13	3.25
E_5	8	4
Average	7	2.35

Similarly, the number of responded high-initial priority events was much larger than low-initial priority events in the SPFR algorithm. On the contrary, the number of all kinds of sensing events that get response from the IoT system was approximately the same in our proposed DPFR algorithm, ensuring that all sensing events could obtain the reasonable computing resources. Thus, DPFR algorithm could better reflect the fairness of IoT system.

6. Implementation and Screenshots

In order to automate the approach presented in this paper, we developed an actual development system for developers to manage the full lifecycle of the collaborative IoT-aware

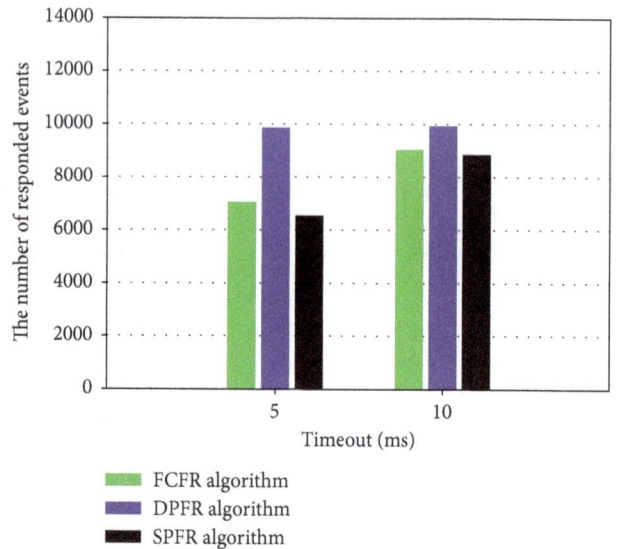

FIGURE 15: The throughput based on different algorithms.

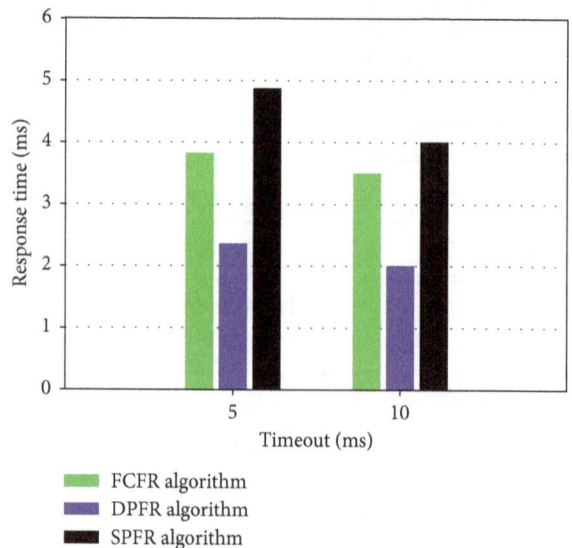

FIGURE 16: The average response time based on different algorithms.

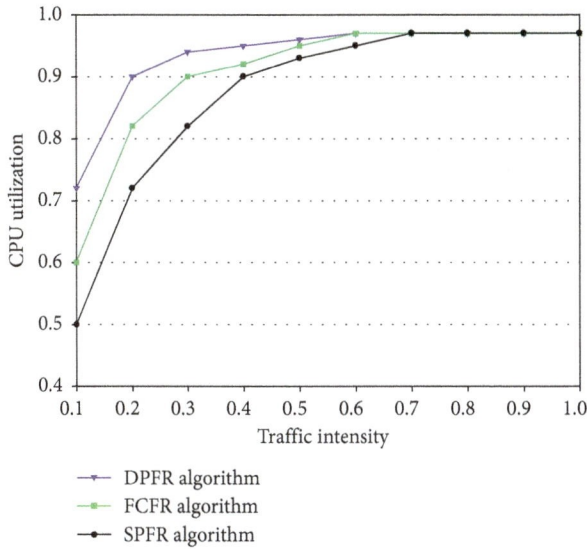

FIGURE 17: CPU Utilization of the system.

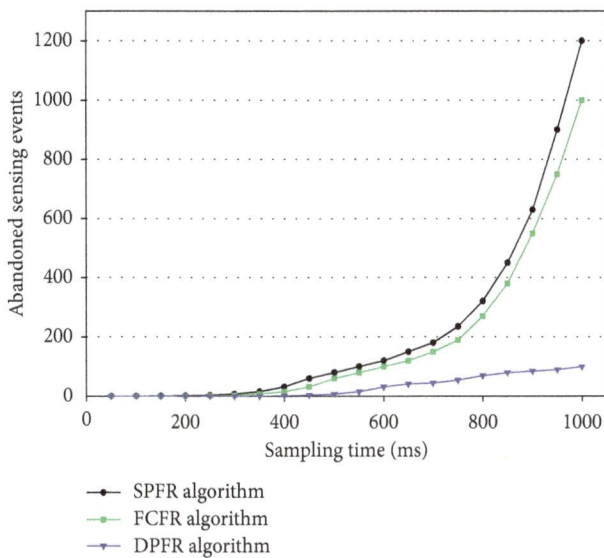

FIGURE 19: The amount of abandoned sensing events (timout = 5 ms).

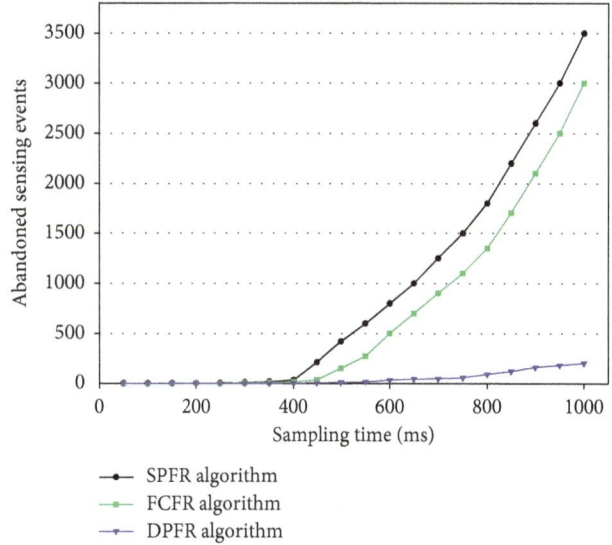

FIGURE 18: The amount of abandoned sensing events (timout = 10 ms).

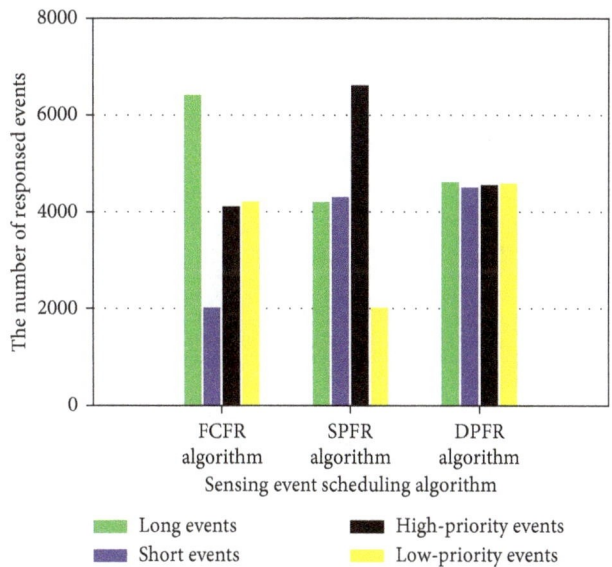

FIGURE 20: The number of IoT system responses to various sensing events.

business process applications. Figure 21 illustrates the comprehensive GUI of the system. It consists of 5 major parts and provides all essential tools for developers. Each part is responsible for a different function. Part 1 is a basic graphic element library, which provides developers draggable process elements. Part 2 is a workspace panel where developers could model the specific business processes, create the corresponding Petri net models, and bind the sensing events to send or receive tasks. Part 3 is a configuration panel where developers configure the properties of process elements, including sensing event topic, timeout value, and the types of interfaces by which they would like to access the raw data. Part 4 is a menu bar where developers could decouple the business processes into fine-grained process fragments, package the IoT application into a war file, and deploy process fragments to distributed servers with

custom buttons. Part 5 is a package explorer where developers could open or browse the element hierarchy of the collaborative IoT-aware business process applications. More information about our work could be found at this YouTube video.

Figure 22 illustrates the screenshots of our proposed system to manage and monitor the collaborative business process towards IoT applications. Figure 22(a) shows the collaborative business processes, where developers could graphically add or remove process elements, dynamically define the process interaction interfaces, and logically validate the logic errors of process model (A demonstrative video could be found at this https://youtu.be/T5HN0AboJ2AYouTube video). Figure 22(b) shows the

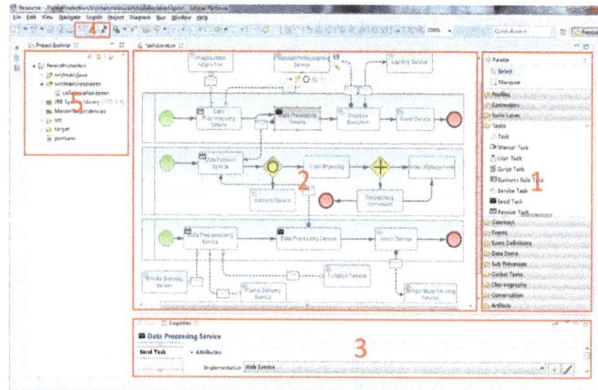

FIGURE 21: GUI of our proposed development system.

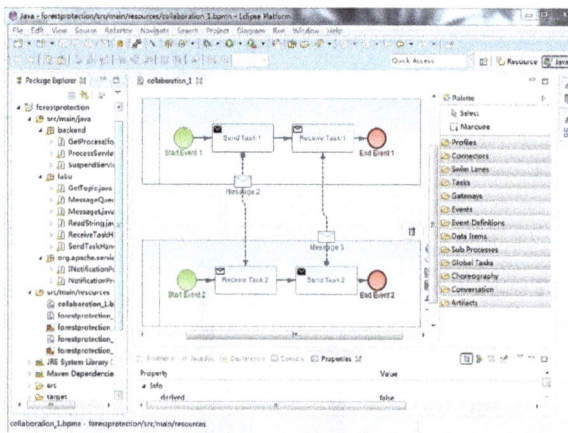

(a)

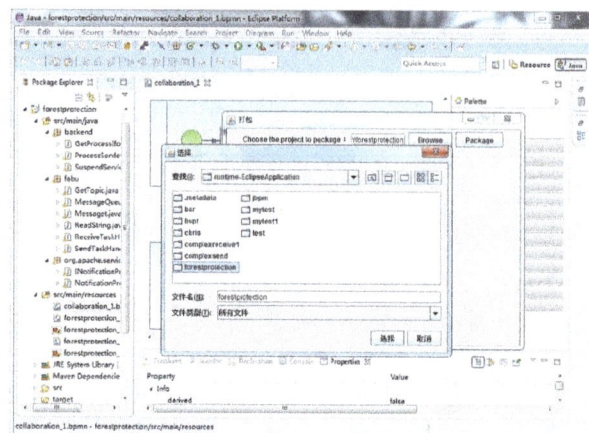

(b)

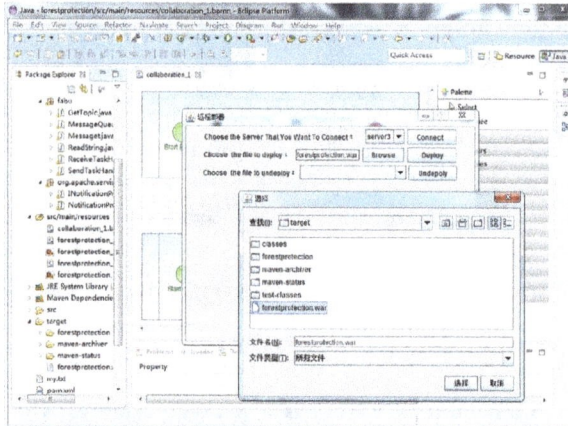

(c)

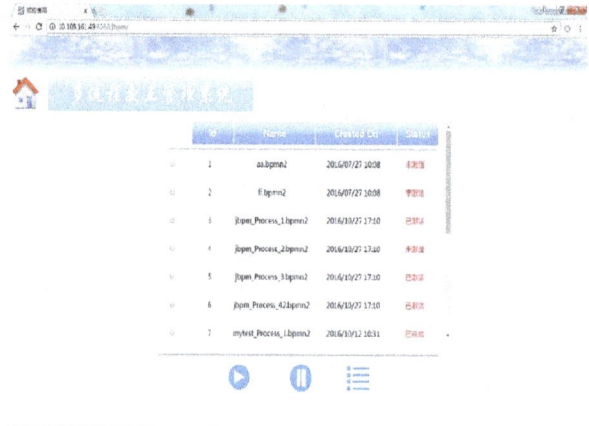

(d)

FIGURE 22: Screenshots of our proposed development system. (a) Creating Business Process. (b) Packaging Business Process. (c) Deploying Business Process. (d) Monitoring Business Process.

way to package a designed IoT-aware business process application. Developers could package the application into a war file through this step. Similarly, Figure 22(c) shows the way to deploy a packaged war file to distributed servers. It is worth noting that unavailable servers have been automatically filtered off by the system. Namely, developers only need to choose the war file which they want to deploy. Figure 22(d) shows the way to monitor the information of

a deployed IoT application, including the name, created time, and execution status.

7. Conclusion

In this paper, we have presented a novel modeling and optimization approach for the collaborative business process towards IoT applications, making it fit in the dynamic IoT sensing environment. In order to reach this goal, we initial extend the traditional Petri nets with sensing event factor. Based on a real collaborative business process towards IoT application in the forest-protection scenario, the formal modeling specification of senor nodes is investigated and their coordination patterns are formally defined. Then we could easily deduce its corresponding sequence diagram from the extended Petri net model. Next, we propose a DPFR-based optimization algorithm to solve the problem of sensing event scheduling. Moreover, we qualitatively and quantitatively analyze each algorithm and conduct a series simulation experiment to compare different algorithms. Finally, the approach presented in this paper has been validated to be valid and implemented through an actual system.

However, our work is still in its infancy and requires more actual applications to prove its applicability. Thus, we plan to leverage our proposed approach to implement advanced collaborative business process towards IoT applications in a larger and more complicated IoT sensing environment.

Acknowledgments

This work has been supported by National Natural Science Foundation of China (Grant nos. 61501048, U1536111, and U1536112), Beijing Natural Science Foundation (Grant no. 4182042), and Fundamental Research Funds for the Central Universities (Grant no. 2017RC12).

References

[1] R. Mirzavand, M. M. Honari, and P. Mousavi, "Direct-conversion sensor for wireless sensing networks," *IEEE Transactions on Industrial*, vol. 64, no. 12, pp. 9675–9682, University of Alberta, Edmonton, Canada, 2017.

[2] M. Y. Cheng, K. C. Chiu, Y. M. Hsieh, I. T. Yang, J. S. Chou, and Y. W. Wu, "BIM integrated smart monitoring technique for building fire prevention and disaster relief," *Automation in Construction*, vol. 84, pp. 14–30, Elsevier Science, New York, NY, USA, 2017.

[3] A. E. Marquez-Chamorro, M. Resinas, A. Ruiz-Cortes, and M. Toro, "Run-time prediction of business process indicators using evolutionary decision rules," *Expert Systems with Applications*, vol. 87, pp. 1–14, Elsevier Science, Seville, Spain, 2017.

[4] S. Appel, P. Kleber, S. Frischbier, T. Freudenreich, and A. Buchmann, "Modeling and execution of event stream processing in business processes," *Information Systems*, vol. 46, pp. 140–156, Elsevier Science, Darmstadt, Germany, 2014.

[5] K. M. You, W. Yang, and R. S. Han, "The video collaborative localization of a Miner's lamp based on wireless multimedia sensor networks for underground coal mines," *Sensors*, vol. 15, no. 10, pp. 25103–25122, 2015.

[6] H. M. N. D. Bandara and A. P. Jayasumana, "Distributed, multi-user, multi-application, and multi-sensor data fusion over named data networks," *Computer Networks*, vol. 57, no. 16, pp. 3235–3248, 2013.

[7] C. U. Lei, W. K. Chong, and K. L. Man, "Utility-oriented placement of actuator nodes with a collaborative serving scheme for facilitated business and working environments," *The Scientific World Journal*, vol. 2014, Article ID 835260, 11 pages, 2014.

[8] C. F. Liu, Q. Li, and X. H. Zhao, "Challenges and opportunities in collaborative business process management: overview of recent advances and introduction to the special issue," *Information System Frontiers*, vol. 11, no. 3, pp. 201–209, 2014.

[9] M. Graiet, A. Mammar, S. Boubaker, and W. Gaaloul, "Towards correct cloud resource allocation in business processes," *IEEE Transactions on Services Computing*, vol. 10, no. 1, pp. 23–36, 2017.

[10] M. J. Tsai and C. S. Wang, "A computing coordination based fuzzy group decision-making (CC-FGDM) for web service oriented architecture," *Expert System With Applications*, vol. 34, no. 4, pp. 2921–2936, 2008.

[11] H. W. Kim, J. H. Park, and Y. S. Jeong, "Efficient resource management scheme for storage processing in cloud infrastructure with Internet of Things," *Wireless Personal Communications*, vol. 91, no. 4, pp. 1635–1651, 2016.

[12] A. A. Alsaffar, H. P. Pham, C. S. Hong, E. N. Huh, and M. Aazam, "An Architecture of IoT service delegation and resource allocation based on collaboration between fog and cloud computing," *Mobile Information Systems*, vol. 2016, Article ID 6123234, 15 pages, 2016.

[13] W. Tan, Y. S. Fan, and M. C. Zhou, "A petri net-based method for compatibility analysis and composition of web services in business process execution language," *IEEE Transactions on Automation Science and Engineering*, vol. 6, no. 2, p. 392, 2009.

[14] A. Chen and D. Buchs, "Towards service-based business process modeling, prototyping and integration," in *Proceedings of International Workshop on Rapid Integration of Software Engineering Techniques*, vol. 3943, pp. 218–233, Heraklion, Greece, September 2006.

[15] W. Y. Yu, C. G. Yan, Z. J. Ding, C. J. Jiang, and M. C. Zhou, "Modeling and validating E-commerce business process based on Petri nets," *IEEE Transactions on Systems, Man, and Cybernetics: Systems*, vol. 44, no. 3, pp. 327–341, 2013.

[16] S. C. Pang, Y. Li, H. He, and C. Lin, "A model for dynamic business processes and process changes," *Chinese Journal of Electronics*, vol. 20, pp. 632–636, 2011.

[17] D. S. Liu, J. M. Wang, S. C. F. Chan, J. G. Sun, and L. Zhang, "Modeling workflow processes with colored Petri nets," *Computers In Industry*, vol. 49, no. 3, pp. 267–281, 2002.

[18] I. D. Mironescu and L. Vinłean, "A task scheduling algorithm for HPC applications using colored stochastic Petri Net models," in *Proceedings of IEEE International Conference on Intelligent Computer Communication and Processing*, pp. 479–486, Cluj-Napoca, Romania, September 2017.

[19] L. G. He, N. Chaudhary, and S. A. Jarvis, "Developing security-aware resource management strategies for workflows," *Future Generation Computer Systems*, vol. 38, pp. 61–68, 2014.

[20] M. A. A. Aziz, M. N. Taib, and N. M. Hussin, "The effects of event selection based on soft constraint violation (ESSCV) in a modified PSO algorithm to solve class scheduling problems," in *Proceedings of International Conference on Computer Applications and Industrial Electronics*, vol. 28, pp. 584–587, Kuala Lumpur, Malaysia, December 2010.

[21] C. Erbas, S. Cerav-Erbas, and A.D. Pimentel, "Static priority scheduling of event triggered real time embedded systems," in *Proceedings of Second ACM and IEEE International Confer-*

ence on Formal Methods and Models for Co-Design, pp. 109–118, San Diego, CA, USA, 2004.

[22] W. van der Aalst, "Loosely coupled interorganizational workflows: Modeling and analyzing workflows crossing organizational boundaries," *Information and Management,* vol. 37, no. 2, pp. 67–75, 2000.

[23] Q. T. Zeng, F. M. Lu, C. Liu, H. Duan, and C. H. Zhou, "Modeling and verification for cross-department collaborative business processes using extended Petri nets," *IEEE Transactions on Systems, Man, and Cybernetics: Systems,* vol. 45, no. 2, pp. 349–362, 2014.

[24] K. Munir, S. Javed, M. Welzl, and M. M. Junaid, "Using an event based priority queue for reliable and opportunistic scheduling of bulk data transfers in grid networks," in *Proceedings of IEEE International Multitopic Conference,* pp. 1–6, Lahore, Pakistan, December 2007.

[25] F. F. Li, C. Liu, G. Yu, and Z. Chen, "A scheduling algorithm of events with uncertain timestamps for CPS," in *Proceedings of International Conference on Big Data Computing and Communications,* pp. 313–319, Chengdu, China, August 2017.

[26] A. Kheldoun, K. Barkaoui, and M. Ioualalen, "Formal verification of complex business processes based on high-level Petri nets," *Information Sciences,* vol. 385-386, pp. 39–54, 2017.

[27] R. Baldoni, M. Contenti, and A. Virgillito, "The evolution of publish/subscribe communication systems," in *Future Directions in Distributed Computing: Research and Position Papers,* vol. 2584, pp. 137–141, Springer, Rome, Italy, 2003.

[28] D. A. Hapsari, A. E. Permanasari, S. Fauziati, and I. Fitriana, "Management information systems development for veterinary hospital patient registration using first in first out algorithm," in *Proceedings of International Conference on Biomedical Engineering,* pp. 1–5, Yogyakarta, Indonesia, October 2016.

[29] T. T. Nguyen and X. T. Tran, "A novel asynchronous first-in-first-out adapting to multi-synchronous network-on-chips," in *Proceedings of International Conference on Advanced Technologies for Communications,* pp. 365–370, Hanoi, Vietnam, October 2014.

[30] H. Yu, S. Ruepp, and M. S. Berger, "Enhanced first-in-first-out-based round-robin multicast scheduling algorithm for input-queued switches," *IET Communications,* vol. 5, no. 8, pp. 1163–1171, 2011.

[31] C. L. Liu, J. Zhu, and H. Y. Liu, "Queue management algorithm for multi-terminal and multi-service models of priority," in *Proceedings of 14th IEEE Annual Consumer Communications and Networking Conference,* pp. 1–5, Las Vegas, NV, USA, January 2017.

[32] A. Maoudj, B. Bouzouia, A. Hentout, A. Kouider, and R. Toumi, "Distributed multi-agent approach based on priority rules and genetic algorithm for tasks scheduling in multi-robot cells," in *Proceedings of IEEE International Conference on Industrial Electronics Society,* pp. 692–697, Florence, Italy, October 2016.

Permissions

List of Contributors

Wernhuar Tarng, Kuo-Liang Ou and Yun-Chen Lu
Institute of Learning Sciences and Technologies, National Tsing Hua University, Hsinchu, Taiwan

Yi-Syuan Shih and Hsin-Hun Liou
Department of Computer Science and Information Engineering, National Central University, Taoyuan, Taiwan

Xin Li and Yan Wang
School of Computer Science and Technology, China University of Mining and Technology, Xuzhou 221116, China

Xin Li and Kourosh Khoshelham
Department of Infrastructure Engineering of the University of Melbourne, Melbourne 3010, Australia

Junwoo Seo and Kyoungmin Kim
Department of Cyber Defense (CYDF), Korea University, Seoul, Republic of Korea

Mookyu Park and Kyungho Lee
Center for Information Security Technologies (CIST), Korea University, Seoul, Republic of Korea

Moosung Park
Agency for Defense Development, Seoul, Republic of Korea

Jinjun Luo, Shilian Wang and Eryang Zhang
School of Electronic Science and Engineering, National University of Defense Technology, Changsha 410073, China

Xin Man
College of Electronic Engineering, Naval University of Engineering, Wuhan 430033, China

Yali Zhang
School of Management, Northwestern Polytechnical University, Xi'an 710072, China

Jun Sun and Ying Wang
College of Business and Entrepreneurship, University of Texas Rio Grande Valley, Edinburg, TX 78539-2999, USA

Zhaojun Yang
School of Economics and Management, Xidian University, Xi'an 710126, China

Abayomi Otebolaku and Gyu Myoung Lee
Department of Computer Science, Faculty of Engineering and Technology, Liverpool John Moores University, Liverpool, UK

Samantha Yasivee Carrizales-Villagómez and Marco Aurelio Nuño-Maganda
Universidad Politécnica de Victoria, Av. Nuevas Tecnologías, Parque Científico y Tecnológico Tecnotam, Km. 5.5 Carretera a Soto la Marina, Ciudad Victoria, TAMPS, Mexico

Javier Rubio-Loyola
Centro de Investigación y de Estudios Avanzados del Instituto Politécnico Nacional, Av. Nuevas Tecnologías, Parque Científico y Tecnológico Tecnotam, Km. 5.5 Carretera a Soto la Marina, Ciudad Victoria, TAMPS, Mexico

MarioNieto-Hidalgo, Francisco Javier Ferrández-Pastor, Rafael J. Valdivieso-Sarabia, Jerónimo Mora-Pascual and Juan Manuel García-Chamizo
Department of Computing Technology, University of Alicante, Campus San Vicente del Raspeig, Alicante, Spain

Jung-Yoon Kim
Graduate School of Game, Gachon University, 1342 Seongnam Daero, Sujeong-Gu, Seongnam-Si, Gyeonggi-Do 461-701, Republic of Korea

Sung-Jong Eun
Health IT Research Center, Gil Medical Center, Gachon University College of Medicine, Incheon, Republic of Korea

Dong Kyun Park
Department of Gastrointestinal Medicine, Gil Medical Center, Gachon University College of Medicine, Incheon, Republic of Korea

Berglind F. Smaradottir
Department of Information and Communication Technology, University of Agder, Grimstad N-4879, Norway

Jarle A. Håland
Kongsgård School Centre, Kristiansand N-4631, Norway

Santiago G. Martinez
Department of Health and Nursing Science, University of Agder, Grimstad N-4879, Norway

Martín Hernández-Ordoñez
Instituto Tecnológico Superior de Alvarado, Alvarado, VER, Mexico
Instituto Tecnológico de Veracruz, Veracruz, VER, Mexico

Marco A. Nuño-Maganda, Carlos A. Calles-Arriaga and Karla E. Bautista Hernández
Universidad Politécnica de Victoria, 87138 Cd. Victoria, TAMPS, Mexico

Omar Montaño-Rivas
Universidad Politécnica de San Luis Potosí, 78363 San Luis Potosí, SLP, Mexico

Hanjin Kim, Heonyeop Shin, Hyeong-su Kim and Won-Tae Kim
Smart CPS Lab, The Department of Computer Science & Engineering, KOREATECH University, Cheonan, Republic of Korea
Department of Special Education, University of Eastern Finland, Joensuu, Finland

Calkin Suero Montero
Department of Computer Science, University of Eastern Finland, Joensuu, Finland

Federica Burini, Nicola Cortesi, Kevin Gotti and Giuseppe Psaila
University of Bergamo, Bergamo, Italy

Sinem Bozkurt Keser and Ahmet Yazici
Department of Computer Engineering, Eskisehir Osmangazi University, Eskisehir, Turkey

Serkan Gunal
Department of Computer Engineering, Anadolu University, Eskisehir, Turkey

Andres Neyem, Juan Diaz-Mosquera and Jose I. Benedetto
Computer Science Department, Pontificia Universidad Católica de Chile, Santiago, Chile

Rakesh Shrestha and Seung Yeob Nam
Department of Information and Communication Engineering, Yeungnam University, 280 Daehak-Ro, Gyeongsan-si, Gyeongsangbuk-do 712-749, Republic of Korea

Yongyang Cheng, Shuai Zhao, Bo Cheng and Junliang Chen
State Key Laboratory of Networking and Switching Technology, Beijing University of Posts and Telecommunications, Beijing 100876, China

Shoulu Hou
Data61, Commonwealth Scientific and Industrial Research Organization, Marsfield, NSW 2122, Australia

Yulong Shi
Middleware Systems Research Group, University of Toronto, Toronto, Canada M5S 2E4

Index